T0229610

DNA LIQUID-CRYSTALLINE DISPERSIONS AND NANOCONSTRUCTIONS

THE LIQUID CRYSTALS BOOK SERIES

Edited by

Virgil Percec

Department of Chemistry
University of Pennsylvania
Philadelphia, PA

The Liquid Crystals book series publishes authoritative accounts of all aspects of the field, ranging from the basic fundamentals to the forefront of research; from the physics of liquid crystals to their chemical and biological properties; and from their self-assembling structures to their applications in devices. The series will provide readers new to liquid crystals with a firm grounding in the subject, while experienced scientists and liquid crystallographers will find that the series is an indispensable resource.

PUBLISHED TITLES

Introduction to Liquid Crystals: Chemistry and Physics
By Peter J. Collings and Michael Hird

The Static and Dynamic Continuum Theory of Liquid Crystals:
A Mathematical Introduction
By Iain W. Stewart

Crystals That Flow: Classic Papers from the History of Liquid Crystals
Compiled with translation and commentary by Timothy J. Sluckin, David A. Dunmur, and Horst Stegemeyer

Nematic and Cholesteric Liquid Crystals: Concepts and Physical Properties
Illustrated by Experiments
By Patrick Oswald and Pawel Pieranski

Alignment Technologies and Applications of Liquid Crystal Devices
By Kohki Takatoh, Masaki Hasegawa, Mitsuhiro Koden, Nobuyuki Itoh, Ray Hasegawa, and Masanori Sakamoto

Adsorption Phenomena and Anchoring Energy in Nematic Liquid Crystals
By Giovanni Barbero and Luiz Roberto Evangelista

Chemistry of Discotic Liquid Crystals: From Monomers to Polymers
By Sandeep Kumar

Cross-Linked Liquid Crystalline Systems: From Rigid Polymer Networks to Elastomers
Edited By Dirk J. Broer, Gregory P. Crawford, and Slobodan Žumer

DNA Liquid-Crystalline Dispersions and Nanoconstructions
By Yuri M. Yevdokimov, V.I. Salyanov, S.V. Semenov, and S.G. Skuridin

THE LIQUID CRYSTALS BOOK SERIES

DNA LIQUID-CRYSTALLINE DISPERSIONS AND NANOCONSTRUCTIONS

Yuri M. Yevdokimov

V.I. Salyanov

S.V. Semenov

S.G. Skuridin

CRC Press

Taylor & Francis Group

Boca Raton London New York

CRC Press is an imprint of the
Taylor & Francis Group, an **informa** business

CRC Press
Taylor & Francis Group
6000 Broken Sound Parkway NW, Suite 300
Boca Raton, FL 33487-2742

First issued in paperback 2019

© 2012 by Taylor & Francis Group, LLC
CRC Press is an imprint of Taylor & Francis Group, an Informa business

No claim to original U.S. Government works

ISBN-13: 978-1-4398-7146-1 (hbk)
ISBN-13: 978-0-367-38214-5 (pbk)

Visit the Taylor & Francis Web site at
http://www.taylorandfrancis.com

and the CRC Press Web site at
http://www.crcpress.com

Contents

SECTION I   The Liquid-Crystalline State of the DNA

SECTION II   DNA LIQUID-CRYSTALLINE FORMS AND THEIR BIOLOGICAL ACTIVITY

SECTION III   DNA LIQUID-CRYSTALLINE
DISPERSIONS IN
NANOTECHNOLOGY
AND BIOSENSORICS

Authors' Preface

Many of the results included in the book *DNA Liquid-Crystalline Dispersions and Nanoconstructions* were received at a time that was especially difficult for Russian science; they were achieved only because colleagues from various institutes of our country and other countries united in our endeavor in this area of science.

We have attempted to write this book in the hope that it will assist readers to see not only the many similarities between the properties of liquid-crystalline state of DNA and the peculiarities of the DNA state in living cells but also the very deep gap existing between properties of isolated DNA molecules in solution and the very complicated problems related to functioning DNA in cellular conditions.

We are greatly indebted to our coworkers and colleagues from the V. A. Engelhardt Institute of Molecular Biology of the Russian Academy of Sciences: Ya. M. Varshavsky, A. S. Tikhonenko, N. M. Akimenko, T. L. Pyatigorskaya, V. A. Kadykov, G. B. Lortkipanidze, A. L. Platonov, and N. S. Badaev, as well as to colleagues from other Russian institutes and universities: A. S. Sonin, V. P. Shibaev, L. M. Blinov, V. A. Belyakov, A. Yu. Grossberg, E. I. Kats, A. T. Dembo, V. G. Pogrebnyak, and V. P. Varlamov. We also acknowledge the assistance of colleagues from other countries: L. Lerman (US), B. Zimm (US), M. Maestre (US), C. Chandrasekhar (India), M. Palumbo (Italy), G. Gottarelli (Italy), F. Spener (Germany), and A. Turner (UK), who have stated various, sometimes very critical, but still well-wishing remarks when discussing certain steps of our investigations, and these have definitely contributed to the improvement of the quality of this book.

We are grateful to B. P. Gottikh (the Russian Academy of Sciences, Biological Department), whose knowledge and long-term scientific experience have contributed to more precise formulation of the general concept of this book.

We want to express our thanks to many students of various colleges and universities for the direct participation in the experiments described, in part, in our book.

For purely technical reasons, we cannot enumerate all of the participants at every stage of our studies, but their names and scientific contribution can be found in the references to the chapter of this book.

We are especially grateful to our wives, who, regardless of our (as well as all Russian scientists') complicated financial circumstances, created the conditions for our productive work related to the presentation of the scientific material included in this book.

Editor's Preface

When considering the known properties of one condensed state of polymeric molecules, namely, the liquid-crystalline state, and, especially, the lyotropic liquid-crystalline state of these molecules, we see a number of peculiarities that apparently have much in common with the properties of double-stranded DNA molecules in biological objects. First, some of the mechanisms of liquid crystals formation are realized almost without any expenditure of energy. Second, the x-ray parameters of the DNA molecules in artificially condensed phases are close enough to the parameters of the DNA molecules in biological objects. Third, there are many differing liquid-crystalline phases, and the transitions between them are regulated not only by the properties of DNA but also by the properties of the medium in which the condensed phases are obtained. Finally, not only the specific structure of liquid-crystalline phase itself but also that of the double-stranded DNA it contains is restored by almost 100% after the removal of the factors, such as temperature, inducing their "denaturation."

Double-stranded native DNA has two peculiarities that determine many of the physical properties of tightly packed DNA. Double-stranded DNA molecules have their own anisotropic properties because they contain both geometric anisometry (as the molecules are helical) and optical anisotropy (caused mainly by the presence of asymmetric carbon atoms in the structure of sugar residues). This is typical of the native DNA feature and cannot be changed in any way; it is an internal property that is inherent to native DNA only, under any condition. In the language of physics and physical chemistry of liquid crystals, this fact means that the fragments (molecules) of native DNA will always tend to so-called cholesteric packing as they approach one another. The double-stranded DNA molecules cannot be "deprived" of this property; they can only be affected by forcing DNA molecules or their complexes to pack into different liquid-crystalline phases. But at the first opportunity, the tendency of DNA molecules to form cholesterics will become apparent and dominate, determining the properties of the condensed phase of DNA.

A number of extra peculiarities of DNA liquid crystals are interesting from the biological point of view. On the one hand, a tailored modification of the DNA secondary structure in a liquid-crystal phase in a relatively broad extent does not result in the distortion of the mode of the spatial packing (i.e., the liquid-crystalline system has a high "structural memory" determined not by the properties of single molecules (fragments) of DNA but by the properties of the whole ensemble of neighboring molecules). On the other hand, at the "moment of formation" of a liquid crystal, when single fragments (molecules) of DNA and their complexes (for instance, with histone proteins) "recognize" one another, the properties of these fragments exert a significant, if not determining, influence on the mode of spatial packing, and relatively small changes in the structure of the single DNA molecule can regulate the spatial structure of the whole liquid-crystalline ensemble of the molecules.

This fact has allowed one of the authors of this book to illustrate a "gap" existing between the properties of DNA molecules freely floating in the laboratory solvent, where the linear responses to the external effects are observed, and the properties of the highly ordered liquid-crystalline state of these molecules, where DNAs have non-linear properties and generate nonlinear responses to external effects. This means that the correct transfer of knowledge received when studying the properties of iso-lated DNA molecules compared to the properties of condensed DNA molecules in biological objects is not a simple problem, and must be solved in the near future. The authors illustrate, also, the existence of a very complicated problem related to the precise descriptions of functioning of DNA molecules under conditions of living cells. However, this book is not attempting an absolute definition of all these prob-lems but rather is a general introduction to the subject.

In addition, the authors demonstrate that the experimental results obtained in studies of the fundamental problems can be used for resolving important and practi-cal questions in medicine and biotechnology—for instance, how to obtain various types of biosensing units for bioanalytical systems.

It can be assumed that readers of this book specializing in various areas of the studies of living systems will find not only parts that can be criticized but also parts that could suggest many more possibilities of applying their own knowledge to explain the peculiarities of the packing and functioning of DNA molecules in bio-logical objects. It can be expected that the input of physicists, chemists, and biolo-gists dealing with the various aspects of living systems and interested, especially, in the biological processes that take place in living cells will bring to this field many new achievements that may be useful for molecular biology, nanobiotechnology, and applied medicine.[*]

Yu. M. Yevdokimov

---

[*] Note regarding this edition: Only minor changes were introduced in the text of the book and to the figures during translation of its contents from Russian to English.

Foreword

The discovery of the spatial structure of the double-stranded DNA molecule is one of the greatest achievements of science. It would not be an exaggeration to say that the DNA double helix is a distinguished symbol of modern biology, as its leading branches in the second half of the twentieth century were molecular biology and molecular genetics.

The deciphering of the replication mechanism of genetic information and the expression of the genome's basic vital processes have not only become the heritage of fundamental science but also determine the development of many areas of medical science, agriculture, and a number of industrial sectors.

DNA has been one of the most intensively studied objects for decades. The results of these researches are described in many thousands of scientific papers. Nevertheless, the simple-looking spatial "image" of complementary polynucleotide chains twisted into a helix still holds many secrets connected with the relationship between the structure and functions of DNA.

The January 2003 issue of *Nature* (Vol. 421), devoted to the fiftieth anniversary of the publication of an article by J. D. Watson and F. H. C. Crick entitled "Molecular structure of nucleic acids," contains a very interesting and remarkable statement by P. Ball: "The double helix is idealized for its aesthetic elegant structure, but the reality of DNA's physical existence is quite different. Most DNA in the cell is compressed into a tangled package that somehow still exposes itself to meticulous gene-regulatory control.... One has the impression of a genome as a book lying open, waiting to be read. However, it is not straightforward. The book is closed up, sealed, and packed away. Moreover, the full story is not merely what is written on the pages; these operations on DNA involve information transfer over many length scales.... We know about molecules; we know about cells and organelles; but the stuff in between is messy and mysterious."

This statement that the functioning of the "DNA package" cannot be understood is based on the iconic model of the Watson and Crick ideal double helix. This also shows that there is a gap in the "transfer" of the properties of isolated DNA molecules to the properties of packed (condensed) DNA form that really exists in a living cell. Moreover, the standard methods of bioinformatics used for the annotation of genomes depicted as linear DNA chains do not reflect the whole set of legitimacies (patterns) that determine the connection between a certain genome and the features of an individual person.

In this respect, the research and the modeling of both physicochemical properties and the biological activity of the condensed state of nucleic acids becomes especially important, considering the fact that this spatial form of DNA molecules can exist in the content of chromosomes.

The idea of collecting into a book the results that demonstrate the existence of a complicated relationship between the peculiarities of the condensed state of nucleic acid molecules and the functioning of these molecules in a cell was presented to the editorial board office of *Technology of Living Systems* journal, since a group of authors had written a number of reviews concerning the issue and published them in this journal in 2007. The reviews have attracted the interest of a worldwide scientific audience.

The book *DNA Liquid-Crystalline Dispersions and Nanoconstructions* is an attempt to summarize the results received by scientists from different countries and laboratories that demonstrate the multiplicity and variability of condensed forms of nucleic acids. Here, in particular, the works of Russian scientists in the field of liquid-crystalline state of nucleic acids are well presented.

The conclusion that follows from the results represented in the book is that the phase exclusion of double-stranded linear or circular DNA molecules or their complexes with polycations induces transition of these molecules into a special liquid-crystalline state that is not at all biologically inert.

Moreover, the fundamental results so far have helped the authors to demonstrate the possible application of nucleic acid liquid crystals for practical purposes, namely, for the creation of nanoconstructions with unique physicochemical properties and sensing units for sensor devices used to detect chemical or biologically active compounds that affect the genetic material of a cell.

The hope is that the book will attract the attention of biologists, physicists, and chemists to this interesting area of science. It will be useful for both students and lecturers, and may contribute to the application of the concept of the condensed DNA for the study of various biological systems and their operational mechanisms.

A. I. Grigoriev

Introduction

Many years before physicists and chemists began to investigate liquid crystals of polymeric macromolecules, biologists had already supposed that some of biomacromolecules in living cells could adopt a specific structural condition that is now called the *liquid-crystalline state*.

The existence of a liquid-crystalline state of biological structures was first mentioned in 1933 at the Faraday Society Meeting, and, later in the 1970s, G. H. Brown and J. J. Wolken initiated a discussion to find out what the connection was between the distinctive properties of biological structures (and biological reactions) and the structure and properties of liquid crystals.

A number of original monographs and translated books were published meanwhile in Russian. The most remarkable of them are *The Physics and Chemistry of Life* (D. Flanagan, Moscow, Russian trans., 1960), *Coacervates and Protoplasm* (K. B. Serebrovskaya, Nauka-Edition, Moscow, 1971), *Periodical Colloid Structures* (I. F. Yefremov, Khimiya-Edition, Leningrad, 1971), *Liquid Crystal State of Polymers* (S. P. Papkov and V. G. Kulichikhin, Khimiya-Edition, Moscow, 1977), *The Physics of Liquid Crystals* (P. G. de Gennes Russian trans., Moscow, 1977), *Comb-Shaped Polymers and Liquid Crystals* (N. A. Plate, V. P. Shibaev, Khimiya-Edition, Moscow, 1980), *Liquid Crystals* (S. Chandrasekhar, Russian trans., Moscow, 1980), *Liquid Crystalline Order in Polymers* (A. Blumstein, Russian trans., Moscow, 1981), and *Liquid Crystals and Biological Structures* (G. H. Brown and J. J. Wolken, Russian trans., Moscow, 1982). The authors of these works touch on the properties of various liquid-crystalline biopolymers to some extent. However, they virtually left aside the issue of the liquid-crystalline state of nucleic acids.

This book was written as an attempt to consider the state of nucleic acid molecules in the cells and trace the possible connections between the peculiarities of this state and the functioning of nucleic acid molecules. We did not have the intention of covering all the issues of the nucleic acid molecules functioning under cell conditions; therefore, in Table I.1 we enumerate in chronological order experimental works of the authors who, in our opinion, made the most significant contribution to the formation of the current concept of the liquid-crystalline state of nucleic acid molecules and their peculiarities. Though this table does not cover all the works in this field, it shows that the works of Russian scientists are not inferior to any of the works of foreign authors, and many of them can be considered pioneers.

Further, there are references in every chapter to special literature that reflect the contributions of scientists from different countries to the solution of various aspects of problems related to the liquid-crystalline state of nucleic acids.

TABLE I.1

Chronology of the Most Significant Works Devoted to the Experimental Research of Peculiarities of Nucleic Acids' Condensed State

Order of Publication	Year	Author(s)	Paper Title	Journal
1.	1961	Robinson K.	Liquid-crystalline structures in polypeptide solutions	*Tetrahedron*, Vol. 13, pp. 219–234
2.	1971	Lerman L.S.	A transition to a compact form of DNA in polymer solutions	*Proc. Natl. Acad. Sci. USA*, Vol. 68, pp. 1886–1890
3.	1972	Evdokimov Yu.M., Platonov A.L., Tikhonenko A.S., Varshavsky Ya.M.	A compact form of double-stranded DNA in solution	*FEBS Lett.*, Vol. 23, pp. 180–184
4.	1973	Akimenko N.M., Dijakova E.B., Evdokimov Yu.M., Frisman E.V., Varshavsky Ya.M.	Viscosimetric study on compact form of DNA in water-salt solutions containing polyethyleneglycol	*FEBS Lett.*, Vol. 38, pp. 61–63
5.	1973	Evdokimov Yu.M., Akimenko N.M., Glukhova N.E., Tikhonenko A.S., Varshavsky Ya.M.	Formation of the compact form of double-stranded DNA in solution in the presence of polyethylene glycol	*Molecular Biology* (Russian edition), Vol. 7, pp. 151–159
6.	1974	Maniatis T., Venable J.H., Lerman L.S.	The structure of ψ DNA	*J. Mol. Biol.*, Vol. 84, pp. 37–64
7.	1976	Gosule L.C., Schellman J.A.	Compact form of DNA induced by spermidine	*Nature*, Vol. 259, pp. 333–335
8.	1977	Evdokimov Yu.M., Pyatigorskaya T.L., Kadikov V.A., Polyvtsev O.F., Doscočil J., Koudelka Ya., Varshavsky Ya.M.	DNA compact form in solution: 12. A formation of a compact form of double-stranded RNA in the presence of poly(ethylene glycol)	*Molecular Biology* (Russian edition), Vol. 11, pp. 891–900
9.	1977	Iizuka I.	Some new findings in the liquid crystals of sodium salt of desoxyribonucleic acid	*Polymer J.*, Vol. 9, pp. 173–180
10.	1977	Sipski M.L., Wagner T.E.	Probing DNA quaternary ordering with circular dichroism spectroscopy studies of equine sperm chromosomal fibers	*Biopolymers*, Vol. 16, pp. 573–582

TABLE I.1 (continued)
Chronology of the Most Significant Works Devoted to the Experimental Research of Peculiarities of Nucleic Acids' Condensed State

Order of Publication	Year	Author(s)	Paper Title	Journal
11.	1978	Skuridin S.G., Kadykov V.A., Shashkov V.S., Evdokimov Yu.M., Varshavsky Ya.M.	The formation of a compact form of DNA in solution induced by interaction with spermidine	*Molecular Biology* (Russian edition), Vol. 12, pp. 413–420
12.	1979	Wilson R.W., Bloomfield V.A.	Counterion-induced condensation of deoxyribonucleic acid	*Biochemistry*, Vol. 18, pp. 2192–2196
13.	1979	Becker M., Misselwitz R., Damaschun H., Damaschun G., Zirver D.	Spermine–DNA complexes build up metastable structures: small-angle X-ray scattering and circular dichroism studies	*Nucl. Acids Res.*, Vol. 7, pp. 1297–1309
14.	1980	Widom J., Baldwin R.L.	Cation-induced toroidal condensation of DNA: studies with $Co^{3+}(NH_3)_6$	*J. Mol. Biol.*, Vol. 144, pp. 431–453
15.	1980	Tinoko I., Bustamante C., Maestre M.F.	The optical activity of nucleic acids and their aggregates	*Annu. Rev. Biophys. Bioeng.*, Vol. 9, pp. 107–146
16.	1981	Potaman V.N., Alexeev D.G., Skuratovskii I.Ya., Rabinovich A.Z., Shlyakhtenko L.S.	Study of DNA films by the CD, X-ray and polarization microscopy techniques	*Nucl. Acids. Res.*, Vol. 9, pp. 55–64
17.	1981	Huey R., Mohr S.C.	Condensed state of nucleic acids: III. $\psi_{(+)}$ and $\psi_{(-)}$ conformational transitions of DNA induced by ethanol and salt	*Biopolymers*, Vol. 20, pp. 2533–2552
18.	1981	Allison S.A., Herr J.C., Schurr J.M.	Structure of viral φ29 DNA condensed by simple triamines: light-scattering and electron microscopy study	*Biopolymers*, Vol. 20, pp. 469–488
19.	1983	Rill R.L., Hilliard P.R., Levy G.C.	Spontaneous ordering of DNA	*J. Biol. Chem.*, Vol. 258, pp. 250–256

continued

TABLE I.1 (continued)
Chronology of the Most Significant Works Devoted to the Experimental Research of Peculiarities of Nucleic Acids' Condensed State

Order of Publication	Year	Author(s)	Paper Title	Journal
20.	1984	Livolant F.	Cholesteric organization of DNA *in vivo* and *in vitro*	*Eur. J. Cell Biol.*, Vol. 33, pp. 300–311
21.	1984	Marx K.A., Ruben G.C.	Studies of DNA organization in hydrated spermidine-condensed DNA toruses and spermidine-DNA fibres	*J. Biomol. Struct. Dynam.*, Vol. 1, pp. 1109–1132
22.	1986	Brandes R., Kearns D.R.	Magnetic ordering of DNA liquid crystals	*Biochemistry*, Vol. 25, pp. 5890–5895
23.	1986	Rill R.L.	Liquid crystalline phases in concentrated aqueous solutions of Na^+ DNA	*Proc. Natl. Acad. Sci. USA*, Vol. 83, pp. 342–346
24.	1986	Livolant F.	Cholesteric liquid crystalline phases given by three helical biological polymers: DNA, PBLG and xanthan: a comparative analysis of their textures	*J. Physique*, Vol. 47, pp. 1605–1616
25.	1986	Livolant F., Bouligand Y.	Liquid crystalline phases given by helical biological polymers (DNA, PBLG and xanthan) Columnar textures	*J. Physique*, Vol. 47, pp. 1813–1827
26.	1987	Livolant F.	Precholesteric liquid crystalline states of DNA	*J. Physique*, Vol. 48, pp. 1051–1066
27.	1987	Strzelecka T.E., Rill R.L.	Solid-state ^{31}P NMR studies of DNA liquid crystalline phases: the isotropic to cholesteric transition	*J. Amer. Chem. Soc.*, Vol. 109, pp. 4513–4518
28.	1987	Baeza I., Gariglio P., Rangel L.M., Chavez P., Cervantes L., Arguello C.	Electron microscopy and biochemical properties of polyamine-compacted DNA	*Biochemistry*, Vol. 26, pp. 6387–6392

TABLE I.1 (continued)
Chronology of the Most Significant Works Devoted to the Experimental Research of Peculiarities of Nucleic Acids' Condensed State

Order of Publication	Year	Author(s)	Paper Title	Journal
29.	1988	Yevdokimov Yu.M., Skuridin S.G., Salyanov V.I.	The liquid-crystalline phases of double-stranded nucleic acids *in vitro* and *in vivo*	*Liq. Crystals*, Vol. 3, pp. 1443–1459
30.	1988	Livolant F., Maestre M.F.	Circular dichroism microscopy of compact forms of DNA and chromatin *in vivo* and *in vitro*: cholesteric liquid-crystalline phases of DNA and single dinoflagellate nuclei	*Biochemistry*, Vol. 27, pp. 3056–3068
31.	1988	Strzelecka T.E., Davidson M.W., Rill R.L.	Multiple liquid crystal phases of DNA at high concentrations	*Nature*, Vol. 331, pp. 457–460
32.	1988	Spada G.P., Brigidi P., Gottarelli G.	The determination of the handedness of cholesteric superhelices formed by DNA fragments	*J. Chem. Soc., Chem. Commun.*, Vol. 14, pp. 953–954
33.	1989	Rill R.L., Livolant F., Aldrich H.C., Davidson M.W.	Electron microscopy of liquid crystalline DNA: direct evidence for cholesteric-like organization of DNA in dinoflagellate chromosomes	*Chromosoma* (Berl.), Vol. 98, pp. 280–286
34.	1989	Livolant F.	Lines in liquid crystalline phases of biopolymers	*J. Phys. France*, Vol. 50, pp. 1729–1741
35.	1989	Livolant F., Levelut A.M., Doucet J., Benoit J.P.	The highly concentrated liquid-crystalline phase of DNA is columnar hexagonal	*Nature*, Vol. 339, pp. 724–726
36.	1989	Torbet J., DiCapua E.	Supercoiled DNA is interwound in liquid crystalline solutions	*EMBO J.*, Vol. 8, pp. 4351–4356
37.	1990	Van Winkle D.H., Davidson M.W., Wan-Xu Chen, Rill R.L.	Cholesteric helical pitch of near persistence length DNA	*Macromolecules*, Vol. 23, pp. 4140–4148

TABLE I.1 (continued)

Chronology of the Most Significant Works Devoted to the Experimental Research of Peculiarities of Nucleic Acids' Condensed State

Order of Publication	Year	Author(s)	Paper Title	Journal
38.	1990	Strzelecka T.E., Rill R.L.	Phase transitions of concentrated DNA solutions in low concentrations of 1:1 supporting electrolyte	*Biopolymers,* Vol. 30, pp. 57–71
39.	1991	Salyanov V.I., Dembo A.T., Yevdokimov Yu.M.	Liquid-crystalline phases of circular superhelical DNA and their modification by the action of nuclease enzymes	*Liq. Crystals,* Vol. 9, pp. 229–238
40.	1991	Livolant F.	Supramolecular organization of double-stranded DNA molecules in the columnar hexagonal liquid crystalline phase: an electron microscopic analysis using freeze-fracture methods	*J. Mol. Biol.,* Vol. 218, pp. 165–181
41.	1991	Rill R.L., Strzelecka T.E., Davidson M.W., van Winkle D.H.	Ordered phases in concentrated DNA solutions	*Physica,* Vol. A 176, pp. 87–116
42.	1991	Livolant F.	Ordered phases of DNA *in vitro* and *in vivo*	*Physica,* Vol. A 176, pp. 117–137
43.	1991	Bloomfield V.	Condensation of DNA by multivalent cations: considerations on mechanism	*Biopolymers,* Vol. 31, pp. 1471–1481
44.	1991	Gottarelli G., Spada G.P., Miranda de Morais M.	The effect of ethidium bromide on the liquid crystalline phases of aqueous DNA	*Chirality,* Vol. 3, pp. 227–232
45.	1991	Baeza I., Ibanez M., Wong C., Chavez P., Gariglio P., Oro J.	Possible prebiotic significance of polyamines in the condensation, protection, encapsulation, and biological properties of DNA	*Orig. Life Evol. Biosph.,* Vol. 21, pp. 225–242
46.	1992	Yevdokimov Yu.M., Skuridin S.G., Lortkipanidze G.B.	Liquid-crystalline dispersions of nucleic acids	*Liq. Crystals,* Vol. 12, pp. 1–16

TABLE I.1 (continued)
Chronology of the Most Significant Works Devoted to the Experimental
Research of Peculiarities of Nucleic Acids' Condensed State

Order of Publication	Year	Author(s)	Paper Title	Journal
47.	1992	Ghirlando R., Wachtel E.J., Arad T., Minsky A.	DNA packaging induced by micellar aggregates: a novel *in vitro* DNA condensation system	*Biochemistry*, Vol. 31, pp. 7110–7119
48.	1992	Rau D.C., Parsegian V.A.	Direct measurement of the intermolecular forces between counterion-condensed DNA double helices	*Biophys. J.*, Vol. 61, pp. 246–259
49.	1992	Durand D., Doucet J., Livolant F.	A study of the structure of highly concentrated phases of DNA by x-ray diffraction	*J. Phys.* (France), Vol. 2, pp. 1769–1783
50.	1993	Leforestier A., Livolant F.	Supramolecular ordering of DNA in the cholesteric liquid crystalline phase: an ultrastructural study	*Biophys. J.*, Vol. 65, pp. 56–72
51.	1994	Leforestier A., Livolant F.	DNA liquid crystalline blue phases: electron microscopy evidence and biological implications	*Liq. Crystals*, Vol. 17, pp. 651–658
52.	1994	Merchant K., Rill R.L.	Isotropic to anisotropic phase transition of extremely long DNA in an aqueous saline solution	*Macromolecules*, Vol. 27, pp. 2365–2370
53.	1994	Gottarelli G., Spada G.P.	Application of CD to the study of some cholesteric mesophases (In: *Circular Dichroism: Principles and Applications*)	New York: VCH, pp. 105–119
54.	1994	Robinov C., Kellenberger E.	The bacterial nucleoid revisited	*Microbiol. Rev.*, Vol. 58, pp. 211–232
55.	1994	Sikorav J.-L., Pelta J., Livolant F.	A liquid crystalline phase in spermidine-condensed DNA	*Biophys. J.*, Vol. 67, pp. 1387–1392

continued

TABLE I.1 (continued)
Chronology of the Most Significant Works Devoted to the Experimental Research of Peculiarities of Nucleic Acids' Condensed State

Order of Publication	Year	Author(s)	Paper Title	Journal
56.	1994	Reich Z., Wachtel E.J., Minsky A.	Liquid-crystalline mesophases of plasmid DNA in bacteria	*Science,* Vol. 264, pp. 1460–1463
57.	1994	Reich Z., Levin-Zaidman S., Gutman S.B., Arad T., Minsky A.	Supercoiling-regulated liquid-crystalline packaging of topologically constrained, nucleosome-free DNA molecules	*Biochemistry,* Vol. 33, pp. 14177–14184
58.	1996	Yevdokimov Yu.M., Salyanov V.I., Gedig E., Spener F.	Formation of polymeric chelate bridges between double-stranded DNA molecules fixed in spatial structure of liquid-crystalline dispersions	*FEBS Lett.,* Vol. 392, pp. 269–273
59.	1996	Livolant F., Leforestier A.	Condensed phases of DNA: structures and phase transitions	*Prog. Polym. Sci.,* Vol. 21, pp. 1115–1164
60.	1996	Pelta J., Livolant F., Sikorav J.-L.	DNA aggregation induced by polyamines and cobalthexamine	*J. Biol. Chem.,* Vol. 271, pp. 5656–5662
61.	1996	Pelta J., Durand D., Doucet J., Livolant F.	DNA mesophases induced by spermidine: structural properties and biological implications	*Biophys. J.,* Vol. 71, pp. 48–63
62.	1996	Leforestier A., Richter K., Livolant F., Dubochet J.	Comparison of slam-freezing and high-pressure freezing effects on the DNA cholesteric liquid crystalline structure	*J. Microsc.,* Vol. 184 (Pt. 1), pp. 4–13
63.	1997	Leforestier A., Livolant F.	Liquid crystalline ordering of nucleosome core particles under macromolecular crowding conditions: evidence for a discotic columnar hexagonal phase	*Biophys. J.,* Vol. 73, pp. 1771–1776

TABLE I.1 (continued)
Chronology of the Most Significant Works Devoted to the Experimental Research of Peculiarities of Nucleic Acids' Condensed State

Order of Publication	Year	Author(s)	Paper Title	Journal
64.	1998	Kassapidou K., Jesse W., van Dijk J.F., van der Maarel J.R.	Liquid crystal formation in DNA fragment solutions	*Biopolymers*, Vol. 46, pp. 31–37
65.	1998	Yevdokimov Yu.M., Salyanov V.I., Lortkipanidze G.B., Gedig E., Spener F., Palumbo M.	Sensing biological effectors through the response of bridged nucleic acids and polynucleotides fixed in liquid-crystalline dispersions	*Biosens. Bioelectron.*, Vol. 13, pp. 279–291
66.	1998	Lin Z., Wang C., Feng X., Liu M., Li J., Bai C.	The observation of the local ordering characteristics of spermidine-condensed DNA: atomic force microscopy and polarization microscopy studies	*Nucl. Acids Res.*, Vol. 26, pp. 3228–3234
67.	1999	Saminathan M., Antony M., Shirahata A., Sigal L.H., Thomas T., Thomas T.J.	Ionic and structural specificity effects of natural and synthetic polyamines on the aggregation and resolubilization of single-, double-, and triple-stranded DNA	*Biochemistry*, Vol. 38, pp. 3821–3830
68.	1999	Wolf S.G., Frenkiel D., Arad T., Finkel S.E., Kolter R., Minsky A.	DNA protection by stress-induced biocrystallization	*Nature*, Vol. 400, pp. 83–85
69.	2000	Yevdokimov Yu.M., Salyanov V.I., Skuridin S.G., Dembo A.T., Platonov Yu.V., Il·ina A.V., Varlamov V.P.	Complexes of the (dsDNA–Chitosan) form cholesteric liquid-crystalline dispersions	*Doklady Academii Nauk* (formerly *Doklady of the USSR Academy of Sciences*), Vol. 374, pp. 696–698 (Russian edition)
70.	2001	Yevdokimov Yu.M., Salyanov V.I., Zakharov M.A.	A novel type of microscopic size chip based on double-stranded nucleic acids	*Lab on a Chip*, Vol. 1, pp. 35–41

continued

TABLE I.1 (continued)
Chronology of the Most Significant Works Devoted to the Experimental Research of Peculiarities of Nucleic Acids' Condensed State

Order of Publication	Year	Author(s)	Paper Title	Journal
71.	2001	Vijayanathan V., Thomas T., Shirahata A., Thomas T.J.	DNA condensation by polyamines: a laser light scattering study of structural effects	*Biochemistry*, Vol. 40, pp. 13644–13651
72.	2001	Bouligand Y., Norris V.	Chromosome separation and segregation in dinoflagellates and bacteria may depend on liquid crystalline states	*Biochimie*, Vol. 83, pp. 187–192
73.	2001	Sartori B.N., Senn A., Leforestier A., Livolant F., Dubochet J.	DNA in human and stallion spermatozoa forms local hexagonal packing with twist and many defects	*J. Struct. Biol.*, Vol. 134, pp. 76–81
74.	2001	Piraccini S., Gottarelli G., Mariani P., Spada G.P.	The chirality of the cholesteric phases of DNA and G-wires: its connection to their molecular structures	*Chirality*, Vol. 6, pp. 3249–3253
75.	2002	Saminathan M., Thomas T., Shirahata A., Pillai C.K., Thomas T.J.	Polyamine structural effects on the induction and stabilization of liquid crystalline DNA: potential applications to DNA packaging, gene therapy and polyamine therapeutics	*Nucl. Acids Res.*, Vol. 30, pp. 3722–3731
76.	2002	Kato T.	Self-assembly of phase-segregated liquid crystal structures	*Science*, Vol. 295, pp. 2414–2418
77.	2002	Zakharova S.S., Jesse W., Backendorf C., van der Maarel J.R.	Liquid crystal formation in supercoiled DNA solutions	*Biophys. J.*, Vol. 83, pp. 1119–1129
78.	2002	Goobes R., Cohen O., Minsky A.	Unique condensation patterns of triplex DNA: physical aspects and physiological implications	*Nucl. Acids Res.*, Vol. 30, pp. 2154–2161

TABLE I.1 (continued)

Chronology of the Most Significant Works Devoted to the Experimental Research of Peculiarities of Nucleic Acids' Condensed State

Order of Publication	Year	Author(s)	Paper Title	Journal
79.	2004	Vijayanathan V., Thomas T., Antony M., Shirahata A., Thomas T.J.	Formation of DNA nanoparticles in the presence of novel polyamine analogues: a laser light scattering and atomic force microscopic study	*Nucl. Acids Res.*, Vol. 32, pp. 127–134
80.	2004	Englander J., Klein E., Brumfeld V., Sharma A.K., Doherty A.J., Minsky A.	DNA toroids: framework for DNA repair in Deinococcus radiodurans and in germinating bacterial spores	*J. Bacteriol.*, Vol. 186, pp. 5973–5977
81.	2004	Gottarelli G., Spada G.P.	The stepwise supramolecular organization of guanosine derivatives	*Chem. Rec.*, Vol. 4, pp. 39–49
82.	2005	Yevdokimov Yu.M., Skuridin S.G., Nechipurenko Yu.D., Zakharov M.A., Salyanov V.I., Kurnosov A.A., Kuznetsov V.D., Nikifirov V.N.	Nanoconstructions based on double-stranded nucleic acids	*Int. J. Biol. Macromol.*, Vol. 36, pp. 103–115
83.	2005	Yevdokimov Yu.M., Salyanov V.I., Kondrashina O.V., Borshevsky V.I., Semenov S.V., Gasanov A.A., Reshetov I.V., Kuznetsov V.D., Nikiforov V.N., Akulinichev S.V., Mordovskoi M.V., Potashev S.I., Skorkin V.M.	Particles of liquid-crystalline dispersions formed by (nucleic acid–rare earth element) complexes as a potential platform for neutron capture therapy	*Int. J. Biol. Macromol.*, Vol. 37, pp. 165–173

continued

TABLE I.1 (continued)

Chronology of the Most Significant Works Devoted to the Experimental Research of Peculiarities of Nucleic Acids' Condensed State

Order of Publication	Year	Author(s)	Paper Title	Journal
84.	2005	Wong J.T., Kwok A.C.	Proliferation of dinoflagellates: blooming or bleaching	*Bioassays,* Vol. 27, pp. 730–740
85.	2006	Sundaresan N., Thomas T., Thomas T.J., Pillai C.K.	Lithium ion induced stabilization of the liquid crystalline DNA	*Macromol. Biosci.,* Vol. 6, pp. 27–32
86.	2006	Smalyukh I.I., Zribi O.V., Butler J.C., Lavrentovich O.D., Wong G.C.	Structure and dynamics of liquid crystalline pattern formation in drying droplets of DNA	*Phys. Rev. Lett.,* Vol. 96, pp. 177801
87.	2006	Safinya C.R., Ewert K., Ahmand A., Evans H.V., Rativ U., Needleman D.J., Lin A.J., Slack N.L., George C., Samuel C.	Cationic liposome-DNA complexes: from liquid crystal science to gene delivery applications	*Philos. Transact. A Math. Phys. Eng. Sci.,* Vol. 364, pp. 2573–2596
88.	2006	Livolant F., Mangenot S., Leforestier A., Dertin A., Frutos M., Raspaud E., Durand D.	Are liquid crystalline properties of nucleosomes involved in chromosome structure and dynamics?	*Philos. Transact. A Math. Phys. Eng. Sci.,* Vol. 364, pp. 2615–2633

PECULIARITIES OF DNA MOLECULES IN SOME BIOLOGICAL OBJECTS

The results obtained in different laboratories up to the present time allow us to enumerate the distinctive properties of the DNA molecules and their complexes with some proteins inside biological objects.

To estimate the amount of DNA required to provide the functioning of different biological objects, let us project the length of their DNAs. According to evaluations [1], the amount of DNA necessary to describe an organism (its genome) is enormous. For instance, even the DNA of a simple bacterium contains 10^6 nucleotide pairs. Eukaryotes, such as mammals and plants, have about 10^{10} nucleotide pairs. A human genome contains $4 \cdot 10^9$ nucleotide pairs. This genome is represented by two copies in each somatic cell so that each cell contains $8 \cdot 10^9$ nucleotide pairs. If all of this DNA were to be laid out in a line, considering that the length of one pair of bases in a B-form of the DNA is 0.34 nm, it would extend about 2.7 m. Given that there are around 10^{13} cells in a human body, there are $3 \cdot 10^{10}$ km of DNA in any person. However, about 90% of this DNA is latent (silent?), and its functions are still uncertain. DNA is spread around every cell in the form of chromosomes. A human cell has 46 chromosomes. Each of them contains approximately $1.7 \cdot 10^8$ nucleotide pairs and is about 6 cm long.

In the smallest human chromosome, the length of a DNA 14 mm long is condensed into a chromosome about 2 μm, with a packing ratio of 7000. Hence, the packing ratio for DNA is enormous.

Therefore, the first question of interest is how to evaluate the local DNA concentration as weight of the DNA per a unit of cell volume for different biological objects (which is also called the *packing density*).

In general, the estimation of the packing density of DNA in different biological objects is nontrivial. In exponentially growing cells of *E. coli*, the DNA concentration reaches 4% of the cell solid material, which is approximately $15 \cdot 10^{-15}$ g [2]. If the DNA is homogeneously dispersed in the cell with a volume of $1.4 \cdot 10^{-12}$ cm³, the packing density of the DNA must be about 10 mg/mL. On average, there are less than 3 genomes in an *E. coli* cell, which conforms to cytological findings (2–4 genomes per cell). Considering DNA concentration in nucleotides and the fact that a volume of a nucleotide is 30–70% of a cell's volume, we can obtain a quantity of 20 to 50 mg/mL.

For the interphase nuclei of hepatocytes, an average DNA concentration is 20–40 mg/mL, considering that there is $6–12 \cdot 10^{-12}$ g of DNA in each cell, and the nucleus volume is $2.8 \cdot 10^{-10}$ cm³. The possible difference of DNA concentrations in heterochromatin and euchromatin is set aside. It is also interesting that the DNA concentration in *E. coli* is comparable with concentration typical of the hepatocytes' interphase nuclei.

DNA can be packed even more densely. For some objects such as bacteriophages, the packing density can be calculated quite accurately. The calculations of the DNA packing density of bacterial viruses are based on the comparison of the bacteriophages' head volume and chemically detected DNA concentration in these objects [3]. The results of such calculations vary from 800 mg/mL for the T4 bacteriophage (calculation for this bacteriophage is the most accurate) to 350 mg/mL for the T3 bacteriophage. Calculation of the local DNA concentration in metaphase chromosomes is much more complicated, and there is hope that it can be estimated by atomic force microscopy.

Some of the data [4] on DNA packing density are given in Table I.2.

That the DNA local packing density in typical eukaryotic cells of animals and plants constantly changes because the diffuse chromatin "condenses" to the

TABLE I.2
DNA Packing Density in Various Objects

Object	Packing Density, mg/mL	
	Calculated	Measured
Bacteriophage T4 head	800	1000–2000
E. coli nucleoid	20–50	100 ± 50
Dinoflagellate chromosome	—	220 ± 80
CV-1 cell metaphase chromosome	—	300–500

metaphase chromosomes is also worth mentioning. It is important that, in the "condensed" DNA state, negative charges of phosphate groups are neutralized by the formation of ion pairs with counterions of different kinds (histones, etc.).

Therefore, the foregoing data show that high DNA packing density (up to several hundreds mg/mL [!]) is an obligatory property of biological objects. It is also typical that the local density changes during various biological processes. Moreover, the DNA molecules in biological objects are not only concentrated but also highly ordered. That is proved by the characteristic maximum (peak) on the small-angle x-ray scattering curves of such objects.

There is another fact that has been attracting the attention of researchers in recent years [5]. The biological medium in which the genome is functioning does not represent by itself an ideal solution. First, the biological milieu contains a number of macromolecules. The concentration of each type of macromolecules is not very high, but taken together, they represent a significant part of the total volume of the medium. Second, the biological medium is "structurated" due to the presence of grids formed by such objects as actin, microtubules, and the areas near the membrane, especially in the case of eukaryotic cells. Therefore, living systems are functioning under conditions of a very high concentration of background macromolecules; that is why a biological medium can be designated as "crowded." Even without any direct interaction between background and investigated biopolymers (test macromolecules), macromolecular crowding can exert significant influence on the tertiary and the quaternary structure of biopolymeric molecules, their aggregate state (even trigger the formation of a particular phase), and the kinetics of the reactions they are involved in.

This means that the divergence between the properties of media typical of *in vivo* and *in vitro* can cause the wrong interpretation of intracellular events, such as DNA compaction, its interaction with proteins, and the rate of biological processes if the properties of the crowded biological medium are taken into account.

It is obvious that the clarification of the mechanism of the genome functioning under the foregoing conditions is a complicated experimental problem. The methods of analysis that would provide a definite solution to this problem have not been developed yet.

When we are able to describe more accurately the state of DNA molecules in biological objects and the properties of biological medium in which the genome functions, we will be able to resolve two questions of interest:

1. Can the DNA molecules be packed *in vitro* so that they match the size of biological objects (nucleus, bacteriophage head, viruses, etc.), which is about 10^{-5} cm?
2. Will the DNA condensed *in vitro* under conditions modeling the biological environment sustain biological activity?

It is important that more or less accurate answers to these questions can be obtained under experimental conditions by simulation of the DNA condensation process in the laboratory. The simulation makes it possible to reduce the obvious complexity of the cell architecture to the relative simplicity of self-assembling DNA structures.

Then, the only question left will be whether the structures formed under simulation conditions take part in realization of the cellular functions.

Interest in biopolymeric liquid crystals is heightened by a number of factors. First, many biopolymeric liquid-crystalline structures are known [6,7], and the transition between different liquid-crystalline forms is triggered by minor changes in the physical and chemical properties of the solvent. Second, liquid-crystalline structures can spread to a distance comparable with the size of a cell [8]. Third, a one (or two)-dimensional order of the liquid-crystalline phase provides quick diffusion of biologically significant compounds throughout [9]. Fourth, after liquid-crystalline packing, the native secondary DNA-structure is retained, and the efficiency of the DNA molecules' rapid renaturation is even increased [8]. The transition to the liquid-crystal state is also interesting in terms of biological morphogenesis, especially for the primitive forms of life where nucleic acids are the principal components of the replicating organisms. Finally, the interest in liquid crystals is caused by a recently discovered possibility of their use in practice, particularly in nanotechnology, to create nanoconstructions with unique physical and chemical properties. Such constructions can be used as carriers of genetic makeup and biologically significant compounds fixed to the DNA molecules. Liquid crystals can also be used as biosensing units for sensory devices. This means that the properties of the liquid crystals formed by the phase exclusion of double-stranded DNA have become the object of intensive theoretical and experimental research [10,11].

There are two complementary areas in the investigation of DNA liquid crystals. The first one is the research of intramolecular condensation of a single high-molar-mass DNA molecule. The second one is the study of intermolecular condensation of the low-molar-mass DNA molecules. The most significant results of the research of condensation of natural nucleic acid and some synthetic polynucleotide molecules under different conditions will be considered in this book.

REFERENCES

1. van Holde, K. Structure of DNA, chromatin and chromosomes. In Internatl. Interdisciplinary Workshop: Structure and function of DNA. A physical approach. Sept. 30, 1996. Abbay du Sainte-Odile, Alsace (France).
2. Kellenberger, E., Carleman, E., Sechaud, J. et al. Consideration on the condensation and the degree of compactness in non-eukariotic DNA-containing plasmas. In *Bacterial Chromatin*, Ed. by C.O. Gualezi and C.L. Pon. Berlin: Springer Verlag, 1986, p. 11–25.
3. Kellenberger, E. About the organization of condensed and decondensed non-eukaryotic DNA and the concept of vegetative DNA (a critical review). *Biophys. Chem.*, 1988, vol. 29, p. 51–62.
4. Kellenberger, E. and Arnold-Schulz-Gahmen, B. Chromatins of low-protein content: Special features of their compaction and condensation. *FEMS Microbiol. Lett.*, 1992, vol. 100, p. 361–370.
5. Zimmerman, S.B. and Minton, A.P. Macromolecular crowding: biophysical, and physiological consequences. *Annu. Rev. Biomol. Struct.*, 1993, vol. 22, p. 27–65.

6. Brown, G.H. and Wolken, J.J. *Liquid Crystals and Biological Structures*. New York: Academic Press, 1979. p. 200.
7. Rizvi, T.Z. Liquid crystalline biopolymers: A new arena for liquid crystal research. *J. Mol. Liquids*, 2003, vol. 106, p. 43–45.
8. Sikorav, J.-L. and Church, G.M. Complementary recognition in condensed DNA: Accelerated DNA denaturation. *J. Mol. Biol.*, 1991, vol. 222, p. 1085–1108.
9. Holyst, R., Blazejczyk, M., Burdzy, K. et al. Reduction of dimensionality in a diffusion search process and kinetics of gene expression. *Physica A*, 2000, vol. 277, p. 71–82.
10. Bouligand, Y. and Norric, V. Chromosome separation and segregation in dinoflagellates and bacteria may depend on liquid crystalline states. *Biochimie*, 2001, vol. 83, p. 187–192.
11. Yevdokimov, Yu.M. Liquid-crystalline forms of nucleic acids. *Herald of the Russian Acad. Sci. (Russian Edition)*, 2003, vol. 73, p. 712–721.

Outline

This book is aimed not only at understanding the information presently available on the condensation of various forms of DNA but also at describing practical applications of the peculiar properties of the liquid-crystalline state of nucleic acids.

The book contains three inherently connected sections:

Section I. The liquid-crystalline state of DNA
Section II. DNA liquid-crystalline forms and their biological activity
Section III. DNA liquid-crystalline dispersions in nanotechnology and biosensorics

In Section I, the main methods used for the condensation of linear high- and low-molecular-mass DNA, and their complexes with polycations as well as circular DNA are considered, and the properties of the structures they adopt are discussed.

According to the results presented in Section II, two essentially different types of condensation of double-stranded (ds) nucleic acids (NA) result in the formation of their liquid-crystalline dispersions (LCD). The first type consists in an «entropy» condensation, that is, a process of phase "exclusion" of rigid, linear ds NA from water–salt solutions added with water-soluble polymers, in particular, PEG. As a result of this process, low-molecular-mass NA tends to be ordered and can form LCD particles at high concentration of added polymer. The theoretically estimated size of an LCD particle is close to 500 nm, each particle containing about 10^4 NA molecules. The infringement of the "critical" conditions for LCD particle formation is accompanied by particle collapse and NA transition from a condensed to an isotropic state. Some characteristic features should be mentioned. First, the polymer does not participate in the composition of the forming dispersion particles. Second, for particles of ds NA LCD, a high (from 160 up to 600 mg/mL!) local NA concentration is typical. Third, the distance (d) between adjacent NA molecules in LCD particles can be adjusted between 2.5 and 5.0 nm by changing the osmotic pressure of the solution. Fourth, NA molecules, because of their inherent geometrical and optical anisotropies, tend to generate mainly cholesteric LCDs (CLCD), the formation of which is accompanied by the appearance of an intense (abnormal) band in the CD spectrum in the absorption region of NA chromophores (nitrogen bases). (In Section I, the circular dichroism [CD] spectra of DNA dispersions, in the form of CLCD particles in a water–salt solution, have been calculated theoretically. Calculated curves have been fitted to the experimental CD spectra generated by the LCD particles under various conditions. The amplitude and the sign of the intense [abnormal] band in the absorption region of NA nitrogen bases vary with the size of particles, the pitch, and the twist sense of the cholesteric structure. The CD spectra of particles of the complexes formed by DNA and intercalating compounds has been similarly described.) Fifth, the packaging of NA molecules in the CLCD particles is not only ordered but also "fluid," so that the NA molecules in each quasi-nematic layer are capable both of rotating about their axis and shifting laterally. Sixth, because of the fluid character of NA packaging in quasi-nematic layers, the different compounds can easily diffuse inside CLCD particles. Finally, the CLCD particles formed in water–polymeric solutions preserve their physicochemical properties under a rather wide interval of conditions.

The second method of ordering low-molecular-mass, rigid, linear ds NA molecules is by "enthalpy" condensation, that is, their phase exclusion from water–salt solutions results from the attraction between NA molecules, the negative charges of phosphate groups being neutralized by positively charged counterions, generally polycations. The relevant stage in the process of enthalpy condensation is the interaction between approaching adjacent molecules of (NA–polycation) complexes.

The biodegradable polymer chitosan (a copolymer, consisting of β-[1→4]-2-amino-2-deoxy-d-glucopyranose and β-(1→4)-2-acetamido-2-deoxy-d-glucopyranose residues) attracts great attention as a polycation exhibiting a wide range of properties. Indeed, the chemical and spatial structures of chitosan enable it to form stable complexes with various compounds. Hence, chitosan is considered in Section I as an example of a polycation capable of interacting with NA. Ds–NA molecules are shown to interact with chitosan to form multiple types of liquid-crystalline dispersions under appropriate conditions (chitosan molecular mass, content of amino groups, distance between amino groups, solution pH, etc.). The dispersions formed differ in their spatial structures, and hence, in the sense and magnitude of abnormal optical activity. The physicochemical properties of NA–chitosan particles are investigated, and the accessibility of these NA LCD with respect to enzyme and drug action is tested. The multiplicity of liquid-crystalline forms of NA–chitosan complexes is explained by the influence of the dipole distribution over the surface NA molecules in the sense of the spatial twist of the cholesteric liquid-crystalline dispersions resulting from these complexes.

It is shown in Section I that the features of enthalpy condensation depend on the surface properties of NA molecules, on the spatial structure of polycation molecules, and on the character of allocation of positive charges. The features of LCD formed by (NA–chitosan) complexes allow an outline of the peculiarities of these dispersions.

1. The phase exclusion of (NA–polycation) complexes and the formation of dispersions occur at "critical" polycation concentrations in solution (that is comparable to the initial NA concentration).
2. The polycation always participates in the dispersion particles.
3. The electrostatic (NA–polycation) energy of interaction provides a fixed distance (d) between adjacent molecules in the dispersion particles. As a rule, this distance is equal to 2.6–2.9 nm, which is characteristic of a hexagonal packaging of NA molecules.
4. Only in some ("lucky") cases is it possible to generate the cholesteric mode of packaging of (NA–polycation) complexes in the dispersion particles, which results in the appearance of an abnormal band in the CD spectrum in the region of NA chromophores' (nitrogen bases') absorption. (In Section I, an attempt is made to provide a theoretical description of the polymorphism of the liquid-crystalline structures formed by [NA–polycation] complexes and the consequent different shapes of their CD spectra.)

Despite noticeable differences between these types of ds NA molecule condensation, it is necessary to underline that the forming particles of NA dispersions share properties of both crystalline bodies and fluids. A high local concentration of NA in combination with an ordered location of adjacent NA molecules in the LCD particles does not restrict diffusion of chemically or biologically important compounds inside the dispersion particles, which provides not only a prompt permeation of these compounds into particles but also a high speed of interaction with ds NA molecules.

Section II is dedicated to a comparison between the state and reactivity of ds NA molecules fixed spatially in the liquid-crystalline dispersions and of the same molecules under intracellular conditions.

The numerous results of various researchers allow one to identify the biologically significant properties of the liquid-crystalline NA dispersions:

1. First, a high-level structural ordering encompassing considerable distances, compatible with a multiplicity of liquid-crystalline forms and opportunities for transitions from one ordered structure to the other
2. A preferred cholesteric liquid-crystalline mode of packaging anisotropic, anisometric, and ds NA molecules (or complexes with polycations) under appropriate conditions
3. A diffusive mobility of NA molecules
4. The ability of ds NA molecules to easily respond to factors such as temperature, properties and composition of medium, availability for enzymes, etc.
5. The evolution of the type of packaging as a result of the chemical or biological "action" initial NA molecules are undergoing
6. The appearance of new reactive sites in circular DNA molecules packaged in the dispersion particles, connected with modifications in the "work" mode of some enzymes
7. In general, conservation of biological potency of NA molecules packaged in a liquid-crystalline state

The listing of these properties of liquid-crystalline NA dispersions under model conditions attracts the reader's attention to the fact that structural ordering, structural polymorphism, and adjustments of properties comprise the idiosyncrasies of living cells.

Indeed, analysis of the fine details and operational features of both genome as a whole and separate genes, characteristic of the normal and malignant development of living cells and of their differentiation, which is very relevant to any living system, suggests that it is necessary to take into account additionally the properties of the liquid-crystalline state of ds NA possibly occurring in living cells. Therefore, carrying out further investigations in this scientific direction is an imperative and interesting issue.

In addition, the discovery of the fundamental principles underlying the formation of NA LCD particles opens the gate for the operational use of these principles in the areas of nanotechnology and biosensorics (Section III).

The results presented in Section III demonstrate that the development of nanotechnology will be connected to application of biological molecules as a background for the creation of nanostructures with different properties. Indeed, the combination of the chemical reactivity of biopolymers and their tendency to form hierarchic nanostructures, as well as the opportunity to commercialize biopolymers, makes biological molecules suitable for important applications in nanotechnology. Therefore, their use in generating artificial nanostructures based on principles provided by nature appears to be quite logical and productive. Moreover, progress in chemical synthesis and biotechnology, which allows one to combine building blocks of different origin, that is, to design "chimeric" molecules containing, for instance, amino acids and synthetic organic chains, opens up fantastic scenarios for designing nanomaterials and nanostructures that in principle do not exist in nature. It is anticipated that, as nanobiotechnology advances, biopolymers will make a transition from the world of biology to that of technology. It means that, in the next few years, we should expect the rise of a new science, namely, nanobiotechnology. Nanodesign based on ds NAs, that is, rationally directed formation of three-dimensional architectures (nanostructures, nanoconstructions, or NaCs) with tailored properties, the "building blocks" of which are ds NA molecules or their complexes with biologically active compounds, is a topic of high current theoretical and experimental interest. The ds NA nanoconstructions are of significant practical importance at least from two points of view. First, nanoconstructions with adjustable spatial parameters can be used in bioelectronics and biosensorics; second, NA nanoconstructions can be used for the controlled delivery of genotoxicants or relevant biologically active compounds into animal cells.

Several strategies are described for designing NA nanostructures, and most of them could be named conventionally as a consecutive or "step-by-step" design based on successive modification of original NA molecules.

Our strategy of creating NaCs containing ds NA molecules, considered in Section III as well, differs in principle from all variants of the step-by-step strategy because our approach makes use of the LCD, rather than single NA molecules, resulting from the phase exclusion of ds NA from aqueous polymer solutions. The main attention is devoted to the formation and properties of nanoconstructions based on ds DNA, fixed in the spatial structure of CLCD dispersions and cross-linked by artificial nanobridges consisting of alternating copper ions and daunomycin molecules. In contrast to the spatial arrangement of the initial NA LCD, the structure of the resulting nanoconstruction is no longer "liquid-crystalline"; rather, it is a "rigid," crystal-like, three-dimensional structure. One can add that NA molecules complexed with polycations, fixed in a spatial LCD structure, and obtained as a result of an enthalpy condensation, represent a totally different way of building nanoconstructions having unique properties. As an illustration of this new approach, the formation of nanoconstructions based on (DNA–chitosan) complexes is considered.

NA-nanoconstructions, independent of the manner of their formation, can find practical application in different fields of research and technology:

1. Nanoconstructions at high DNA and "guest" concentration may be used as "carriers" for genetic material or as a "reservoir" for various biologically active compounds embedded in the composition of these structures.

2. Nanoconstructions based on CLCDs of ds NA or (DNA–chitosan) complexes can be used as sensing units for optical biosensors, which makes it possible to determine the presence of relevant biologically active or chemical compounds in physiological fluids.

3. Nanoconstructions with controlled physicochemical properties immobilized in synthetic polymeric films (hydrogels) can be used in technical applications (for example, as molecular sieves or optical filters).

Hence, nucleic acids in nanotechnology, despite the existence of miscellaneous strategies of nanodesign based on these molecules, can be applied to build nanoconstructions having predefined properties, which are useful from the practical point of view.

As an example of the practical application of NA-nanoconstructions in such a branch of nanotechnology as biosensorics, the features of the various approaches to create sensing units based on NA molecules (or NA complexes with different compounds), as well as NA molecules in LCD particles, are considered in Section III. Here, at the beginning, the main principles used for creating biosensing units based on single- and double-stranded NA are reviewed. A broad range of analytical abilities of these units is illustrated. Multifunctional biosensing units are capable of detecting various chemically and biologically active compounds (antitumor drugs, genotoxicans, etc.) influencing the DNA secondary structure. Examples of operational use of such biosensing units for the analysis of biologically relevant compounds are given.

An interdisciplinary approach, which takes into account achievements in the field of building ds DNA–based nanoconstructions and the current successes in the field of physical chemistry of polymers, microelectronics, and optics, open a gate for the production of conceptually new types of biosensors. The manufacture of "rigid" biosensing units based on DNA nanoconstructions having high operational stability reflects one of the recent achievements in this branch of biosensorics. The combination of a portable dichrometer with a panel of DNA biosensing units of various types represents the background for new bioanalytical systems with important applications in medicine, biotechnology, and so forth. The NA molecules ordered in LCD particles can be utilized as building blocks in nanobiotechnology, and the ability of the dispersion particles to respond quickly to the action of various chemical/physical factors allows one to utilize these particles as effective sensing units in biosensor devices.

The data presented in this book show that double-stranded nucleic acids, spatially organized in a liquid-crystalline structure, represent an important polyfunctional tool for molecular biology and nanobiotechnology. The possibility of programmed and controlled variations in the properties of these molecules and, hence, in the characteristics of their liquid-crystalline dispersions, provides wide options for the formation of biologically active three-dimensional structures with unique, widely applicable properties.

Section I

The Liquid-Crystalline
State of the DNA

1 The Condensed State of the High-Molecular-Mass Double-Stranded DNA

1.1 THE DNA CONDENSATION AND AGGREGATION

First of all, it would be helpful to define several terms that will be used throughout this book. Following Reference [1], the term *condensation* will be used to define the process of the *intramolecular collapse* (*compaction*) of a single high-molecular-mass, double-stranded nucleic acid molecule, as well as the *intermolecular aggregation* of neighboring molecules. The collapse of a single high-molecular-mass, double-stranded nucleic acid molecule results in the formation of a new structure with a fixed size and morphology. The high segment density is typical of this structure, which means that it has a small radius of gyration, R_g. An *aggregate* means a precipitate that contains a number of nucleic acid molecules disordered in space [2]. Further, nucleic acid *aggregation* is different from *condensation* of the double-stranded molecules. *Condensation* begins with molecules not shorter than 150 Å (about 50 base pairs), while *aggregation* happens even with single-stranded DNA molecules that have only 20 base pairs [3].

The behavior of any polymer, including nucleic acid, in different solvents can be briefly described. In a nonfriendly environment (in a "poor" solvent), the segments of a molecule tend to associate with other segments to reduce their interaction with the solvent. This association can be either intramolecular, which results in the formation of a collapsed structure of a single polymeric molecule, or intermolecular, which leads to an aggregate formation. As the result of intermolecular association, the radius of gyration, R_g, of the collapsed structure rapidly decreases in comparison to the radius typical of the aggregate and the polymer statistical coil in a "good" solvent. Though the single molecules' collapse and aggregation are two different events, they are both based on the reduction of the free energy of the polymer–polymer and the solvent–solvent contacts in comparison to the polymer–solvent contacts' energy. Both intramolecular condensation and formation of the aggregate obtain structures with high segment density and low free energy because of the decreased number of contacts between the polymer and the solvent and the increased number of polymer–polymer and solvent–solvent contacts. In the first case, the high segment density is caused by the solution volume reduction occupied by a single polymeric molecule (decrease in R_g value), and, in the second case, by the increased number of molecules in a volume unit. It is obvious that the higher the polymer concentration, the higher

3

its aggregation level must be. More accurate analysis requires detailed information on the conformational properties of both the polymeric molecules and the solution.

It is known that the polyelectrolyte molecules (which include most of the biopolymers with charged groups) can exist in solutions in different conformations, and the free energy of any conformation depends on the efficiency of the charged polyelectrolyte groups' screening with counterions. According to theoretical calculations, the repulsion of the charged groups in the molecule changes the free energy to a relatively low value for extended conformation and a relatively high value for compact conformation. Though different theories provide different data on the relationship between the free energy level and the extension, all of them lead to the conclusion that a polyelectrolyte macromolecule with charged groups must have a more extended conformation than a noncharged macromolecule. This means that, in a solution with low ionic strength— that is, where the screening of the polyelectrolyte in an electrolyte-charged group is in the low range—its molecules must transform to the most extended conformation.

A DNA molecule in a solution is a typical example of a charged polyelectrolyte whose structure is conditioned by the ionization of negatively charged phosphate groups (the phosphoric acid residues) in the content of its structure [4].

The B-form of DNA has its inner part with a radius of about 9Å formed by nucleotide pairs and two sugar-phosphate chains, twisted in a spiral mode around the inner part. As a result, a well-known double helix with the pitch, P, about of 34 Å is formed. In this structure, there are two phosphate groups per each base pair, so there are 10 nucleotide pairs per each pitch (turn) of the helix. The distance between the base pairs alongside the DNA axis is 3.4 Å; therefore, there is one elementary charge per each 1.7 Å. An average rotation angle of the adjacent base pairs is about 36°, and the average distance between the adjacent charges on the DNA surface is around 7 Å. This distance is much shorter than the pitch, P, value, and is comparable to the value of the Debye screening length under physiological conditions (under physiological conditions, the DNA molecule is surrounded by the ionic atmosphere, and the parameter, λ_D, of the Debye screening length is within 5–10 Å). At a distance shorter than λ_D near the DNA surface, there is nonlinear screening of the negative charges of the DNA phosphate groups. When the distance between the surfaces of neighboring DNA molecules are close to λ_D or longer, the Debye–Hükkel theory (based on the Poisson–Boltzmann equation) is a reasonable approximation to describe the ionic atmosphere around the DNA. Nevertheless, the attraction between the molecules is not entirely considered by this theory.

Since the DNA molecule carries a large number of negatively charged phosphate groups, let us try to evaluate the DNA properties in a water–salt solution using a polyphosphate molecule as the DNA model [5].

1.2   POLYPHOSPHATES AS A SIMPLIFIED DNA MODEL

First of all, it is interesting to watch the alteration of the polyphosphates' spatial structure under the changing ionic composition and ionic strength of the solution as a model. In a polyphosphate chain

$$
\begin{array}{ccc}
O^- & O^- & O^- \\
| & | & | \\
-O-P-O-P-O-P-O- \\
| & | & | \\
O & O & O
\end{array}
$$

the charged groups are quite close to each other (the P–O bond is about of 1.62 Å long [6]), so the change in the ionic strength of the solution exerts a great influence on the electrostatic interaction pattern among these groups, which substantially determines the conformation of the polyphosphate molecules in the solution.

In U. P. Strauss's work, it was shown [7,8] that, without cation excess in the solution, the quantity of the specific viscosity, η, of the polyphosphates changes as $M^{1.9}$ (M is the molecular mass of the polyphosphate molecule). This points up the fact that, under such conditions, the polyphosphate molecules' shape is similar to that of a "rod" (extended chain), characterized by the lowest electrostatic interaction energy and the maximum number of contacts with the water molecules, as mentioned earlier. The type of relationship between M and η changes as the cation concentration grows. For instance, when NaBr concentration reaches 0.4 M, the polyphosphate molecules' behavior becomes similar to that of compact polymer coils with a deeply reduced possibility of contacts between the phosphate groups and water, which is indicated by the fact that, under such conditions, η changes as $M^{0.5}$ [7,8].

Therefore, the common polyelectrolyte property of shaping more compactly after the screening of charged groups can be fully attributed to polyphosphates.

A bigger increase of the cation concentration in the solution results in a sharp decrease in the polyphosphate chains' solubility, as well as their aggregation and the formation of a poorly soluble precipitate (the "salting-out" effect). This typical polyphosphate property is caused by the incompatibility with water of the screened phosphate group of a polyphosphate chain, unlike the situation with a negatively charged group [9–11].

Another issue of interest is the effectiveness of the binding of cations to phosphate groups of a polyphosphate chain. The dilatometric research of the interaction between the alkali metal cations and polyphosphate molecules [12] has shown that the screening of phosphate groups with these cations results in the translocation of water molecules from both the hydrate shells surrounding the phosphate groups and the hydrate shells of the metal cations. This fact points out that the interaction between alkali metal cations and polyphosphates' phosphate groups differs from the common process of ionic atmosphere formation [9,10,13], because this interaction has the obvious peculiarities of specific cation binding. The results of potentiometry [14] show that, at the interaction with polyphosphates, alkali metal cations have the coordination number 2; this corresponds to the formation of a six-member complex with one cation per two phosphate groups, and the lifetime of cation in such complex is about 10^{-2} seconds [15].

According to the data received during the research on the electrophoretic mobility of polyphosphates in solutions containing different alkali metal cations [16,17], the constants of interaction between alkali metal cations and polyphosphates reduces during the transition from Li^+ ($K = 2.7$ M^{-1}) to Cs^+ ($K = 1.11$ M^{-1}). Though the screening of phosphate groups in polyphosphate molecules can be realized with any alkali metal cations, nevertheless—according to some data—in the solutions containing Na^+ cations, polyphosphate molecules have more high level of compactness than in solutions with other alkali metal cations [18].

It is also known that the alkali metal cation concentration at which the polyphosphate starts to salt out is minimal for Na^+ [19].

These facts prove that the tendency of the screened phosphate groups to lose contact with water molecules is exhibited in the most noticeable way if the screening is conducted with Na^+ ions.

These results show that not only the effects caused by the presence of an ionic atmosphere but also the effects attributed to the specific binding of alkali metal ions should be considered when examining the reasons for the salting-out (precipitation) of polyphosphate molecules.

It is obvious that the foregoing data for polyphosphates cannot be directly attributed to the DNA because the DNA phosphate groups are located more distant from one another and are further separated by CH_2-groups of sugar residues. Moreover, the peculiarities of the DNA double-stranded helical secondary structure make these molecules more rigid in comparison to the polyphosphate chains. Nevertheless, the DNA behavior in water–salt solutions is, in many respects, similar to that of the polyphosphates, which is confirmed by the fact that the interaction constants of the alkali metal cations with the phosphate groups of polymers of both types are quite close [20,21]. This means that, under certain conditions, DNA molecules as well as polyphosphate molecules must be characterized by the salting-out effect, that is, the formation of poorly soluble precipitate from DNA molecules. So it can be assumed that, for the DNA molecules, the mechanisms causing the formation of the spatial structures with minimal number of contacts with the solvent show little difference from the mechanisms previously described, which cause similar processes in the case of polyphosphate molecules. By analogy with polyphosphates, we can assume that one of the reasons that support the stability of an expanded conformation of the DNA double-stranded molecules in water–salt solutions is the tendency of the phosphate groups to form as many contacts with water molecules as possible. The screening of a significant number of phosphate groups in DNA molecules with alkali metal cations must cause, at least, a decrease in the DNA molecules' solubility for the polyphosphates as well. Under such conditions, DNA molecules will form a structure with minimal number of contacts with water molecules, that is, at least the DNA molecules will tend to form aggregates.

We can conclude from these results that, to condense the DNA, it is necessary to overcome the repulsive interaction of DNA molecules through the screening of the phosphate groups' negative charges. (At this stage of the study of DNA behavior, it is not essential to question the particular reasons that determine the approach of the adjacent DNA molecules or the particular molecular mass of these molecules.)

The analysis of the DNA condensation (precipitation) can be described in the framework of the concept of how cations neutralize the negative charges of phosphate groups [22,23]. It should be noted once again that the necessary condition for this process is the neutralization of the DNA phosphate groups' negative charges. (DNA molecules carrying negatively charged phosphate groups cannot be precipitated under any standard condition.) This means that the precipitation process is initiated by electrostatic forces; the process of the DNA phosphate groups' negative charges being neutralized by counterions to a certain extent can be examined (in the simplest case) by considering the Manning theory regarding the condensation of counterions [24].

The approach proposed by G. Manning is based on a concept according to which the counterions condense on the polyelectrolyte to reduce the linear charge density to some limited value. Though this theory is quite approximate (in particular, the DNA molecule is modeled as a line of point charges [25]), it makes certain practical and valuable calculations possible.

For a solution containing only one polyelectrolyte and one kind of counterion with the charge, Z, the polymer linear charge density can be reduced by the factor

$$r = 1 - (|Z|\varsigma)^{-1}, \tag{1.1}$$

where $\varsigma = q^2{}_p/\varepsilon kTb$, q_p – proton charge; ε is the solution dielectric constant, and b is the average linear distance between the polyelectrolyte charges in the absence of fixed ions.

For DNA in water solution, $b = 1.7$ Å, $\zeta = 4.2$. After all the constants are substituted in the equation, it shows that the linear density of the negative charges in the DNA molecule in the solution containing Na^+ can be reduced only by 76% in comparison to the initial phosphate groups' charge.

The situation is more complicated for the composition of counterions with a different valency. In this case, there is competition between ions during their condensation, and the charge density is determined not only by the ions concentration but also by their valency [26].

The total share, Θ_T, of the DNA phosphate groups' negative charges neutralized by counterions is $\Theta_T = \Theta_1 + N\Theta_N$, where Θ_1 is attributed to univalent cations, and $N\Theta_N$ is attributed to N-valent cations, which under the excess of 1:1 electrolytes, obey the following equations:

$$\ln(\Theta_1/10^3 V_p C_1) + 1 = -2\,\xi\,(1-\Theta_1-N\Theta_N)\,\ln\,(1-e^{-kb}); \tag{1.2}$$

$$\ln(\Theta_N/C_N) = \ln(10^3 V_p/e) + N\,\ln\,(e\Theta_1/10^3 V_p C_1), \tag{1.3}$$

where C_1, C_N are total concentrations of free univalent and N-valent cations in the solution, respectively, and the following apply: V_p—cylindrical volume around the DNA where the cations can be considered fixed; ($V_p = 6.43 \cdot 10^{-4}$ m^3 per mole of nucleotides for DNA under 20°C); k—Debye screening parameter determined

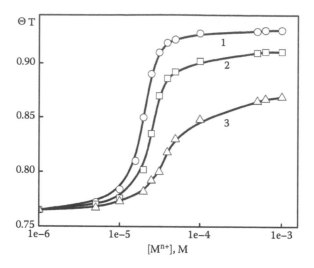

FIGURE 1.1 Dependence of the total share of DNA phosphate groups neutralized by counterions carrying different charges (n) on the cation concentration in the solution ($N = 2, 3,$ and 4): $1 - n = 4$; $2 - n = 3$; $3 - n = 2$; $C_{DNA} = 100\ \mu M$; $C_{Na^+} = 1\ mM$.

as $k = 8\pi L_{AV} 10^3 (q^2/4\pi\varepsilon_0\varepsilon_r kT) I$; N—Avogadro constant; I—solution ionic strength; $\xi = q^2/(4\pi\varepsilon_0\varepsilon_r kTb)$—charge density; $q = 4.8 \cdot 10^{-10}$ CGS units $= 1/6 \cdot 10^{-19}$ C—electron charge; ε_0—dielectric constant of vacuum; ε_r—solution dielectric constant (80 for water under 20°C); κ—Boltzmann constant; T—absolute temperature, K; and b—distance between the phosphate charges (for $b = 3.4/2 = 1,7\text{Å}$; $\xi = 4.2$).

Figure 1.1 shows the total share of DNA phosphate groups neutralized by counterions as a function of N-valent cations, which is calculated from the foregoing equations under the conditions of precipitation experiments [26]. It is evident that the solubility of the DNA molecules changes as an S-shaped curve as the polyvalent cations concentration increases. Polyvalent cations, interacting with the DNA molecule, initiate the formation of DNA aggregates, which easily precipitate as a result of low-speed centrifugation.

Using the curves given, we can understand why the trivalent spermidine[3+] ion concentration necessary to precipitate 50% of the DNA is much higher than the concentration of the spermine[4+] ion, considering that the DNA precipitation takes place when 89–90% of the molecule charges are neutralized [24].

It has to be noted that the formalism developed by Manning describes the counterion fixation on a polyelectrolyte chain as delocalized and nonspecific fixation of point charges. It is also worth mentioning that this point of view does not consider the nature and the properties of alkali metal cations that neutralize the DNA phosphate group charges and influence the properties of DNA molecules quite differently [25]. Unlike alkali metal cations that are fixed nonspecifically, some metal salt cations have site-specific fixation. For example, cation Tb^{3+} forms a chelate complex with a phosphate group and a guanine N-7 atom [27], which is accompanied not only by screening of phosphate groups' negative charges but also by noticeable changes in the parameters of the DNA secondary structure.

Modification of two parameters in Equation 1.1 to Equation 1.3 makes it possible to consider the influence such additives as (1) dielectic constant ε_r and (2) distance **b** between the charges along the chain on the efficiency of DNA condensation. Both parameters reduce the value of parameter ξ but oppositely change the screening parameter e^{-kb}.

The charge density parameter ξ of polyion is a function to the dielectric constant of the solvent. At the given concentration of cations, the extent of neutralized phosphate groups of DNA steadily decreases as the dielectric constant increases. As a result, the fraction of charges that can be neutralized by the ion of a certain valency will be different for various solvents. (The dielectric constant of water is 80 at 20°C.) For the DNA, the increase of ε_r from 80 to 100 corresponds to the decrease of ξ from 4.18 to 3.35. As the limit of the equation ($C_1 \rightarrow 0$) for polyion in a 1:1 salt solution is $\Theta_1 = 1-1/\xi$, the share of neutralized DNA phosphate group charges in an Na solution when this dielectric constant decreases from 0.76 to 0.70. So the DNA molecule in a solution with a high dielectric constant (more than water) carries more charged phosphate groups, and, as a result, it is more difficult to precipitate this molecule.

This statement is also confirmed by the results of experiments performed in 50% methanol solution where $\varepsilon_r = 60$. According to the theory of ion condensation, bivalent cations can neutralize only 88% of DNA charges in water, but they can neutralize up to 91% of charges in 50% methanol. This means that the decrease in the solution dielectric constant will cause the diminishing of the cation concentration necessary to neutralize enough phosphate group negative charges to induce DNA precipitation (condensation).

Therefore, the main part of DNA phosphate group charges neutralized by the cation fixation needed for the condensation is approximately the same in all cases. DNA molecules can be condensed in a water–salt solution when the linear charge density of the macromolecule is reduced by 89–90%.

Similarly, the change in the **b** value (i.e., the distance between the charges along the chain) is accompanied by the alteration of the extent of the DNA neutralization (Figure 1.2, where the inset represents the same dependences for **b** equal to 1.7 and 1.9 Å; $\varepsilon_r = 80$).

The foregoing evaluations show that many factors can influence DNA condensation and its neutralization level. This means that the Manning theory is approximated in many respects. It divides ions only by their charges, so all the ions are regarded as simple point charges. However, it should be noted that, when the counterions reach the DNA surface, such details as discrete nature and helical charge distribution, as well as presence of different grooves on the DNA surface, become important. Nevertheless, Manning's theory does not consider the possibility of site-specific cation binding. All the ions were considered as if they could move freely inside the fixed volume so that their hydration would not change during the binding.

One can add that, though there are different descriptions of the DNA behavior in water–salt solutions with high ionic strength [28], a theory considering all the factors just mentioned does not exist. Therefore, we can draw a conclusion from the analysis given, though it is quite approximated, that the conditions of DNA condensation are

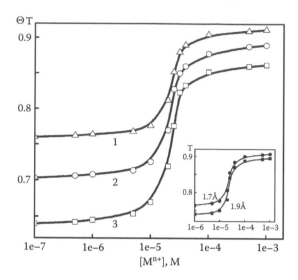

FIGURE 1.2 Dependence of the integral share of DNA phosphate groups neutralized by counterions upon the trivalent cations concentration in the solution.

1. Neutralization has occurred with about 90% of the negative charges of the phosphate groups with positively charged counterions (cations or polycations) [29].
2. There is a high local concentration of the adjacent segments of the DNA molecules.

When the counterions reach the DNA surface, such details of DNA molecules as discrete nature and helical charge distribution, as well as the peculiarities of the grooves on the DNA surface, become important. This means that the interaction between segments (molecules) of the DNA necessary for condensation depends not only on the number of cations around the DNA (common "linear" theories are based on this quantity) but also on the location of these cations near the DNA surface. Moreover, polycations (particularly polyamines) can connect adjacent phosphates of one chain or different chains; this process cannot only result in the formation of different condensed DNA forms but also finely adjust the properties of the forming structures.

1.3   MODELS OF HIGH-MOLECULAR-MASS DNA CONDENSATION IN WATER–POLYMERIC SOLUTIONS

It is obvious that two different approaches to the realization of DNA condensation are possible. According to the first approach, DNA molecules should be placed into a water–salt solution with properties providing the conditions for the phase exclusion of these molecules; according to the second approach, even in a common water–salt solution, the attraction between DNA molecules can be achieved with the help of polycations that neutralize the negative charges of the DNA phosphate groups. This means that the solvent plays a significant role in the process of DNA condensation.

Indeed, according to P. Flory [30], generally a homogeneous solution cannot be made with a high concentration of two polymers that are very soluble in this solution. Therefore, the solution containing DNA and any other water-soluble polymer that does not form complexes with DNA can be expected to spontaneously divide into two phases under certain conditions: one of the phases will be DNA-enriched, while the other will be enriched with the polymer added to the solution (i.e., the problem of DNA local concentration increase can be solved by the phase exclusion of these molecules).

The polymer often used for the phase exclusion of other polymers (proteins and nucleic acids) is poly(ethylene glycol), PEG [31]. This polymer has unique properties:

Preparations of this polymer with various molecular masses are available in significant amounts.
The polymer is highly soluble in water and water–salt solutions.
The solutions of PEG have almost neutral pH and insignificant UV-absorption; they are optically isotropic.
Only the terminal OH-groups are chemically active.
The PEG molecules themselves do not form stoichiometric complexes with nucleic acids.

The main peculiarity of PEG is the alteration of the structural elements such as $-O-$ and $-CH_2-$ groups in the polymeric chain. This alteration results in the disturbance of the water structure, which, in turn, induces the formation of a more complicated spatial structure (network) containing both water molecules and the structural element mentioned before [32]. The hydrodynamic properties of water- and water–salt–PEG-containing solutions indicate their complicated structure [33–35].

The curves in Figure 1.3 show the dependence of viscosity in the PEG-containing water solutions on the PEG molecular mass (M_{PEG}) at 20°C. The parts of the curves corresponding to low- and high-molecular mass are direct lines with different slope angles. As the PEG concentration increases, the slope angle tangent of the direct curve parts corresponding to low-molecular mass increases, tending to 1, while for those corresponding to high-molecular mass it is always equaled to ~ 3.2, regardless of the PEG concentration.

A "transition" area (**Tr**) between the regions marked as "I" and "II" is a typical feature of PEG solutions. Experimental points of the transition area are situated on straight lines with slope angle tangents within range from 1 to 3.2. The transition area is limited, with thick lines connecting the cross-points of straight lines within **Tr** and the respective lines in regions I and II (one can see that the transition area can be achieved by increasing the PEG concentration in the solution or reducing the PEG molecular mass under constant polymer concentration). The molecular mass interval for the given PEG concentration can be described by the average quantity M^{tr}_{PEG} ("transition" molecular mass), and the PEG concentration corresponding to the lower limit of the transition area can be called a "transition" concentration (C^{tr}_{PEG}). According to theoretical concepts [36–38], the transition area exists under such PEG concentrations, which allow a spatial polymer network formation.

The relation between C^{tr}_{PEG} and M^{tr}_{PEG} is described by the curve in Figure 1.4.

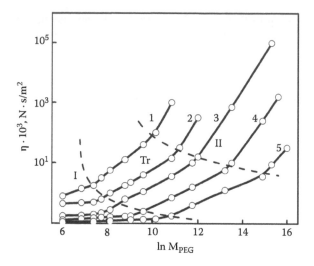

FIGURE 1.3 Dependence of viscosity of PEG-containing water solutions on molecular mass of PEG (20°C).

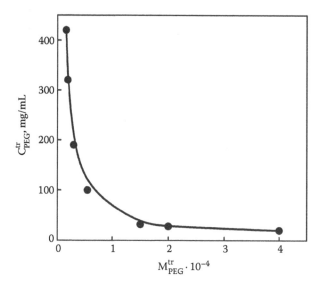

FIGURE 1.4 Dependence of C^{tr}_{PEG} on M^{tr}_{PEG} in water.

The formation of a spatial polymer network in any solvent can be detected by the constant value of the product of the polymer molecular mass and its volume fraction, v (i.e., $M^{tr}_{PEG} \cdot v$). According to theory [38,39], the numerical value of this quantity is determined by the polymer properties (stiffness of the polymeric chain and its thickness) and does not depend on the temperature. The fact that, in PEG solutions, the slope angle tangent for the straight line fragment of the curves that describe the relation between η and **lnM** in the area of low-molecular mass tends to 1, and for the area of high-molecular mass it equals 3.2, which is close to the theoretical value 3.4. The fact that the value of the product $(\mathbf{M^{tr}_{PEG}} \cdot v)$ for different PEG preparations is about 600 means that the theoretical concept [36–38] of the molecular structure of polymer solutions can be used to describe the properties of water solutions of PEG. Therefore, the data describing the relation between PEG molar mass (Figure 1.3) allow one to assume that, within the transitional area corresponding to certain PEG concentrations and molecular mass combinations, the PEG molecules form a specific polymer network [33–35].

The spatial polymeric network formation results in the effective exclusion of macromolecules, such as biopolymers, from this structure, which causes phase separation [32]. It is interesting that the polymeric network has to stabilize the spatial macromolecular structure that has the minimal volume [40].

As the product $\mathbf{M^{tr}_{PEG}} \cdot v$ for the water solutions equals 600, one can assume that the minimal PEG molecular mass that allows phase exclusion (condensation) of DNA in the limiting case must be about 600 (if the PEG volume fraction equals 1) [34,40]. In PEG solutions with molecular mass lower than the limiting value, the condensation of DNA molecules is impossible [34,40].

The sterical exclusion of macromolecular molecules is followed by the increase of their chemical activity. The increase of macromolecules' chemical activity in a polymer solution changes many of their properties, such as solubility, reactivity, and osmotic pressure of the system [41,42]. Here, one can stress that PEG solutions are capable of decreasing the solubility of biopolymer molecules (nucleic acids, proteins, etc.).

A simple consideration speaks in favor of this assumption. Let us assume that the chemical potential μ of the solid phase is constant in the PEG concentration range where the biopolymer precipitates. Under this condition, we can conclude from the invariability of the biopolymer chemical potential μ that

$$\mathbf{a_0} = \gamma_0 \mathbf{S_0} = \gamma_p \mathbf{S_p} = \mathbf{constant} \tag{1.4}$$

where $\mathbf{a_0}$ is biopolymer activity calculated from the equation $\mu = \mu_0 + \mathbf{RT}\ln\mathbf{a_0}$, and $\mathbf{S_0}$ and γ_0 are the solubility and the activity coefficient of the polymer, respectively, at any PEG concentration.

Let us convert the logarithms on both sides of Equation 1.4:

$$\log\mathbf{S_p} = \log\mathbf{a_0} - \log\gamma_p \tag{1.5}$$

A common expression for the biopolymer activity ratio in the PEG environment looks like

$$\log\gamma_p = \log\gamma_0 + \mathbf{A_{11}S_p} + \mathbf{A_{12}[PEG]} + \mathbf{B_{11}S_p^2} + \mathbf{B_{12}[PEG]^2} + \mathbf{C_1S_p[PEG]} + ... \tag{1.6}$$

where γ_{p0} is the biopolymer activity coefficient in a diluted solution; $\mathbf{[PEG]}$ is the PEG concentration; and $\mathbf{A}$, $\mathbf{B}$, and $\mathbf{C}$ are constants.

This equation is simplified for a diluted solution because, when $\gamma_{p0} \rightarrow 1$, the quantity $\lg\gamma_{p0} \approx 0$. All the terms of degree 2 or higher can be neglected because the interactions of polymer–polymer or PEG–PEG types are insignificant in diluted solutions. Similarly, the excluded polymer volume effect (term $A_{11}S_p$) is insignificant at low concentrations. Moreover, cross terms contribute only slightly to the γ_p quantity because high [**PEG**] values are experimentally related to low **Sp** values and vice versa. The term A_{12}[**PEG**] is dominant because it "hides" the excluded volume effect responsible for the biopolymer molecules' aggregation caused by PEG.

Therefore, at low biopolymer concentration, Equation 1.6 is simplified to

$$\lg\gamma_p \approx A_{12}[\textbf{PEG}] \tag{1.7}$$

The replacement of Equation 1.4 by Equation 1.6 results in an equation such as

$$\lg S_p = \lg a_0 - A_{12}[\textbf{PEG}] \tag{1.8}$$

Therefore, at low biopolymer concentrations (for which solubility can be determined within a concentration range where the activity is close to solubility), the relationship of the **lg** (the biopolymer solubility) to the PEG concentration must be linear with a slope equal to A_{12} value and intercepted with a segment "**y**" equal to the biopolymer activity in balance with the solid (precipitate) phase obtained in the PEG solution. Experimental data show that this concept is applicable to the description of exclusion (precipitation) of many proteins in PEG-containing solutions [41,42].

Besides biopolymer phase formation, water molecules in PEG solutions can be redistributed, which results in the dehydration of biopolymers; it is also possible that the solution dielectric properties change [32]. Considering the water–organic solution as a continuous medium with a dielectric constant **D**, the change in ion solubility when it is transferred from water to this solution can be determined with an equation:

$$\ln(S_2/S_{2,w}) = (A/RT)(1/D_0 - 1/D) \tag{1.9}$$

where **A** is a constant, and D_0 is the dielectric constant of water.

Equation 1.9 shows that biopolymer solubility should reduce when an organic solvent with dielectric constant lower than water is added to the solution. So, adding PEG to water solutions can change their dielectric properties and affect the biopolymer molecules' solubility [43]. The fact that organic solvents such as ethanol, acetone, and so forth, at high concentrations are used for the phase exclusion (precipitation) of nucleic acids supports this assumption.

Taking into account the peculiarities of water solutions of PEG, it is interesting to observe whether polyphosphates can be precipitated in PEG solutions and, hence, used as a model to describe DNA behavior, as was noticed previously.

The data given in the Figure 1.5 describe the results of the formation of the polyphosphate condensed phase, which scatters the UV-irradiation (the quantity C^{cr}_{PEG}, i.e., the "critical" PEG concentration in a polyphosphate solution at which an "apparent" optical density, induced by the light scattering due to formation of the particles

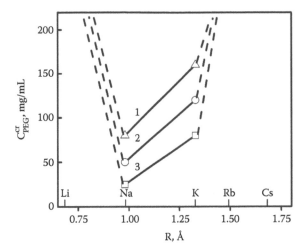

FIGURE 1.5 Dependence of the "critical" PEG concentration on the cation radius at the formation of condensed polyphosphate particles.

of polyphosphate dispersion, appears in the absorption spectrum of this solution [5]) (polyphosphates molecular mass $\approx$ 22,000 Da) under the increasing PEG concentration (PEG molecular mass = 20,000 Da). The broken lines in the right and the left parts of the curves show that, for LiCl, RbCl, and CsCl solutions, the values of C^{cr}_{PEG} are larger than 200 mg/mL, and it is not yet possible to watch the light scattering reflecting the formation of condensed polyphosphate particles at a PEG concentration less than 200 mg/mL.

As noticed before, among the alkali metals, the sodium salts of polyphosphates have the lowest water solubility. It can be concluded from the Figure 1.5 that the transition from water solutions to PEG solutions confirms this observation. Among all the alkali metal salts used (NaCl, LiCl, KCl, RbCl, and CsCl), the polyphosphate phase evolves most efficiently (the lowest C^{cr}_{PEG} value) in NaCl solutions (the difference between the various cations is reduced as the ionic strength of the solution grows).

Therefore, polyphosphate condensation can be induced in water–salt solutions; the efficiency of this process depends on the nature of cations used to neutralize the negative charges of the phosphate groups. It can be assumed that PEG solutions can induce phase exclusion of DNA. One can expect, also, that a concentration of PEG resulting in the transition of DNA molecules into a condensed state can be lower for the Na+-containing solutions than for solutions of the cations of other alkali metals.

It has to be noted that phase exclusion in its simplest case reflects the competition for the solution "space." This principle presents a background for theoretical studies of the peculiarities of macromolecule condensation, including DNA molecules. It is also obvious that the complicated structure of PEG solutions affect different aspects of biopolymer molecules' behavior, specifically DNA, which makes a theoretical description of the process of condensation in such solutions extremely difficult. It can be assumed that all theories represent only attempts to approximate a description of this process and consider only specific aspects of the DNA molecules' behavior in water solutions of PEG. Specifically, the role of cations is not considered as a rule, and the theories describe the behavior of DNA with negative charges neutralized by counterions.

Here, one can briefly repeat the description of the behavior of any polymer, including nucleic acids in different solvents. In a nonfriendly environment (a "poor" solvent), the molecule segments tend to associate with other segments to reduce interaction with the solvent. This association can be either intramolecular, which results in the formation of a collapsed structure of a single polymer molecule, or intermolecular, which leads to the aggregate formation. As the result of intramolecular association (collapse, condensation), the gyration radius of the collapsed structure rapidly decreases in comparison to the radius typical of the aggregate and the polymer statistical coil in a "good" solvent. Though the single molecule collapse and the aggregation of neighboring molecules are two different processes, they are both based on the decrease in the free energy of the polymer–polymer and the solvent–solvent contacts in comparison to the polymer–solvent contacts' energy. Both the intramolecular condensation and formation of the aggregated form cause the obtaining of structures with high segment density and low free energy because of the decreased number of contacts between the polymer and the solvent and the increased number of polymer–polymer and solvent–solvent contacts. In the first case, high segment density is caused by the solution volume reduction achieved by a single polymer molecule (smaller gyration radius) and in the second case by the increased number of molecules in a volume unit.

The intramolecular condensation of the high-molecular-mass double-stranded DNA molecule in water–salt–PEG-containing solution was experimentally discovered by L.S. Lerman in 1971 [44]. He showed the increase in the high-molecular-mass DNA sedimentation coefficient of the phage T4 from 62 to 1035 Svedberg units in a water–salt solution of PEG with a certain concentration. (A similar result was achieved for DNA of the phage T7.) The term ψ-*condensation* is often used to describe the phase exclusion of DNA molecules in water–salt polymer solutions (ψ is an acronym from the words **p**olymer–**s**alt-**i**nduced, **psi**) [45].

In 1973, a sharp decrease was registered in the characteristic viscosity ($\eta/\mathbf{C_{DNA}}$) of the high-molecular-mass DNA molecule of the SD phage under the increase of PEG concentration in a water–salt solution, despite the increase of integral viscosity of the solution [46].

Such a combination of sharp changes in different hydrodynamic properties of the high-molecular-mass DNA molecule at the "critical" PEG concentration in a water–salt solution definitely pointed at the change in the spatial shape of this molecule (i.e., the DNA molecule collapse and the formation of structure with small hydrodynamic radius).

The reason for the collapse of rigid macromolecules such as DNA in a polymeric solution of the flexible coil molecules (i.e., the possibility of the existence of different DNA spatial shapes in polymer solutions) was considered first of all in the framework of the lattice model. The basis for this model was formulated in the works of P. Flory [30].

The Flory lattice model makes some practically important evaluations possible. The total mixing of free energy for the polymer and solvent is calculated by Flory:

$$\Delta G = \Delta G_{ext} + n_2 \, \Delta G_{int}$$

$$\Delta G/kT = n_1 \ln v_1 + n_2 \ln v_2 + \chi n_1 v_2 +$$

$$+ n_2 \, [N \, \{(\chi - 1) + B_2 \, \omega/(2^{3/2} \, \alpha^3) + B_3 \, \omega^2/(2 \cdot 3^{5/2} \alpha^6)\} + 3/2 \, (\alpha^2 - 1) - \ln \alpha^3 \,] \quad (1.10)$$

where n_1 is the number of solvent molecules; v_1 is the solvent volume fraction; the parameters n_2 and v_2 are the same for the polymer; N is the ratio of the molecular volume of polymer to the solvent (i.e., $N = V_P/V_1$); the expansion parameter, α, is the ratio of R_g to the unperturbed radius of gyration, R_{g0} (i.e., to the gyration radius when the net excluded volume effect is zero); the quantities are $B_2 = 1/2 - \chi$; $B_3 = 1 + 12\chi^2/q - 16\chi^3/q2$; $\omega = [9/\pi \, (\, h_0^2)]^{3/2} \, V_P \, (h_0^2 = $ unperturbed end-to-end length of the polymer molecule); and q is the lattice coordination number.

The experimental parameter χ, which is called the *interaction parameter* or the *Flory–Huggins parameter,* is proportional to the difference in the energy of contacts between unlike and like species and does not consider the molecular peculiarities of such interaction. Mathematically, it can be written as h_{ij}, the energy associated with a contact between species i and species j, as

$$\chi = h_{12} - 1/2(h_{11} + h_{22}) \quad (1.11)$$

where 1 and 2 correspond to solvent and polymer segment, respectively.

The increase in h_{ij} is equivalent to the increase in the repulsive energy or the decrease in the attractive energy between the i and j. Parameter χ depends on the molecular mass and concentration of the polymer. The elastic character of the polymer chain is reflected by its terms, which are functions of the α parameter.

It is obvious that the value of χ changes as a result of any change in h_{12}, h_{11}, or h_{22}. The analysis of the Equation 1.11 shows that χ increases (the DNA becomes less soluble) when the value of h_{12} increases or when the values of h_{11} and h_{12} decrease. For example, the neutralization of negative charges of the DNA phosphate groups with polycations reduces the repulsion between DNA segments, hence decreasing the value of the segment–segment interaction parameter, h_{22}. An increase in the segment–solvent interaction parameter, h_{12} , can be produced by adding to the solution such "nonsolvents" as the neutral polymer PEG or ethanol, since such a change in medium makes the segment interaction with the solvent less favorable. (In principle, this means that the proposed consideration is applicable to all DNA solutions because χ is an overall measure of favorability of the interaction of the DNA segments with the solvent relative to the other interactions.)

In the framework of this approximated lattice model, B.H. Zimm and K.B. Post [1,47,48] described, though in a simplified way, the DNA behavior in a solution. Based on the P. Flory model, it can be considered that the attraction between the DNA segments leads to a phase exclusion when the experimental parameter χ is greater than half. If such segment interaction occurs within a single DNA molecule, this results in the formation of a collapsed (compact) DNA structure; if the segments of adjacent DNA molecules take part in such interaction, the molecules aggregate and precipitate. The probability of these processes happening depends on the DNA concentration. The value of free energy, ΔG, in mixing the polymer with the solvent described in Equation 1.10 helps to evaluate the phase equilibrium of DNA solutions.

The chemical potentials μ_1 and μ_2 of the two-components of the mixture can be calculated by differentiation of ΔG with respect to n_1 and n_2, respectively:

$$(\mu_2 - \mu_1^0)/kT = \ln v_2 - (N - 1)(1 - v_2) + \chi N (1 - v_2)^2 +$$

$$+ N \{(\chi - 1) + B_2\, \omega/(2^{3/2}\, \alpha^3) + B_3 \omega^2/(2 \cdot 3^{5/2}\, \alpha^6)\} + 3/2\, (\alpha^2 - 1) - \ln\alpha^3 \quad (1.12)$$

$$(\mu_1 - \mu_1^0)/kT = \ln(1 - v_2) + (1 - 1/N)\, v_2 + \chi\, v_2^2 \quad (1.13)$$

where μ_1^0 and μ_2^0 are the chemical potentials of the pure phases.

With the help of these equations, the phase diagram was plotted to describe the behavior of DNA molecules in a solution (i.e., a curve reflecting the dependence of χ on the DNA concentration was received) [1].

In Figure 1.6, there are such curves for DNA molecules with different concentration: the DNA of molecular mass = $10 \cdot 10^6$ Da; $(h_0^2)^{1/2} = 6.2\ 10^{-5}$ cm; $N = 2.5 \cdot 10^3$; for DNA of molecular mass = $37 \cdot 10^6$ Da; $(h_0^2)^{1/2} = 1.3\ 10^{-4}$ cm; $N = 9.3\ 10^3$; for DNA of molecular mass = $124 \cdot 10^6$ Da; $(h_0^2)^{1/2} = 2.5\ 10^{-4}$ cm; $N = 3.1\ 10^4$.

For each molecular mass, the boundaries separate three regions: (1) the area of the initial extended random DNA coil in solution, (2) the area of the collapsed (compact) DNA in dilute solution, which evolves as a result of intramolecular condensation of a single DNA molecule, and (3) the area of the concentrated precipitate, which evolves

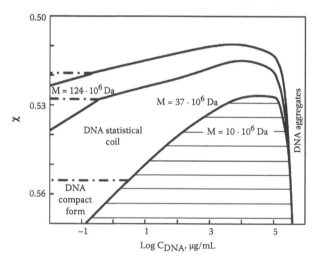

FIGURE 1.6 Phase diagrams of solutions of different DNA molecular mass.

as a result of intermolecular aggregation of DNA molecules. The horizontal line on the side of a diluted solution determines the value χ_{col} at which the free energies of the extended and collapsed coil are equal.

The shaded area marks the two-phase area with the compositions of the coexisting being the values that lie on the curve. The maximum in the coexistence curve indicates the solution composition and χ value critical for phase separation of the two-component system. This χ was designed as χ_{sep}. For χ values less than χ_{sep}, v_2 is a single-valued function of μ_1 and μ_2, and the solution is a homogeneous mixture of solvent and DNA extended coil at all concentrations. For solvent conditions where like contacts are favored, χ value is greater than χ_{sep}, and v_2 is the function of μ_1 and μ_2; here, a region of immiscibility appears. At high DNA concentration, the phase exclusion results in aggregation of these molecules. On the other hand, at low DNA concentration, as χ is increased, there is a first transition "extended coil–monomolecular compact particle," with progression through the collapse area, eventually leading again to aggregation. With decreasing DNA concentration, the collapse area widens in a vertical direction, indicating a larger range in χ values within which the DNA is compacted before aggregation happens. The DNA concentration and χ value at which the phase separation occurs are functions of the DNA molecular mass.

At high molecular masses, the phase separation occurs at low χ values and low DNA concentrations. Figure 1.6 shows that the DNA aggregation dominates at a high DNA concentration (i.e., the higher the DNA concentration, the higher the aggregation degree). When the DNA concentration decreases, the area of the DNA compact shape moves on the χ axis, which corresponds to high χ values at which a collapse can occur (χ is an analogue to temperature, though it actually depends on the solution composition).

The relation between the DNA molecular mass and the concentration at which the compact DNA shape forms is important; as the DNA molecular mass reduces, the region of the DNA compact shape moves to the higher concentration areas. The calculation shows that the area where the DNA compact shape can exist lies within the limits of 2–3 µm/mL for the molecular mass of $10 \cdot 10^6$ Da, and for the molecular mass of $37 \cdot 10^6$ Da, the value is only 0.6 µm/mL. This result is important in a practical sense because it proves that the DNA compact form does not exist at high DNA concentration; this conclusion has to be considered when analyzing the DNA shapes formed under different conditions. This analysis also shows that the DNA compact shape is a thermodynamically stable formation [47].

It has recently become clear that, unlike the traditional concept of the continuous nature of the DNA condensation process, intramolecular collapse is a phase transition of the first kind, "all or nothing" [49,50].

For solutions with complex composition—for instance, solutions containing other polymers and sodium salts—the χ value reflects the "average" properties of the solvent. The DNA phase separation in any "poor" solvent can be described, regardless of whether it was initiated by other polymers, polyamines, or spirits. In all cases, the phase separation is the result of the predominance of "like" contacts to "unlike" contacts.

The polymer concentration, which is "critical" for phase separation, is calculated from the stability equation [51], which looks like

$$[(1/m_2)(1/m_3)] - (ln/m_3)^2 = 0 \qquad (1.14)$$

$$m_2 = m_3/\varepsilon^2 \qquad (1.15)$$

where $m_2 = b/CX(r_h + r_p)^4$; X is the number of nucleotides in a DNA molecule but includes the constants in the expression for $\varepsilon = 0.17\pi CN(r_h + r_p)10^{-24}$; r_h is the double spiral radius; r_p is the effective polymer coil radius, nm; N is the Avogadro number; C is the nucleotide concentration per liter, there being 0.17 nm per one nucleotide; and m_2 and m_3 are the molar concentrations of PEG and DNA, respectively.

The "critical" polymer concentration needed to induce the DNA phase division at a relatively high molar weight can be calculated with the help of r_p, determined by viscosimetric measurements. In this case, r_h = 1.2 nm, r_p = 3.2 nm, C = 10^{-5}, X = 10^5, and the molecular mass of PEG = 6,800 Da. Thus, the m_2 expressed as the weight/volume of the solution equals 145 mg/mL [51].

A simplified calculation can also determine the concentration of the DNA condensed phase in equilibrium with the PEG solution (C_{PEG} = 145 mg/mL). The result is $1.8 \cdot 10^{-5}$ M or 610 mg/mL of DNA. It can be noted that the precise number is insignificant; this evaluation is important because it points at an extremely high concentration of DNA in the condensed phase, which does not conform to the idea of a flexible linear double-stranded DNA form typical of a common diluted solution.

It has to be noted that the theory used is very simplified, and the approximations made are quite rough. The evaluations ignore the DNA polymer nature, and the PEG macromolecules are considered as structureless particles. Moreover, the work [1] is concerned with the DNA behavior in a PEG melt with only a small amount of solvent, which does not permit evaluation of the fine peculiarities of the situation. Studies by many authors contain the development of the Flory theory for the description of DNA condensation in PEG solutions (see, for instance, References 52 and 53]).

1.4   GROSBERG MODEL OF HIGH-MOLECULAR-MASS DNA CONDENSATION

Another issue of interest is high-molecular-mass DNA condensation within the framework of the model proposed by A. Grosberg et al. [54, 55]. The authors note that the DNA compaction is impossible in solutions with low ionic strength because it is prevented by the repulsion of negative charges of phosphate groups. That is why this model describes the compaction of the noncharged DNA molecule without considering the interaction between the counterions and the DNA molecules. The DNA compact form is usually identified with the "globular state," while the noncompacted form is identified with the "coiled state." Therefore, it is necessary to consider the spatial structure of a single, rigid, long macromolecule and the conditions for its existence in a solution of shorter and more flexible chains (PEG) in a "good" solvent.

In order for the PEG solution to be diluted (i.e., for the PEG coils not to cross), the PEG concentration must be low in comparison with the concentration within a single coil:

$$\ll c^*a^3 \approx N^{1-3\upsilon}(a^3/B_p)^{3(2\upsilon - 1)}, \tag{1.16}$$

$$[1-3p \approx 4/5; \; 3(2p - 1) \approx 3/5]$$

where c is the number of PEG monomers within a volume unit; N is the PEG index of polymerization; a is the size of the PEG monomer; B_p is the second virial coefficient of the interaction between PEG monomers; and υ is the critical index for the correlation length for a swollen coil, $\upsilon \approx 3/5$.

Let us assume that a DNA globule occupies a volume, V, and the density of DNA monomers (segments) within that volume is n. The PEG concentration, C^{in}, within the volume, V, must differ from the PEG concentration in the surrounding solution, C^{out}, due to interaction between the DNA and PEG monomers. C^{in} and C^{out} are connected by the condition that the chemical potentials μ^{out} of PEG outside the globule are equal to μ^{in} (i.e., the chemical potential of PEG within the volume, V. Variations in the elimination of PEG and low-molecular-mass solvent from the DNA globule are one of the main causes of DNA condensation.

Having considered the dilute solution of PEG, the solution outside the volume, V, can be represented as an ideal gas in coils, with its chemical potential per coil as

$$\mu^{out} = T \ln(C^{out}/N) \tag{1.17}$$

where C^{out}/N is the concentration of coils.

Within the volume, V, the PEG solution is diluted, too, but here PEG coils interact with DNA segments; hence, we have

$$\mu^{in} = T \ln(C^{in}/N) + TB^C{}_{PD}(C)n \tag{1.18}$$

where $B^C{}_{PD}$ is the second virial coefficient of the interaction between the segment of the DNA and the coil of PEG. (See Figure 1.7.)

Based on the condition $\mu^{in} = \mu^{out}$, we can conclude that at a low n,

$$C^{in} \approx C^{out}(1 - B^C{}_{PD} n) \tag{1.19}$$

This expression does not contain the ideal gas term, $T\ln n$, which originates from the translation entropy, because the DNA monomers form a chain and therefore cannot independently move macroscopic distances (in comparison to the size of a monomer).

The expression for the chemical potential, $\mu^*(n)$, of the DNA segment looks like

$$1/T \cdot \mu^* \approx B_D n + B^C{}_{PD}C^{in}/N + 2C^C{}_{PDD}C^{in}/N - B^C{}_{PD}C^{out}/N \tag{1.20}$$

The first term describes interaction between DNA monomers, and B_D, their virial coefficient; the second and third terms represent the interaction of DNA monomers with a PEG coil. This interaction is marked by the superscript, C, for coil. C^C_{PDD} is the third virial coefficient of interaction between a PEG coil and two DNA segments. Higher powers of C and n can be neglected.

Comparing Equation 1.19 and 1.20, it is easy to calculate that, at low n,

$$\mu^*(n) \approx T\, B_{eff}\, n$$

where

$$B_{eff} = B_D - [(B^C_{PD})2 - 2\, C^{(C)}_{PDD}]C^{out}/N \qquad (1.21)$$

The second term of Equation 1.21 describes (at low n densities) the required effective DNA segments' interaction. This term is positive because the effective interaction is an attraction.

Can effective attraction between DNA monomers, through PEG chains, become stronger than their direct mutual repulsion due to the excluded volume effects?

To answer this question, B^C_{PD} and C^C_{PDD} (i.e., the virial coefficients of the interaction of the coil and one or two long, straight rods) have to be evaluated (see Figure 1.7). If we project a coil on the plane perpendicular to the rod, the surface projection on the plane will be of the order of $(a\, N^\nu)^2$; the density of the links on the projection is of the order of $N/a^2N2\nu$, and we see that is very small for a swollen coil ($\nu > 1/2$).

Consequently, the probability of an interaction of the rod with more than one coil monomer is of the order of $N^{2\nu}\, [(N/N^{2\nu})]^2/N \approx N^{2-2\nu}$, negligibly small in comparison. So, we obtain

$$B^C_{PD} = N\, B_{PD}\; ;\; 2C^C_{PDD} = 1/2N^2(B_{PD})^2 \qquad (1.22)$$

where B_{PD} is the second virial coefficient of the PEG and DNA segments' interaction.

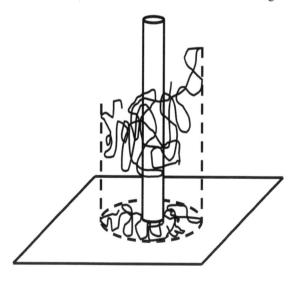

FIGURE 1.7 The calculation of virial coefficient of interaction between the coil and the rods (cylinders).

Replacing Equation 1.22 for 1.21, we can calculate

$$B_{eff} = B_D(1 - B_{PD}^2/B_D \, Nc) \tag{1.23}$$

According to Equations 1.17 and 1.23, under the condition

$$N > (B_{PD}^2/B_P B_D)^{-1/(2-3v)} \, (a^3/B_P)^2, \, (1/2 - 3 \, v) \approx 5 \tag{1.24}$$

In the area of the diluted solution, there is such concentration as

$$C_0(N) = (1/N)(B_D/B_{PD}^2) \tag{1.25}$$

at which $B_{eff} = 0$. If $C > C_0$, the quantity $B_{eff} < 0$ (i.e., the efficient attraction) dominates a direct repulsion. (C and N parameters describing the PEG solution determine the environment for the DNA molecule; their role is similar to the role of temperature in the common polymer theory.)

Let us examine two DNA segments in a PEG solution. Because of repulsion between the PEG and DNA segments, an area around the first DNA segment appears to have a lowered density of PEG monomers that are able to repeal the second DNA segment. This area plays the role of "potential hole" for the second DNA segment. The size of this hole is of the order of correlation length in solution—that is, for a diluted solution, it is of the order of the size of a coil.

For the attraction mechanism described here, long-range correlations in a PEG solution are very important. Regarding PEG chains as unstructured particles of a corresponding volume, the authors [47,56] could not explore the right mechanism of DNA segments in the attraction and compaction of this molecule.

Therefore, even the simplest theory not only provides a reasonable explanation for the high-molecular-mass DNA condensation in PEG solutions but also makes it possible to determine some practically important criteria of this process (specifically, the existence of inverse proportional dependence between the "critical" PEG concentration and PEG molecular mass confirmed in the theoretical works of other authors [52,53].

The condition (1.24) shows that DNA compaction cannot occur in a PEG solution with a low molecular mass—that is, this conclusion of the theory corresponds to the statement that follows from the consideration of the molecular structure of PEG-containing solutions, according to which, at the PEG "critical" molecular mass of 600 Da, DNA compaction does not occur. From the standpoint of the quality interpretation of mechanism of the DNA compaction described before, this dependence can be explained in two ways:

1. The displacement of PEG coils is caused by the failure of the mixing entropy. The smaller the amount of displaced particles, that is, segments, the smaller the mixing entropy failure. Consequently, the longer the chains, the stronger is the displacement effect.
2. As the PEG chain's length increases, the radius of effective attraction of DNA segments grows.

A rigorous expression for the free energy of the high-molecular-mass DNA single molecule (illustrated by phage T2 DNA of molecular mass of $1.24 \cdot 10^8$ Da), considering the influence of PEG solution ionic strength with variable molecular masses, was outlined in Reference 57.

The theory of DNA collapse in PEG solutions considers the following contributions to DNA free energy: (1) the free energy of sugar-phosphate chain elastic expansion or contraction; (2) the free energy of DNA segments' interaction with each other and with the solvent molecules; (3) the free energy of counterions translation causing extra-osmotic pressure inside the DNA coil; and (4) the free energy of DNA–PEG interaction, which commonly affects the external compression osmotic pressure of PEG on the DNA macromolecules. The transition between the statistical coil and the DNA compact form is mainly controlled by contributions (3) and (4), that is, the balance of the expanding osmotic pressure caused by counterions and the compression osmotic pressure caused by PEG. When the last osmotic pressure exceeds the first, DNA collapse occurs. The common physical explanation of DNA condensation in PEG solutions is based on the concept of the contacts between DNA and PEG as thermodynamically unfavorable. This means that, as the PEG concentration in the solution increases, the quality of the solvent for DNA worsens, which in its turn leads to strengthening of the interaction between the adjacent segments of the same DNA molecule. At a certain "critical" PEG concentration in the solution, its osmotic pressure becomes strong enough to initiate phase separation and fix the DNA coil in a new "compressed" state. DNA compression is caused by segregation of DNA-stiff segments (chains) and extraction and segregation of the flexible PEG chains. It can be expected to happen if the PEG chains in the external solvent are linked with one another and form a network. (In other words, in a diluted solution the compression osmotic pressure is not strong enough.)

There is a contradiction between the two statements following from the foregoing theoretical considerations. On the one hand, according to simple evaluations based on the Manning theory, the DNA molecules can be neutralized only by 76% in NaCl solutions, and, consequently, under such conditions, DNA molecule condensation is impossible. On the other hand, effective condensation of polyphosphate in water solutions of PEG shows that such solutions have the conditions providing the extent of neutralization of phosphate groups' negative charges necessary for the condensation. This shows that the active PEG role in the condensation process consists not only in the DNA molecules' exclusion but also in the increase in their activity coefficients—that is, in the decrease of DNA molecule solubility (which is necessary for the formation condensed form of these molecules). Meanwhile, the latter process must depend on the nature of cations used for the neutralization of the DNA negative charges, which is not considered in any theoretical works mentioned previously.

The question of the particular structure of the condensed form of high-molecular-mass DNA molecules formed under the condition of phase exclusion corresponding with physiological conditions (i.e., ionic strength 0.3; pH 7.0; room temperature), began to be analyzed by different authors almost simultaneously with the development of the condensation theories. The first electron-microscopic images of the DNA particles received under conditions of phase exclusion were published in

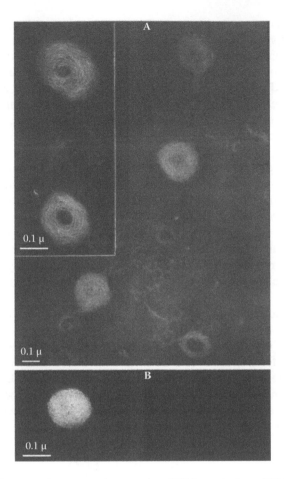

FIGURE 1.8   Electron-microscope photographs of T4 bacteriophage DNA toroid-like particles (a) and a particle formed from denatured DNA (b).

Reference 58. The electron-microscope investigation has shown that, when condensed, single double-stranded DNA molecules form toroid-like particles (Figure 1.8) with external diameters around 1,000 Å. The fact that such particles were immobilized on the surface of a special film without any noticeable distortion of the particles' shape is evidence in favor of the relatively high stability of these particles, which is apparently caused by their low solubility after the formation. (See Figure 1.8.)

A qualitative explanation of the toroidal DNA structures' stability was given in A. Grossberg's work [59] and analyzed by V. Golo et al. [60]. This explanation can be summed up as follows: because of DNA chain stiffness, the DNA molecule has no points where a curve could be fixed. To minimize energy loss caused by the curve, DNA reels in on itself; as a result, an opening (a hole) with a diameter of 200–300 Å evolves in the middle of the DNA. The packing density of the DNA molecule in a toroidal particle is relatively high.

An important question is, how many DNA molecules actually enter into the composition of one toroidal particle? To answer this question (i.e., to evaluate whether the toroid contains only one molecule), simple calculations can be made [61].

A DNA macromolecule self-volume (i.e., the total volume occupied by the DNA chain) is $V_0 = \pi L r_h^2$, where L is the DNA contour length, and r_h is the double-stranded DNA helix radius ($r_h \sim 10$ Å). The contour length, L, of the DNA depends on the N number of base pairs – $L \sim 3.4\,N$ (Å). The self-volume, V_0, of the DNA molecule is about $0.46 \cdot 10^8$ Å^3 (for the λ phage DNA, 48,000 base pairs) and $1.7\ 46 \cdot 10^8$ Å^3 (for the T4 DNA, 166,000 base pairs).

This volume can be compared with the volume, V_1, of toroid-like particles observed in the electron-microscopic photographs. The evaluation of R external radius of toroid and the r radius of its section in the electron photographs show that the value of V_1 toroidal particle volume ($V_1 \sim 2\pi^2 R r^3$) is approximately $0.54 \cdot 10^8$ Å^3 for λ phage DNA and $2.6 \cdot 10^8$ Å for T4 phage DNA. If we take into account the possible errors in V_1 and V_0 evaluations attributed to the inaccuracy of r_h double-stranded helix radius and r value calculation, it can be concluded that every toroidal particle observed in the electron photographs is really formed by one high-molecular-mass DNA molecule.

Short DNA molecules with less that 150 base pairs never form toroidal particles. (See Figure 1.9.)

With the help of a fluorescence microscope, the relation between the length and diameter of the forming DNA structure was evaluated and the phase diagram of the DNA state in PEG solutions was drawn (Figure 1.9). According to the figure, in a PEG solution with molar weight 8,200 Da, the transition of the DNA statistical coil to the toroid state (first-order transition) occurs at a PEG concentration of about 250 mg/mL. For longer PEG molecules, the DNA collapse takes place at a lower

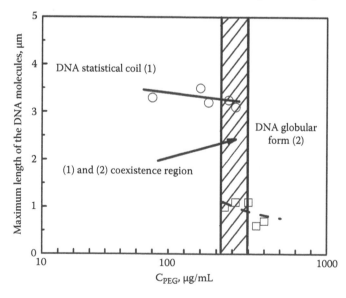

FIGURE 1.9 The dependence of high-molecular-mass DNA length on PEG concentration.

PEG concentration. Meanwhile, the "critical" PEG concentration depends on the NaCl concentration in the solution.

The double-stranded DNA molecules' transition from a coil to the toroid state is fully reversible; PEG-containing solution's dilution leads to the restoration of the DNA statistical coil shape.

The high-molecular-mass DNA molecule collapse also takes place when high concentrations of such polyanions as polyasparagine or polyglutamine acids (PG), which, like PEG, do not form complexes with DNA molecules, are added to the solution [62,63].

In particular, at a PG concentration lower than 0.8 M, the DNA molecules of T4dC bacteriophage exist as statistical coils at a PG concentration ~ 0.8–1.1 M. There are both coil and compact shapes; at higher PG concentrations, DNA molecules form compact structures.

1.5   SUMMARY

Therefore, DNA molecules of high molecular mass form single compact toroidal particles with a diameter of around 1,000 Å because of the phase exclusion of these molecules from water–polymer solutions (intramolecular condensation, compaction, see Figure 1.8). It is theoretically shown that, under phase exclusion, the formation of other compact DNA structures—in particular, shells or dense globules—is also possible.

The factors that determine the formation of one or another spatial structure of the high-molecular-mass DNA molecules at their compaction are now subject to intensive theoretical research [61,64,65].

It can be said that, as a result of phase exclusion, the high-molecular-mass DNA molecules form a dispersed phase, each particle of which consists of a single DNA molecule.

Hence, at the current moment, there is a definite positive answer to the question, can single high-molecular-mass DNA molecules (with a molecular mass over $20 \cdot 10^6$ Da) exist in a compact state in experimental conditions corresponding to actual physiological conditions? Besides, the DNA molecules' secondary structure does not change during the DNA compaction process, and its parameters correspond to the B-form. From this point of view, the DNA compact particles have much in common with viral particles or bacteriophages.

Experimental investigations on the single high-molecular-mass DNA molecules' condensation both in vivo and in vitro are difficult because the only method of direct observation is the electron or atomic-force microscope; meanwhile, the experiments have to be held with the use of solutions with the DNA concentration not higher than 1 µg/mL under conditions precluding both the DNA intermolecular aggregation and the distortion of the secondary structure of these molecules. This means that, on the whole, there is only a small amount of research [66–68], including work by Russian scientists, that contain direct proof of particles formation from single DNA molecules with high molecular mass that have a toroid shape.

Another field of research on the peculiarities of the DNA condensed state, notably the research on double-stranded DNA molecules condensation with molecular mass not more than $3 \cdot 10^6$ Da, has been developing collaterally and almost independently.

REFERENCES

1. Post, C.B. and Zimm, B.H. Theory of DNA condensation: Collapse versus aggregation. *Biopolymers*, 1982, vol. 21, p. 2123–2137.
2. Flory, P.L. Molecular configuration and states of aggregation of biopolymers. *Ciba Found. Symp., Polymerization in Biological Systems*, 1972, vol. 7, p. 109–124.
3. Saminathan, M., Antony, A., Shirahata, A., et al. Ionic and structural specificity effects of natural and synthetic polyamines on the aggregation and resolubilization of single-, double-, and triple-stranded DNA. *Biochemistry*, 1999, vol. 38, p. 3821–3830.
4. Zenger, V. *Principles of Structural Organization of Nucleic Acids*. Ed. by B.K. Vainshtain (Russian Edition). Moscow: Mir, 1987. p. 584.
5. Evdokimov, Yu.M., Salyanov, V.I., and Varshavsky, Ya.M. A compact form of DNA in solution. III. Influence of the ion composition of the solution on compactization process of the double-stranded DNA in the presence of PEG. *Mol. Biol. (Russian Edition)*, 1975, vol. 9, p. 563–573.
6. Flory, P.J. *Statistical Mechanics of Chain Molecules*. New York: Interscience Publishers, 1969. p. 432.
7. Strauss, U.P., Smith, E.H., and Wineman, P.L. Polyphosphates as polyelectrolytes. I. Light scattering and viscosity of sodium polyphosphates in electrolyte solutions. *J. Amer. Chem. Soc.*, 1953, vol. 75, p. 3935–3940.
8. Strauss, U.P. and Wineman, P.L. Dimensions and interactions of long-chain polyphosphates in sodium bromide solutions. *J. Amer. Chem. Soc.*, 1958, vol. 80, p. 2366–2371.
9. van Waser, J.R. and Holst, K. Structure and properties of the condensed phosphates. I. Some general consideration about phosphoric acids. *J. Amer. Chem. Soc.*, 1950, vol. 72, p. 639–644.
10. van Waser, J.R. Structure and properties of the condensed phosphates. II. A theory of molecular structure of sodium phosphate glasses. *J. Amer. Chem. Soc.*, 1950, vol. 72, p. 644–647.
11. Strauss, U.P., Woodside, D., and Wineman, P.L. Counterion binding by polyelectrolytes. I. Exploratory electrophoresis, solubility and viscosity studies of the interaction between polyphosphates and several univalent cations. *J. Phys. Chem.*, 1957, vol. 61, p. 1353–1356.
12. Strauss, U.P. and Leung, Y.P. Volume changes as a criterion for site binding of counterions by polyelectrolytes. *J. Amer. Chem. Soc.*, 1965, vol. 87, p. 1476–1480.
13. van Waser, J.R. Structure and properties of the condensed phosphates. III. Solubility fractionation and other solubility studies. *J. Amer. Chem. Soc.*, 1950, vol. 72, p. 647–655.
14. van Wazer, J.R. and Campanella, D.A. Structure and properties of the condensed phosphates. IV. Complex ion formation in polyphosphate solutions. *J. Amer. Chem. Soc.*, 1950, vol. 72, p. 655–663.
15. Crutchfield, M.M. and Irani, R.R. A P^{31} magnetic resonance study of complexing between Li^+, Ca^{2+}, and Mg^{2+} ions and the lower condensed phosphate polyanions. *J. Amer. Chem. Soc.*, 1965, vol. 87, p. 2815–2820.
16. Strauss, U.P. and Bluestone, S. Counterion binding by polyelectrolytes. II. The determination of binding of univalent cations by long-chain polyphosphates from conductivity and electrophoresis data. *J. Amer. Chem. Soc.*, 1959, vol. 81, p. 5592–5295.
17. Strauss, U.P. and Ross, P.D. Counterion binding by polyelectrolytes. III. Stability constants for the binding of univalent cations by PO_3-groups of polyphosphates from electrophoresis measurements. *J. Amer. Chem. Soc.*, 1959, vol. 81, p. 5295–5262.
18. Ross, P.D. and Strauss, U.P. Counterion binding by polyelectrolytes.V. The effect of binding of univalent cations by polyphosphates on the intrinsic viscosity. *J. Amer. Chem. Soc.*, 1960, vol. 82, p. 1311–1314.

19. Strauss, U.P. and Ander, P.J. Molecular dimensions and interactions of lithium polyphosphate in aqueous lithium bromide solutions. *Phys. Chem.*, 1962, vol. 66, p. 2235–2239.

20. Ross, P.D. and Scruggs, R.L. Electrophoresis of DNA. III. The effect univalent electrolytes on the mobility of DNA. *Biopolymers*, 1961, vol. 2, p. 231–236.

21. Strauss, U.P., Helfgott, C.H., and Pink, H.H. Interactions of polyelectrolytes with simple electrolytes. II. Donnan equilibria obtained with DNA in solutions of 1-1 electrolytes. *J. Phys. Chem.*, 1967, vol. 71, p. 2550–2556.

22. Marquet, R., Houssier, C., and Fredericq, E. An electro-optical study of the mechanism of DNA condensation induced by spermine. *Biochim. Biophys. Acta.*, 1985, vol. 825, p. 365–374.

23. Wilson, R.W. and Bloomfield, V.A. Counterion-induced condensation of deoxyribonucleic acid. *Biochemistry*, 1979, vol. 18, p. 2192–2196.

24. Manning, G.S. The molecular theory of polyelectrolyte solutions with application to the electrostatic properties of polynucleotides. *Q. Rev. Biophys.*, 1978, vol. 11, p. 179–246.

25. Grosberg, A.Yu., Nguyen, T.T., and Shklovskii, B.I. The physics of charge inversion in chemical and biological systems. *Rev. Modern Phys.*, 2002, vol. 74, p. 329–345.

26. Flock, S., Labarbe, R., and Houssier, C. Osmotic effectors and DNA secondary structure: Effect of glycine on precipitation of DNA by multivalent cations. *J. Biomol. Struct. Dynam.*, 1995, vol. 13, p. 87–102.

27. Gersanovski, D., Colson, P., Houssier, C., and Fredericq, E. Terbium(3+) as a probe of nucleic acids structure. Does it alter the DNA conformation in solution? *Biochim. Biophys. Acta.*, 1985, vol. 824, p. 313–323.

28. Kassapidou, K., Jesse, W., Dijk van, J.A.P.P., and Maarel van der, J.R.C. Liquid crystal formation in DNA fragment solution. *Biopolymers*, 1998, vol. 46, p. 31–37.

29. Allahyarov, E., Gomper, G., and Lowen, H. DNA condensation and redissolution: Interaction between overcharged molecules. *J. Phys.: Condens. Matter*, 2005, vol. 17, p. 1827–1840.

30. Flory, P.J. Statistical thermodynamics of polymer solutions. In *Principles of Polymer Chemistry*. Ed. by P. Flory. Ithaca, New York: Cornell University Press, 1953. p. 495–594.

31. Albertson, P.-Å. *Partition of Cell Particles and Macromolecules*. New York: John Wiley & Sons, 1960. p. 231.

32. McPherson, A. Use of polyethylene glycol in the crystallization of macromolecules. In *Methods in Enzymology*, Vol. 114/Ed. by Wyckoff, H.W., Hirs, C.H.W., and Timasheff, S.N. New York: Academic Press, 1985. p.120–125.

33. Pogrebnyak, V.G. and Toranic, A.I. Study of physico-chemical properties of polyethylene glycol water solutions. In *Physical Hydrodynamics (Russian Edition)*. Kiev-Donetsk: Visha shkola, 1977. p. 104–111.

34. Salyanov, V.I., Pogrebnyak, V.G., Skuridin, S.G., et al. On the relation between the molecular organization of the solution of poly(ethylene glycol)-water and the compactization of the double-stranded DNA molecules. *Mol. Biol. (Russian Edition)*, 1978, vol. 12, p. 485–495.

35. Pogrebnyak, V.G. and Toranic, A.I. Viscosity of polymer solutions. *Polym. Sci.*, Series A (Russian Edition), 1979, vol. 21, p. 302–306.

36. Pogrebnyak, V.G., Ivanyuta, Yu..F., Pogrebnyak, L.A., et al. Viscosity and state chart diagrams of polyethylene oxide solutions. *Bull. USSR Institutions of Higher Education* (Russian Edition), Series Chemistry and Chemical Technology, 1986, vol. 29, p. 93–96.

37. Vinogradov, P.V. and Malkin, A.Ya. *Rheology of Polymers* (Russian Edition). Moscow: Khimiya, 1977. p. 440.

38. Privalko, V.I., Lipatov, Yu.C., Lobodina, A.I., et al. Viscosity and molecular properties of narrow fractions of melted polyoxyethylene. *Polym. Sci.*, Series A (Russian Edition), 1974, vol. 16, p. 2771–2780.

39. Mikhailov, V.P. *Fundamentals of Physics and Chemistry of Polymers*. Ed. by V.N. Kulichikhin (Russian Edition). Moscow: Vysshaya shkola, 1977. p. 242.

40. Laurent, T.C., Preston, B.N., and Carlson, B. Conformational transition of polynucleotides in polymer media. *Eur. J. Biochem.*, 1974, vol. 43, p. 231–235.

41. Laurent, T. Enzyme reactions in polymer media. *Eur. J. Biochem.*, 1971, vol. 21, p. 498–506.

42. Middaugh, C.R., Tisel, W.A., Haire, R.N., and Rosenberg, A. Determination of apparent thermodynamic activities of saturated protein solutions. *J. Biol. Chem.*, 1979, vol. 254, p. 367–370.

43. Arakawa, T. and Timasheff, S. Theory of protein solubility. In *Methods in Enzymology*, Vol. 114/Ed. by Wyckoff, H.W., Hirs, C.H.W., and Timasheff, S.N. New York: Academic Press, 1985. p. 49–79.

44. Lerman, L.S. A transition to a compact form of DNA in polymer solutions. *Proc. Natl. Acad. Sci.*, 1971, vol. 68, p. 1886–1890.

45. Lerman, L.S. Intercalability, the ψ transition, and the state of DNA in nature. *Proc. Mol. Biol. Subcell. Biol.*, 1971, vol. 2, p. 382–391.

46. Akimenko, N.M., Dijakova, E.B., Evdokimov, Yu.M., et al. Viscosimetric study on compact form of DNA in water-salt solutions containing polyethyleneglycol. *FEBS Lett.*, 1973, vol. 38, p. 61–63.

47. Post, C.B. and Zimm, B.H. Internal condensation of a single DNA molecule. *Biopolymers*, 1979, vol. 18, p. 1487–1501.

48. Post, C.B. and Zimm, B.H. DNA condensation and how it relates to phase equilibrium in solution. *Biophys. J.*, 1980, vol. 32, p. 448–450.

49. Yoshikawa, K., Yoshikawa, Y., and Kanbe, T. All-or-none folding transition in giant mammalian DNA. *Chem. Phys. Lett.*, 2002, vol. 354, p. 354–359.

50. Yamasaki, Y. and Yoshikawa, K. Higher order structure of DNA controlled by redox state of Fe^{2+}/Fe^{3+}. *J. Am. Chem. Soc.*, 1997, vol. 119, p. 10573–10578.

51. Lerman, L.S. The polymer and salt-induced condensation of DNA. In *The Physico-Chemical Properties of Nucleic Acids*, vol. 3. 1973, p. 59–76.

52. Naghzadeh, J. and Massih, A.R. Concentration-dependent collapse of a large polymer. *Phys. Rev. Lett.*, 1978, vol. 40, p. 1299–1302.

53. Tanaka, F. Concentration-dependent collapse of large polymer in a solution of incompatible polymers. *J. Chem. Phys.*, 1983, vol. 78, p. 2788–2794.

54. Grosberg, A.Yu., Erukhimovich, I.Ya., and Shakhnovich, E.I. On DNA compactization in diluted polymeric solutions. *Biophysics* (Russian Edition), 1981, vol. 26, p. 415–420.

55. Grosberg, A.Yu., Erukhimovich, I.Ya., and Shakhnovich, E.I. On the theory of DNA compactization in polymeric solution. *Biophysics* (Russian Edition), 1981, vol. 26, p. 897–905.

56. Frisch, H.L. and Fescian, S. DNA phase transitions: The ψ transition of single coils. *J. Polym. Sci., Polym. Lett. Ed.*, 1978, vol. 17, p. 309–315.

57. Vasilevskaya, V.V., Khokhlov, A.R., Matsuzawa, Y., et al. Collapse of single DNA molecule in poly(ethylene glycol) solutions. *J. Chem. Phys.*, 1995, vol. 102, p. 6595–6602.

58. Evdokimov, Yu.M., Platonov, A.L., Tikhonenko, A.S., et al. A compact form of double-stranded DNA in solution. *FEBS Lett.*, 1972, vol. 23, p. 180–184.

59. Grosberg, A.Yu. and Zhestkov, A.V. On the theory of toroidal compact form of DNA in polymeric solution. *Biophysics* (Russian Edition), 1985, vol. 30, p. 233–238.

60. Golo, V., Yevdokimov, Yu.M., and Kats, E.I. Toroidal structures due to anisotropy of DNA-like molecules. *J. Biomol. Struct. Dynamics*, 1988, vol. 15, p. 750–754.

61. Vasilevskaya, V.V., Khokhlov, A.R., Kidoaki, S., et al. Structure of collapsed persistent macromolecule: Toroids vs. spherical globule. *Biopolymers*, 1997, vol. 41, p. 51–60.
62. Laemmli, U.K., Paulson, J.K., and Hitchins, V. Maturation of the head of bacteriophage T4. V. A possible DNA-packing mechanism: In vitro cleavage of the head proteins and the structure of the core of the polyhead. *J. Supramol. Struct.*, 1974, vol. 2, p. 276–301.
63. Ichiba, Y. and Yoshikawa, K. Single chain observation on collapse transition in giant DNA induced by negatively charged polymer. *Biochem. Biophys. Res. Commun.*, 1998, vol. 242, p. 441–445.
64. Golo, V.L., Yevdokimov, Yu.M., and Kats, E.I. Effect of electroelastic anisotropy of DNA-like molecules on their tertiary structure. *JETP*, 1997, vol. 85, p. 1180–1186.
65. Miller, I.C.B., Keentok, M., Pereira, G.G., and Williams, D.R.M. Semiflexible polymer condensates in poor solvents: Toroids versus spherical geometries. *Phys. Rev.*, 2005, E71, p. 031802-1–031802-8.
66. Evdokimov, Yu.M., Akimenko, N.M., Glukhova, N.E., et al. Formation of the compact form of double-stranded DNA in solution in the presence of polyethylene glycol. *Mol. Biol.* (Russian Edition), 1973, vol. 7, p. 151–159.
67. Minagawa, K., Matsuzawa, Y., Yoshikawa, K., et al. Direct observation of the coil-globule transition in DNA molecule. *Biopolymers*, 1994, vol. 34, p 555–558.
68. Laemmli, U.K. Characterization of DNA condensates induced by poly(ethyleneglycol) and polylysine. *Proc. Natl. Acad. Sci.*, 1975, vol. 72, p. 4288–4292.

2 Liquid-Crystalline Phases of the Low-Molecular-Mass Double-Stranded DNA Molecules

2.1 ORDERING OF LOW-MOLECULAR-MASS DOUBLE-STRANDED DNAS

The mutual impermeability of rigid (stiff) polymeric molecules with a molecular mass not higher than $5 \cdot 10^6$ Da underlies the so-called entropy ordering of these molecules, causing the formation of the liquid-crystalline phases of these polymers. The "moving" power of this process is the change in the entropy of the system.

The principles of the entropy-driven ordering of achiral, rod-like molecules have been known since L. Onsager's research [1]. Because of mutual impermeability, the volume excluded from a single molecule's center (regarded as a hard sphere) to another similar molecule's center is a sphere with a radius equal to the sum of the two hard spheres' radii. This means that, for a diluted solution, the excluded volume is eight times larger than the volume occupied by one spherical molecule. For molecules with a prolate shape, the excluded volume can be much larger, depending on the neighboring molecules' relative orientation. In 1949, L. Onsager showed theoretically that the interaction between the molecules' excluded volumes is enough to induce orientation ordering of the prolate-shaped molecules. Such transition is a function of the axial relation of these molecules and leads to the transition of an "isotropic phase–ordered phase," determined only by a change in the entropy of the system. This is the first-order transition. The transition depends on the relation of length, **L,** to diameter, **D,** of the molecule; it is realized at the axial relation **L/D** › 5, and its efficiency depends on how much the molecules' attraction and repulsion (in comparison to their interaction to the solvent) determines the extent of interaction between the excluded volumes of these molecules [2].

The higher the axial ration of the molecules, the lower the concentration at which the spontaneous ordering of molecules occurs (i.e., the lower the volume fraction at which the orientation ordering of a molecule relative to another molecule becomes entropically efficient). The result achieved by L. Onsager was qualified in many further studies based on more precise models considering high concentration, flexibility, and heterogeneity of molecules [3–9].

It is obvious that, in the case of the rigid, linear double-stranded DNA molecules, the formation of ordered phases (i.e., their condensation that occurs when the concentration of these molecules in a water–salt solution increases) is an intermolecular process that results in the formation of the DNA phase (precipitate).

The research on the properties of phases formed by DNA molecules of low molecular masses is significantly facilitated (in comparison to the research on the toroidal structures of high-molecular DNA) by not only experimental methods for observation of this process but also exactly determined parameters that allow one to classify the phase.

2.2 BRIEF CONCEPT OF TYPES OF LIQUID-CRYSTALLINE PHASES

Before we discuss the question of the existence of nucleic acids' liquid-crystalline state, let us briefly examine the peculiarities of liquid-crystalline phases formed by stiff (rigid) linear molecules (Figure 2.1). The term *liquid crystal* seems inherently controversial, though it actually identifies the most specific properties of a certain substance state.

The classification of all types of liquid crystals was introduced at the beginning of the last century (in 1922) by the French physicist G. Friedel and has undergone only a few insignificant changes up to the present.

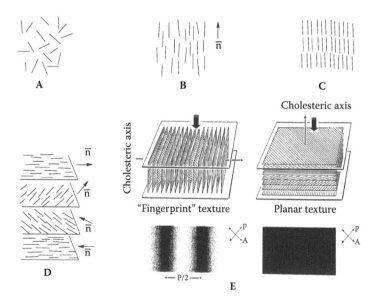

FIGURE 2.1 Scheme illustrating the location of molecules in an isotropic liquid (A), nematic liquid crystal (B), one of the types of smectic liquid crystals (C), and cholesteric liquid crystal (D); Scheme illustrating the formation of different types of structures typical of liquid crystals at the observation of the properties of liquid crystal thin layers by a polarizing microscope in a direction marked with the bold arrow (E).

There are two distinctive types of liquid crystals—thermotropic (i.e., formed under the change in the temperature of solid substance) and lyotropic (i.e., formed by the change in the properties of solution) liquid crystals.

Nematic liquid crystals (Figure 2.1B). In the case of nematic liquid crystals, the molecules are ordered and situated almost parallel to each other (a high degree of long-range orientational order is typical of the nematics), though the long-range translational order is absent because the mass centers of the molecules are situated chaotically (isotropically). The molecules can diffuse in any direction and rotate around the axis. (The picture on the cover of this book—a row of gondolas in a Venetian canal—gives an idea of the structure of a nematic liquid crystal. Each gondola rocks on the water independently from the neighboring gondolas, but there is a certain order in the orientation of gondolas, based on their shape. Meanwhile, though gondolas can float in any direction, the direction of their movement is usually determined by the channel bed [its "director"].)

A thin layer of liquid crystal is observable with the help of a polarizing microscope, revealing the "texture" characteristic of the liquid crystal. The texture is determined by the macroscopic orientation of molecules in the sample; it helps to determine the molecular structure of liquid crystal. For a nematic liquid crystal, the texture is a combination of movable "threads," which prompted the name "nematic" for this type of liquid crystal. (The word comes from the Greek "νημά," meaning "thread.")

Smectic liquid crystals (Figure 2.1C). In this structure (with some exceptions), the molecules form separate equidistant layers (on the order of the molecule's length), divided by some distance. There are several different ways the molecules are arranged in layers. Sometimes the molecules can be situated chaotically inside the layer, though the layers themselves are still located at the same distance from one another. Nevertheless, as a rule, the molecules are located with their long axes almost parallel to one another and perpendicular to the planes of the layers. Both the orientation and translation long-range order in the arrangement of the molecules is typical of smectic liquid crystals. The molecules can diffuse in two directions: on the layer's plane or rotating around the axis. Under certain conditions, layers of molecules can easily move and slide relative to one another as though there were a lubricant between them, which inspired these crystals' name, originating from the Greek "smegma," or "σμηγμά," meaning "lubricant," or "soap").

Cholesteric liquid crystals (Figure 2.1D). These are a specific type of nematic liquid crystals. In this structure, the molecules are placed in the same layers typical of nematics (quasi-nematic layers); in each quasi-nematic layer, the average orientation of molecules is determined by the vector, **n,** identified as the "pilot." Because of the anisotropy of optical and geometrical properties of the molecules, each quasi-nematic layer rotates to a certain angle relative to the previous layer, which leads to the rotation of the "pilot" to a certain angle and to the formation of a spatially twisted helical structure called "cholesteric." (The term *cholesteric* is historical by nature; it means a class of liquid crystals comprising predominantly cholesterol derivatives. A phase with such properties was discovered more than a hundred years ago [in 1888] by Austrian chemist K. Reinitzer.) Many compounds can form cholesteric liquid crystal phases or just "cholesterics" without being cholesterics by their chemical nature. Figure 2.1D illustrates the rotation of adjacent quasi-nematic layers

followed by the spatial rotation of the layer's "pilot." The spiral structure of cholesterics causes specific optical properties in these liquid crystals; notably, by observing
a cholesteric's thin layer at a certain angle, a change in color can be seen. Further,
if the cholesteric's axis in a thin layer of the phase is situated as in Figure 2.1E, a
polarizing microscope texture can be noticed in the direction marked with the bold
arrow, appearing as a combination of dark and light lines. This is known as the
"fingerprint texture." It carries information on the parameters of cholesterics' liquid
crystal helical structure. A very strong rotation of the polarization plane is typical of
cholesterics; the value of the polarization-plane-specific rotation reaches hundreds of
thousands of degrees per millimeter, while for common optically active crystals this
value is rarely higher than 1,000 degree/mm.

Consequently, the most important properties of the liquid crystal types considered are that the molecules forming these crystals maintain their molecular mobility
typical of the molecules in an isotropic liquid and, moreover, the molecules appear to
be structurally ordered in one (as nematics or cholesterics) or two (as smectics) (but
not in three, as real crystals) dimensions. That is why the combination of properties
typical of liquid and crystalline states of the substance allows one to use the term
liquid crystals to define this specific state of matter.

A liquid crystal can be conceived of as a condensed liquid with spontaneous
anisotropy. This means that not only is this long range of the molecules' orientation
typical of liquid crystals but the substance in the liquid crystal state can also flow
and even form separate drops.

Liquid crystals easily react to external actions such as light, mechanic pressure,
change in temperature, magnetic field, and so forth, as well as changes in environmental composition. The properties of different types of liquid crystals, including
those not mentioned here, are given in detail in special literature [10,11].

Thereby, the liquid-crystalline state is a specific state in which a substance has
the properties of both a liquid and solid matter; in this state, many properties are
intermediate between liquid and crystal properties.

2.3   LIQUID-CRYSTALLINE PHASES OF LOW-MOLECULAR-MASS DOUBLE-STRANDED DNA MOLECULES

In 1961, K. Robinson [12] studied the behavior of a thin layer of calf thymus DNA
solution during its drying in the air. With the help of a polarizing microscope, he was
observing the texture, that is, the picture of a thin layer typical of the explored object.
In the concentrated DNA solution, Robinson apparently first noticed the appearance of a "fingerprint" texture. This texture was similar to the texture of thin layers of poly-γ-benzyl-glutamate solution in organic solvents. Being concerned about
the possibility of a protein admixture in the solution, he cautiously formulated an
assumption according to which DNA molecules in concentrated solutions can exist
in the liquid-crystalline state.

Later microscopic research as well as studies of x-ray scattering supplemental
to Robinson's work were conducted by E. Iizuka [13], Y. Bouligand [14], G. Maret
[15], R. Rill [16,17], and other authors who have not only confirmed Robinson's

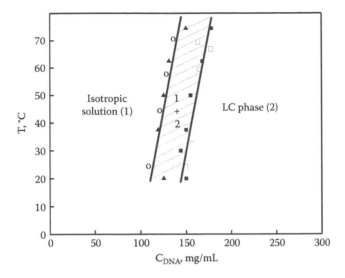

FIGURE 2.2   Phase diagram of the state of low-molecular-mass DNA molecules (from isotropic [1] to liquid-crystalline [2] state). Various marks correspond to the data received by different authors.

assumption but also proved the multiplicity of liquid-crystalline phases formed by double-stranded DNA molecules with length about 500 Å evolving under the conditions of phase exclusion in different situations.

These results conform well to the concept of "entropy ordering" of DNA molecules. Fragments of double-stranded DNA with length about 500 Å behave as rigid rods in water–salt solutions. As the average value of effective diameter, **D**, of DNA molecule in water–salt solutions of moderate ionic strength (~ 0.1 M NaCl) is close to 50 Å, the axial ration (**L/D**) for a molecule of length **L** ~500 Å is close to 10.

According to theoretical calculations, rigid molecules with such parameters tend to form an ordered phase when concentrated (Figure 2.2). The DNA phase transition is satisfactory described within the framework of the P. Flory theory for rigid synthetic polymers. Consideration of DNA molecule flexibility leads only to the improvement of the transition description quality. In the case of the application of L. Onsager ideology for the description of DNA molecule transition, this molecule is considered as a single, rigid double-stranded structure without any breaks in the sugar-phosphate chains.

Here, attention needs to be paid to the unique feature of fragments with lengths shorter than that corresponding to the **L/D** Onsager criteria, containing a few pairs of nitrogen bases (for instance, 6–10 pairs), which has never been taken into consideration when analyzing the peculiarities of DNA entropy condensation. The fact is that the nitrogen base pairs at the end of such fragment are capable of the so-called vertical stacking interaction with the nitrogen bases of adjacent fragments. Because of this vertical stacking interaction at certain solvent properties, an "elongated" columnar structure containing alternating short fragments without a joint, rigid

TABLE 2.1

Low-Molecular-Mass Double-Stranded DNA Phases and Their Parameters

Isotropic Phase	Cholesteric Phase	Hexagonal Phase	Orthorhombic Phase
C_{DNA}, mg/mL			
100	400	700	1000
Average Spacing between DNA Molecules			
~51 Å	~30 Å	~29 Å–~23 Å	Unit cell parameters: a = 24.09 Å b = 39.33 Å

sugar-phosphate chain can be formed. It is possible that such columnar structures, initially consisting of only short DNA fragments, can also form liquid crystals.

Based on the summary of different authors' research, the main phases formed by the double-stranded DNA molecules during the concentration of solutions can be determined [18,19] (Table 2.1).

Analyses of Figure 2.2 and Table 2.1 show that, at a certain "critical" concentration (C^{cr}), double-stranded DNA molecules spontaneously condense with the formation of phases that possess a maximum (peak) in small-angle x-ray scattering curves whose value can change from 51 Å to 30 Å.

There are two distinctive features of the DNA phase formed under its concentration in a solution higher than the C^{cr} value. First, the presence of only one maximum small-angle x-ray scattering curve of the formed phase means that the DNA molecules in the phase are ordered, but there is no three-dimensional order in the arrangement of DNA molecules. Hence, *the phase has the properties of an one-dimensional crystal*. Second, the phase maintains fluidity, that is, the adjacent DNA molecules maintain some diffusion degrees of freedom, which means that *the phase has the properties of a liquid*. Hence, the combination of two different features makes it possible to use the term *lyotropic liquid-crystalline phase* to define the evolving phase because of the increase in the DNA concentration in a solution. The term will be used further in this work in this sense.

"Critical" concentrations of the transition from isotropic to two-phase and from two-phase to completely anisotropic (liquid crystal) phase can be evaluated theoretically; these depend much on the axial ration of DNA. The C^{cr} value is inversely proportional to the DNA molecule length—the shorter the molecules, the higher C^{cr}. For instance, at the DNA length of 147 base pairs, the C^{cr} value is about 170 mg/mL, and at the length of 437 base pairs it decreases to 90 mg/mL. The C^{cr} value depends on the ionic strength of solutions—the higher the ionic strength, the lower C^{cr}. The C^{cr} value does not depend on DNA solution temperature (from 20 to 60°C), which confirms the conception of the entropy ordering of DNA molecules.

Because of anisotropy (geometrical and optical) typical of double-stranded DNA molecules, these molecules tend to be ordered when concentrated so that the

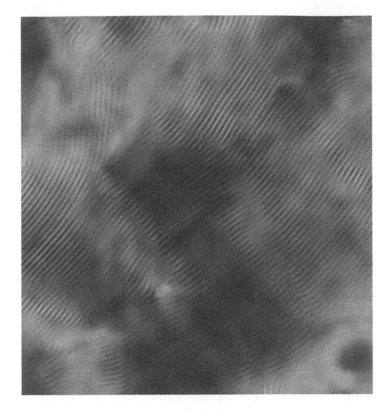

FIGURE 2.3 Texture of a thin layer of a cholesteric liquid-crystalline phase formed by double-stranded DNA molecules of low molecular mass.

liquid-crystalline formation phase would be spatially twisted. This leads to the formation of the cholesteric phase. Because of the helical twisting of quasi-nematic layers of the adjacent DNA molecules, the fingerprint texture observable under analysis of thin layers of the DNA phase by the polarizing microscope represents an excellent confirmation of the formation of the DNA cholesterics phase (Figure 2.3, where the DNA molecular mass is ~7–8·10^5 Da; polarized light).

Cholesteric packing of DNA molecules in the phase was confirmed by electron microscopy; combined with cryofixation, this phase is kept in its native state.

(In some works, a poorly characterized, mostly dynamic, precholesteric DNA phase was observed between the isotropic and the cholesteric phase; the properties of this phase have not been fully described by the current moment.)

The value of pitch, **P**, of the spatial twist of the cholesteric phase or the so-called DNA cholesteric is 2–3 μm. The **P** value increases with the increase in DNA concentration (that is, at the transition from the cholesteric to the hexagonal phase (Table 2.1), the helical structure of the DNA cholesteric phase unwinds. Compare the results given in Table 2.1 of this chapter to those in Table I.2 (Introduction). This comparison shows that the DNA packing density in liquid-crystalline phases formed in the model conditions corresponds to the DNA packing density in biological objects.

Obtaining very concentrated DNA solutions with exactly fixed physicochemical properties is a complicated experimental problem, and the very process of liquid-crystalline phase formation in such a system can last for weeks and even for months. Under such conditions, the question of the role of different factors affecting the formation of certain types of phases is subject to debate.

It is also important to note the following fact: there is no more or less full and consistent theory of liquid-crystalline phase formation from rigid molecules like double-stranded DNA, which is caused by the complicity of the structure of these molecules and the nature of their interaction with each other and with components of the solution. There is an approach that has established a certain reputation in the theory of low-molecular-mass liquid crystals, which accepts that the structure of the liquid-crystalline phase is determined by the competition of dispersion attraction and steric repulsion, though attempts to apply this approach to the description of the DNA phases' properties cannot be considered successful for two reasons. First, for the description of experimental data, many fitting parameters are introduced into the theory, although some of them are hard to calculate or could not be directly measured (for instance, the spatial correlations in interaxial molecule displacements). Second, to describe different experimental data, noticeably different sets of fitting parameters have to be used. The circumstance mentioned above is not casual. It is caused by the fact that, for the liquid crystals formed by low-molecular-mass compounds, the solvent plays mostly a passive role, working only as a "carrier" (medium) of electrostatic or dispersive interactions. As for the DNA molecules forming liquid-crystalline phases, the solvent plays a more complicated role. There is an assumption according to which DNA molecule packing is caused not only by DNA molecules' anisotropic properties but also by the properties of water molecules positioned between and nearby DNA molecules and approaching one another at the phase exclusion. The role of water in the process of stabilization and "reflection" of the DNA structure and its conformation has been subject to theoretical and experimental research for a long period of time. Interactions between DNA molecules and the solvent (which are responsible for the formation of a liquid crystal) are not local by nature because they are always physically related to the spatial dispersion of dielectric properties. In the case of low-molecular-mass compounds, this nonlocality is proportional to a/λ (a, molecular size; λ, wave length) and can be insignificant; however, for DNA segments with lengths of about 100 base pairs strongly interacting with water molecules, this nonlocality becomes most noticeable.

A significant issue is the sense of the cholesteric DNA spatial structure twist. Different theories give different answers to the question regarding the nature of the relationship between the direction of the initial spiral twist of a polymeric molecule and the sense of twist of cholesteric liquid crystal spatial structures formed by the molecules of the polymer (biopolymer). In some theories of the protein or DNA molecule cholesteric ordering, there is a point according to which the direction of the initial molecule spiral twist determines the sense of the spatial twist of a cholesteric structure. The steric (geometrical) model of the packing of low-molecular-mass compounds contains a concept of "ridges and grooves" on the molecule's surface; spiral molecules pack so that the ridges of a molecule enter the grooves of another molecule. Within the framework of this model, the angle of the spatial twist of adjacent layers in a liquid crystal's structure is quite large because of the angle formed by ridges and grooves in relation to the axis of initial polymer molecule spiral. Nevertheless, in the case of cholesteric DNA liquid crystals, the experimental angle of the helical twist in the molecules' adjacent layers is not larger than 1°, which is significantly different from the angle (~50°) formed by spiral grooves in relation to the nucleic acid molecule helical axis.

The goniometric effect observed is that slices of *Dinoflagellate chromosomes* and DNA cholesteric liquid crystals were used to determine the sense of the spatial twist of these structures [20]. The research has shown that *right-handed* double-stranded DNA molecules at a concentration of around 200 mg/mL form a cholesteric with a *left-handed* twist of the spatial structure [21]. The **P** value of the cholesteric structure formed by DNA molecules determined under these conditions is 2–3 μm.

2.4   SUMMARY

Therefore, the unquestionable fact is that the low-molecular-mass, rigid double-stranded DNA molecules form multiples of ordered liquid-crystalline phases as a result of intermolecular condensation (phase exclusion). Meanwhile the parameters of the DNA molecules' secondary structure virtually do not change during condensation.

Concerning the practical application of the DNA liquid-crystalline phases, the following fact must be considered: The time of transition to the equilibrium state reaches several weeks or months, and the dependence of phase formation on many factors make it difficult to take into account and sharply restricts the possibilities. Nevertheless, the existence of information on the conditions for formation, types, and physicochemical properties of DNA lyotropic liquid-crystalline phases makes it possible to move on to the next stage of research on this model system, notably to studies on DNA liquid-crystalline dispersions.

REFERENCES

1. Onsager, L. The effect of shape on the interaction of colloidal particles. *Ann. N.Y. Acad. Sci.*, 1949, vol. 51, p. 627–659.
2. Herzfeld, J. Entropically driven order in crowded solutions: From liquid crystals to cell biology. *Acc. Chem. Res.*, 1996, vol. 29, p. 31–37.
3. Khokhlov, A.R. and Semenov, A.N. Liquid-crystalline ordering in the solutions of long persistent chains. *Physica*, 1981, vol. 108 A, p. 546–556.
4. Khokhlov, A.R. and Semenov, A.N. Liquid-crystalline ordering in the solutions of partially flexible macromolecules. *Physica*, 1982, vol. 112 A, p. 606–614.
5. Lekkerkerker, H.N.W., Coulon, Ph., and van der Haegen, V. On the isotropic—liquid crystal phase separation in solution of rodlike particles of different lengths. *J. Chem. Phys.*, 1984, vol. 90, p. 3427–3433.
6. Nyrkova, I.A. and Khokhlov, A.R. Liquid-crystalline ordering in polyelectrolyte solutions. *Biophysics* (Russian Edition), 1986, vol. 31, p. 771–776.

7. Odjik, T. Theory of lyotropic polymer liquid crystals. *Macromolecules*, 1986, vol. 19, p. 2313–2329.
8. Wissenburg, P., Odijk, T., Circel, P., et al. Multimolecular aggregation of mononucleosomal DNA in concentrated isotropic solutions. *Macromolecules*, 1995, vol. 28, p. 2315–2328.
9. Fraden, S. and Kamien, R.D. Self-assembly in vivo. *Biophys. J.*, 2000, vol. 78, p. 2189–2190.
10. Belyakov, V.A. and Sonin, A.S. *Optics of Cholesteric Liquid Crystals* (Russian Edition). Moscow: Nauka, 1982. p. 360.
11. Sonin, A.S. *Introduction in Physics of Liquid Crystals* (Russian Edition). Moscow: Vysshaya shkola, 1983. p. 320.
12. Robinson, C. Liquid crystalline structure of polypeptide solutions. *Tetrahedron*, 1961, vol. 13, p. 210–219.
13. Iizuka, E. Some new findings in liquid crystals of sodium salt of deoxyribonucleic acid. *Polym. J.*, 1977, vol. 9, p. 173–180.
14. Bouligand, Y. Liquid crystalline order in biological materials. In *Liquid Crystalline Order in Polymers*. Ed. by A. Blumstein. New York: Academic Press, 1978, p. 262–297.
15. Senechal, E., Maret, G., and Dransfeld, K. Long-range order of nucleic acids in aqueous solutions. *Int. J. Biol. Macromol.*, 1980, vol. 2, p. 256–262.
16. Strelecka, T.E., Davidson, M.W., and Rill, R.L. Multiple liquid crystal phases of DNA at high concentrations. *Nature*, 1988, vol. 331, p. 457–460.
17. Rill, R.L., Strzelecka, T.E., Davidson, M.W., et al. Ordered phases in concentrated DNA solutions. *Physica A*, 1991, vol. 176, p. 87–116.
18. Yevdokimov, Yu.M., Skuridin, S.G., and Salyanov, V.I. The liquid-crystalline phases of double-stranded nucleic acids in vitro and in vivo. *Liq. Crystals*, 1988, vol. 3, p. 1443–1459.
19. Livolant, F. and Leforestier, A. Condensed phases of DNA: Structure and phase transitions. *Prog. Polym. Sci.*, 1996, vol. 21, p. 1115–1164.
20. Livolant, F., Giraud, M., and Bouligand, Y. A goniometric effect observed in sections of twisted fibrous materials. *Biol. Cellulaire*, 1978, vol. 31, p. 159–168.
21. Livolant, F. Ordered phases of DNA in vivo and in vitro. *Physica A*, 1991, vol. 176, p. 117–137.

3 Dispersions of Low-Molecular-Mass Double-Stranded DNA Molecules

3.1 LOW-MOLECULAR-MASS DOUBLE-STRANDED DNA DISPERSIONS IN WATER–POLYMER SOLUTIONS

Much attention has been paid to research on the phase exclusion of double-stranded low-molecular-mass nucleic acids (both natural molecules and synthetic polynucleotides). This is caused by adding the acids to solutions containing polymers that take up the volume of the solutions and induce entropy ordering (entropy condensation). Phase exclusion of nucleic acids from polymer-containing solutions can result in the formation of dispersions of these molecules, providing the ordered packing of adjacent nucleic acid molecules typical of the dispersion particles.

The nature of the properties of DNA dispersions is interesting from at least two points of view. From the theoretical point of view, the interest in dispersions is caused by the fact that the physicochemical properties of dispersion particles of size 100–1,000 Å can significantly differ from the properties typical of continuous solid phases. The difference in the properties evolves as a result of the "size effect" [1]. Such an effect is caused by the free energy attributed to the surface tension of particles and the existence of possible defects in molecule packing in the particles. The physicochemical properties of liquid-crystalline dispersions of nucleic acids are of biological interest because such particles open a gateway to describing the properties of *Protozoan* chromosomes and DNA containing viruses, which are isolated systems of microscopic size with an ordered but labile packing.

3.2 FORMATION OF DNA DISPERSIONS IN PEG-CONTAINING SOLUTIONS

The formation of nucleic acids (specifically DNA) dispersions is caused as the result of phase exclusion when their water–salt solutions are mixed with water–salt solutions of some synthetic, water-soluble, chemically neutral synthetic polymers, for instance, poly(ethylene glycol), PEG [2]. The scheme in the Figure 3.1 shows that the efficiency of DNA dispersion formation is affected by a number of variables.

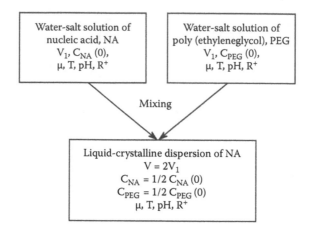

FIGURE 3.1   Scheme of formation of dispersion from double-stranded nucleic acid molecules.

The parameters of dispersion particles can be evaluated only by theoretical calculations based on the results of measurement of the particles' optical and hydrodynamic properties or by the application of certain theories.

The easiest way to demonstrate a dispersion formation is registration of the absorption spectrum. Starting from a particular PEG concentration in solution, the optical density in the absorption region of the nucleic acid nitrogen bases decreases, while at wavelengths above 320 nm (where neither DNA nor PEG molecules absorb) it increases (Figure 3.2). The increase in optical density in this area of the spectrum is caused by light scattering due to formation of the DNA dispersion (an "apparent" optical density, A_{app}). The registration of absorption spectra makes it possible to evaluate the "critical" PEG concentration, C^{cr}_{PEG} (i.e., the concentration at which the DNA dispersion begins to

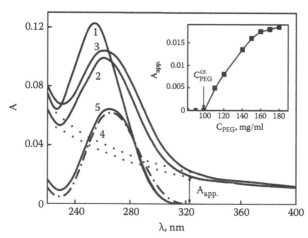

FIGURE 3.2   Change in DNA absorption spectrum at the dispersion formation in PEG solutions.

form). Figure 3.2 shows the dependence A_{app} ($\lambda = 340$ nm) on PEG concentration (PEG molecular mass = 4,000 Da; 0.3 M NaCl) used for the estimation of the C^{cr}_{PEG} value).

The diameter, D, of the particles of formed DNA dispersion has been calculated using the relationship

$$A_{app} = K\lambda^{-n}, \tag{3.1}$$

where λ is the wavelength, and parameter n is related under certain conditions to the diameter, D, of the particles.

For DNA dispersions, D is equal to about 10^3 Å. This tends to rise with DNA concentration and decreases as the PEG concentration is increased. For instance, $D = 4.5 \times 10^3$ Å at $C_{PEG} = 100$ mg/mL, and $D = 3.8 \times 10^3$ Å at $C_{PEG} = 300$ mg/mL.

The dependence of D on PEG concentration in the solution is expressed by the equation

$$D \text{ [}\mu\text{m]} = (K_1/C_{PEG}) x + K_2 \tag{3.2}$$

where: $x \sim 0.2$; $K_2 \sim 4$; C_{PEG} in mg/mL.

The radius, R, of DNA particles has been calculated in terms of the Grosberg theory that describes the condensation of the DNA molecules in polymer-containing solutions. The theoretical value of R ($R = 2.9 \times 10^3$ Å at $C_{PEG} = 170$ mg/mL) is the same order of magnitude as that D-value found experimentally.

The translational diffusion coefficient, D_T, has been found for DNA dispersions ($C_{PEG} = 170$ mg/mL; PEG molecular mass = 4,000; 0.3 M NaCl) using laser correlation spectroscopy: $D_T = 14 \times 10^{-10}$ cm^2/s as $C_{DNA} \rightarrow 0$. The received D_T value corresponds to a spherical particle with a diameter $D = 3.7 \cdot 10^3$ Å.

The DNA molecules' phase exclusion leads to the formation of dispersion, whose particles sediment at low-speed centrifugation (Figure 3.3). (In this figure, the C^{cr}_{PEG} value corresponding to the beginning of the scattering of UV-radiation by particles of DNA dispersion is marked with a dotted line). With the help of an analytical centrifuge, the value of the sedimentation coefficient, S, of DNA particles can be determined: $S = 14.3 \times 10^3$ Svedberg units as $C_{DNA} \rightarrow 0$ (PEG molecular mass equals to 4,000 Da; 0.3 M NaCl).

Evaluation of the molecular mass of the DNA dispersion particles based on the obtained values of D_T and S shows that molecular mass of a single particle is close to $5 \cdot 10^{10}$ Da. Because the molecular mass for a single DNA molecule used for the dispersion formation is ~10^6 Da, it can be considered that there are ~10^4 DNA molecules [3] in one DNA dispersion particle formed at $C_{PEG} = 170$ mg/mL.

Thereby, the combination of results achieved with the help of different physicochemical methods and the application of theoretical concept about the correlation between molecular mass and shape of particles being close to sphere shows that stiff, linear, double-stranded DNA molecules of low molecular mass in PEG-containing solutions can form particles of microscopic size ($D \sim 10^3$ Å); there are ~10^4 DNA molecules in each particle. The explanation of such a size of DNA dispersion particles

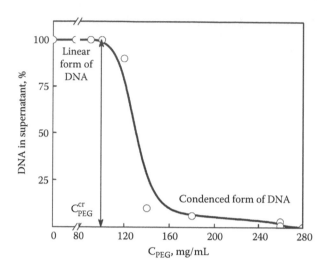

FIGURE 3.3 Relative concentration of DNA linear form in PEG-containing water–salt solutions after low-speed centrifugation.

in PEG-containing solutions considers the contributions of both kinetic and thermo-dynamic factors [3].

By definition, the "critical" PEG concentration, C^{cr}_{PEG}, is a concentration at which PEG coils collide [4]. The PEG coils' collision that occurs when PEG concentration increases is an important factor contributing to the phase separation (otherwise, the "compressing" osmotic pressure of PEG is insufficient for the condensation of DNA molecules [4,5]).

The order of polymeric segments (monomers) concentration, c^*, at which the collision occurs and which must correspond to experimentally observed C^{cr}_{PEG}, can be evaluated. The c^* is approximately equal to the average concentration of segments in a PEG polymer coil; that is,

$$c^* = N/4/3\pi R^3_g \qquad (3.3)$$

where N and R_g are the number of monomers in a PEG molecule and the gyration radius of this polymeric molecule, respectively.

The R_g value was evaluated as follows. The average distance between the ends of the polymer chain is calculated by the equation

$$R^2 = C_\infty N (l^2_{C-C} + 2l^2_{C-O}) \qquad (3.4)$$

where C_∞ is a constant related to the effect of "short-range force," which is equal to 4.1 for PEG in water; l_{C-C} and l_{C-O} are the length of $C-C$ and $C-O$ bonds equal to 1.54 and 1.43 Å, respectively; and N is the number of monomers in polymer chain equals to 91 for PEG with molecular mass 4000 Da.

Mean-square radius of gyration R^2_g is calculated as $R^2_g = R^2/6$ for a flexible-chain polymer, without considering the excluded volume effect. Using the received

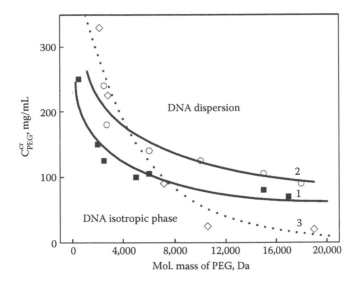

FIGURE 3.4 Dependence of C^{cr}_{PEG} on PEG molecular mass.

R_g value and Equation 3.3, c* was calculated as equal to 4.48 M monomer/L or about 190 mg/mL of PEG.

The data given in Figure 3.4 show that, first, DNA dispersion forms at C^{cr}_{PEG} corresponding to c* only of the order; second, C^{cr}_{PEG} is inversely proportional to the molecular mass of PEG preparations [6,7]. This proves the approximate nature of evaluation and that the effect of such parameters as solution ionic strength, hydrate shell around PEG molecules, and so forth, on the PEG structure is not considered as the evaluation. The PEG molecular mass at which DNA dispersion forms must be above 600 Da [6].

Two principal results, namely, the existence of dependence between the "critical" PEG concentration and PEG molecular mass and the formation of DNA dispersion when the PEG molecular mass is above the "critical" value, were used as background to the theory proposed by A. Grosberg to describe DNA molecules' condensation [8]. In this theory, PEG molecules not only leave the volume of the formed DNA dispersion particles but also exert a compression effect on the particles (which causes the dependence on C_{PEG}).

The fact that depending on conditions C^{cr}_{PEG} varies from 50 to 200 mg/mL means that the formation of DNA particles occurs under the conditions of "crowding" of the solution volume.

3.3 SUMMARY

Therefore, stiff, linear, double-stranded nucleic acid (and synthetic polynucleotide) molecules can form dispersions in PEG-containing solutions. The solution properties, such as concentration and PEG molecular mass, and so forth, can affect the efficiency of such dispersions' formation.

The particles of the low-molecular-mass DNA dispersions are "microscopic droplets of concentrated DNA solution," which cannot be "taken in hand" or "directly seen." A "liquid" mode of packing these molecules in the particles of dispersions prevents their immobilization on the surface of a membrane filter. Hence, the foregoing evaluations are approximated and based on the application of calculation methods regarding DNA dispersion particles as spherical structures. Consequently, the problem of the possibility of "seeing" these particles and directly measuring their size has not been solved yet. The question of how the nucleic acids molecules are packed in dispersion particles is also still open.

REFERENCES

1. Adamczyk, A. Phase transition in freely suspended smectic droplets. Cotton-Mouton technique, architecture of droplets and formation of nematoids. *Mol. Cryst. Liq. Cryst.*, 1989, vol. 170, p. 53–69.
2. Yevdokimov, Yu.M., Skuridin, S.G., and Lortkipanidze, G.B. Liquid-crystalline dispersions of nucleic acids. *Liq. Crystals*, 1992, vol. 12, p. 1–16.
3. Yevdokimov, Yu.M. Liquid-crystalline dispersions of nucleic acids. Bull. *USSR Acad. Sci., Phys.* (Russian Edition), 1991, vol. 55, p. 1804–1816.
4. Salyanov, V.I., Pogrebnyak, V.G., Skuridin, S.G. et al., On the relation between the molecular organization of the solution of poly(ethylene glycol)-water and the compactization of the double-stranded DNA molecules. *Mol. Biol.* (Russian Edition), 1974, vol. 8, p. 396–405.
5. Minagava, K., Matsuzawa, Y., Yoshikawa, K., et al. Direct observation of coil-globule transition in DNA molecules. *Biopolymers*, 1994, vol. 34, p. 5556–558.
6. Evdokimov, Yu.M., Akimenko, N.M., Glukhova, N.E., and Varshavsky, Ya.M. DNA compact form in solution. 1. Patterns of absorption spectra of polyribonucleotides and DNA in PEG-containing water-salt solutions. *Mol. Biol.* (Russian Edition), 1978, vol. 12, p. 485–495.
7. Tanaka, F. Concentration-dependent collapse of a large polymer in a solution of incompatible polymers. J. Chem. Phys., 1983, vol. 78, p. 2788–2794.
8. Grosberg, A.Yu., Erukhimovich, I.Ya., and Shakhnovich, E.I. On the theory of DNA compactization in polymeric solution. *Biophysics* (Russian Edition), 1981, vol. 26, p. 897–905.

4 Circular Dichroism of Nucleic Acid Dispersions

4.1 CIRCULAR DICHROISM AS A METHOD OF PROOF OF CHOLESTERIC PACKING OF NUCLEIC ACID MOLECULES IN DISPERSION PARTICLES AND ANALYSIS OF THEIR PROPERTIES

The mode of packing of low-molecular-mass double-stranded DNA molecules in liquid-crystalline phases can be determined by analyzing the phases' textures. To answer the question of how these DNA molecules are packed in particles of dispersions, some extra research needs to be performed.

The physicochemical methods used earlier (Chapter 3) do not make it possible to determine the character of DNA molecules and their complexes with chemical or biological compounds packed in particles of dispersions. As noted in Chapter 3, the possibility of "size effect" can lead to noticeable differences between packing the DNA molecules in particles of dispersions and their packing in solid phases. This difference is especially noticeable in the case of liquid-crystalline dispersions, and the issue of the molecules' packing mode in this case needs extra testing. The existence of "size effect" can cause the absence of the packing typical of liquid-crystalline phases in case of dispersions. That is why, for double-stranded DNA molecules of low molecular mass (forming dispersions at phase exclusion from water-polymeric solutions) whose particles are microscopic-sized, it is necessary to determine the character of DNA packing in these particles experimentally. In this connection, the methods that make it possible to answer the question given earlier become especially important. One of the most significant methods is the method of circular dichroism (CD).

The experimentally measured CD spectra of an initial, linear, double-stranded DNA molecule (B-form, curve 1), and a dispersion formed by phase exclusion of the DNA molecules from water–salt–PEG-containing solution (curve 2), together with a thin layer of DNA cholesteric liquid-crystalline phase (20 µm-cell, curve 3) formed by the same DNA in a PEG-containing water–salt solution ($C_{PEG} > C^{cr}_{PEG}$) are compared in Figure 4.1. (where 1 is the left ordinate; 2, 3 is the right ordinate; $C_{PEG} = 170$ mg/mL; 0.3 M NaCl; 0.01 M Na-phosphate buffer, pH 6.8) [1]. The comparison of curves 1 and 2 shows that the formation of the DNA cholesteric phase results in the appearance of an intense negative band in CD spectrum in the region where the DNA nitrogen bases absorb. The shape of the band in the CD spectrum of the DNA cholesteric phase is similar to the shape of the band in the absorption spectrum, though the maximum of this band is noticeably "red" shifted ($\lambda \sim 300$ nm).

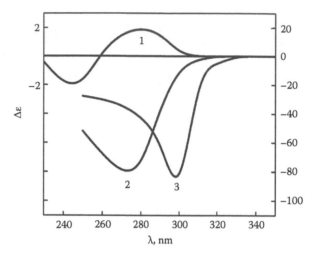

FIGURE 4.1 The CD spectra of B-form of double-stranded DNA water–salt solution (1 Left ordinate), DNA dispersion (2), and a thin layer of cholesteric liquid-crystalline phase of DNA (3).

In addition, in the case of the thin film of the DNA phase, it is difficult to obtain a perfect alignment of the cholesteric phase in respect to the light beam. Both the wavelength, which corresponds to the maximum of negative band in CD spectrum, and the amplitude of the intensive band depend on the geometry of the layer, on its position in respect to the circularly polarized light beam, on the thickness of the layer, and on homogeneity of the layer structure at the point of polarized light penetration. These parameters influence the real magnitude of amplitude of the intense band in the CD spectrum.

Furthermore, the circular dichroism evaluated for the negative band in the CD spectrum of the cholesteric phase is greater than that of the molar circular dichroism ($\Delta\varepsilon$, about 2 units), which reflects the properties of chromophores (nitrogen bases absorbing in the UV-region of the spectra) within the initial linear DNA molecule.

The question then is how to interpret the CD spectrum measured for the DNA dispersion. To answer this, a few facts have to be considered:

(1) Negative band. In the CD spectrum of dispersion, there is a negative band in the region of absorption of nitrogen bases (Figure 4.1), whose amplitude, as well as the amplitude of intense band typical of the thin layer of DNA cholesteric phase, is much higher than that of DNA nitrogen bases.

(2) Pitch of cholesteric spatial structure. The studies on cholesteric phases formed by low-molecular-mass DNA molecules (see Chapter 2) show that the pitch, **P**, of the cholesteric spatial structure varies within 2000–3000 nm, and therefore the so-called region of selective reflection is located far from the band of DNA nitrogen bases' absorption ($\lambda \sim 260$ nm). An extremely sharp increase in the band amplitude in the CD spectrum observed at formation of the DNA dispersion, which is noticeably above the CD signal for isolated DNA molecules in a solution, could not be explained by the diffraction effect (Bormann effect) that evolves when the regions of absorption and selective reflection of the cholesteric phase collide. This conclusion is confirmed by the experiments in measuring the local value of circular dichroism for relatively small sites (domains) in the film of DNA cholesteric phases with the help of the CD-microscope [4].

The condition of circularly polarized light selective reflection looks like

$$m\lambda = P \sin \alpha, \tag{4.1}$$

where m is the degree of diffraction reflection, λ is the wavelength in the environment ($\lambda_{medium} = \lambda_{vacuum}/n$), n is the refractive index of liquid crystal; and α is the angle between the light beam and the plane of the cholesteric phase.

For a sample of the cholesteric phase with a left-handed twist of the spatial structure, the light beam with a left circular polarization will be selectively reflected, while the light with right circular polarization will penetrate the sample without reflection [6]. In the case of the cholesteric structure with a right-handed spatial structure twist, the situation will be opposite to the one described.

For domains in the film with a planar orientation of the cholesteric phase located on its periphery and whose directions of cholesteric axes coincide with the CD-microscope optical axis and, consequently, $\sin \alpha = 1$, an intense band in CD spectrum was observed at $\lambda \sim 270$ nm. That is why the intense band in the CD spectrum of the cholesteric DNA phase in the region of the absorption of nitrogen bases cannot be explained by the oblique incidence of the light. The existence of an intense band is caused by collective effects evolving as the light penetrates the cholesteric phase with both dense helical packing of adjacent DNA molecules and ordered position of the molecules' absorbing dipoles. In the central part of the film of the cholesteric phase, the CD signal has the same form as for more perfect peripheral domains but smaller amplitude. This result shows that there is a large number of domains with different orientations of axes of cholesteric structures in this area of the phase. In this case, the averaging by axial orientations relatively to the polarized radiation beam direction leads to the decrease of amplitude of the intense band in the CD spectrum.

The negative sign of the band in the CD spectrum of the cholesteric phase means that the adjacent DNA molecules in this phase have a left-handed helical twist. This conclusion is based on the following simple assumption. The circular dichroism can be described by the equation

$$CD = A_L - A_R \tag{4.2}$$

where A_L and A_R is the absorption of left-polarized and right-polarized light, respectively.

The intensity of the penetrating light and the absorbance, A, are related by the classical Lambert–Beer law:

$$I = 10^{-A} I_0 \tag{4.3}$$

For example, if an electromagnetic wave with right circular polarization spreads along the cholesteric spiral axis, the momentary allocation of electric field vector directions forms a left-handed-twisted spiral corresponding to the directions of absorbing dipoles of the transition in the molecule, more than for the left-handed circular polarization. That is why, in this case, the absorbing elements conform to the relation $\mathbf{A_R} > \mathbf{A_L}$ and the circular dichroism has a negative sign. Let us note that this consideration is acceptable if the transition dipoles are mostly parallel to the molecule's axle, but if they form a certain angle in respect to the cholesteric axle, the observed picture becomes more complicated [7].

The relation between the sign of the band in the CD spectrum of a cholesteric DNA phase and the direction of the cholesteric helical twist was confirmed by the experiments on thin layers of DNA cholesteric phase when the change in the system's spatial twist by turning the glass plates, between which the phase was placed, caused the change in the circular dichroism sign [8]. The conclusion about the left-handed spatial twist of a cholesteric helix formed by right-handed twisted DNA molecules based on the analysis of CD spectra of liquid-crystalline phase coincides with the conclusion made earlier when analyzing the properties of this phase with the help of other methods.

(3) Intense band in CD spectrum of cholesteric DNA. The existence of the intense band in the CD spectrum of the cholesteric DNA phase is caused by abnormal absorption that evolves as the light is scattered in the phase with the adjacent DNA molecules' dense packing, but not by the change in the secondary structure of these molecules. This conclusion is based on the following assumptions:

The common expression of the CD signal looks like

$$\mathbf{CD} = (\mathbf{a_L} - \mathbf{a_R}) + (\mathbf{S_L} - \mathbf{S_R}) \tag{4.4}$$

where the term $(\mathbf{a_L} - \mathbf{a_R})$ is the contribution of differential absorption (in the region of chromophores absorption) to the CD signal, and the term $(\mathbf{S_L} - \mathbf{S_R})$ is attributed to the differential scattering in the region of wavelengths corresponding to the pitch of the cholesteric helical structure. For the left-handed twist, the sign $(\mathbf{S_L} - \mathbf{S_R})$ is positive because the helically twisted structure mostly scatters the circularly polarized light with the same direction of circular polarization [6].

The differential absorption is described with the equation

$$\mathbf{a_L} - \mathbf{a_R} = (\varepsilon_L - \varepsilon_R)\mathbf{Cl} + (\psi_L - \psi_R) \tag{4.5}$$

The term attributed to the differential absorption $(\varepsilon_L - \varepsilon_R)\mathbf{Cl}$ corresponds to a molecular circular dichroism of any type (A, B, or Z) of isolated linear double-stranded DNA molecules. The term $(\psi_L - \psi_R)$ corresponds to the far order of nitrogen bases' ordering in the cholesteric structure. The value of the term $(\psi_L - \psi_R)$ is determined by both the nitrogen bases' packing density and the cholesteric structure ordering extent. That is why, in such system as the DNA cholesteric phase, which is characterized not only by ordered arrangement of adjacent molecules but also a high packing density, and, consequently, high degree of ordering and high packing density of nitrogen bases, the value of this contribution can be several degrees higher than $(\varepsilon_L - \varepsilon_R)\mathbf{Cl}$ by absolute value, causing the existence of an intense band in the CD spectra.

To analyze the optical properties of the DNA dispersions, it is reasonable to consider the problem of interaction between the circularly polarized electromagnetic wave and the model system consisting of transition dipoles placed within the cubic lattice points and cholesterically ordered [9–11]. The approach based on the approximation of discrete dipoles was developed in studies from References 12 and 13.

Theoretical consideration [9–11] shows that the initial mechanism causing the increase in optical activity includes the collective excitation of the whole liquid-crystalline ensemble, which becomes possible because of the dipoles' long-range binding. The proper modes of dipole moments' oscillation evolve in a dispersion particle similarly to the normal modes evolving in dielectric conductors in response to the effect of the external field. Long-range binding becomes strong enough for the abnormal optical properties to appear only in the case of the three-dimensional ordering of the molecules with a density about one chromophore in one cubic nanometer and at the particles' size above or equal to a quarter of wavelength on the absorption maximum. As the data given in Chapter 2 show, the DNA dispersions conform to these requirements.

A phenomenological theory based on the theory of the optical properties of imperfect absorbing cholesteric liquid crystals [2,14], which makes it possible to examine the dependence of abnormal optical activity of DNA dispersions on the size of particles and the pitch of their helical structure [15], was applied to the calculation of the CD spectra. It is also assumed that the separate particles are small enough to justify the application of the kinematic approximation of the theory of diffraction to describe the optical properties of individual particles. An approach based on such assumptions was previously applied to imperfect nonabsorbing liquid crystals [16]. The calculation of the optical properties of the perfect absorbing cholesteric liquid crystals in the case of light propagation along the cholesteric axis has also been performed in References 17 and 18. This theory allows one to describe and predict many optical peculiarities of particles of double-stranded (ds) DNA dispersions formed in PEG-containing solutions. Here, the particles of DNA dispersions are considered polycrystalline objects with random distribution and the orientation of individual particles possessing their own absorption in the UV-region of the spectra due to the presence of nitrogen bases ("chromophores") in the content of DNA molecules.

The theory in Reference 15 takes into account both the layered structure of the packed double-stranded DNA molecules and the microscopic data shown above regarding these particles of dispersions. In particular, experimental data enable one to treat the particles of dispersions, formed by DNA molecules of low molecular mass (about 10^6 Da), as spheres of diameter, D, for which, because of the inherent rigidity of the secondary DNA structure, the liquid-crystalline ordering is specific.

It is known that the modification of the effective dielectric constant of a medium (ε_0) due to light scattering is proportional to the forward-scattering amplitude of an isolated scattering object measured at the frequency of the propagating light (ω_o). Therefore [16],

$$\varepsilon_{eff} = \varepsilon_0 + \frac{4\pi c^2}{V_0 \omega^2} \overline{\psi}(0) \tag{4.6}$$

where V_0 is the volume of a single dispersion particle, and $\overline{\psi}(0)$ is the forward-scattering amplitude averaged over all orientations of the particles, that is, over all possible directions of the reciprocal lattice vector τ.

Taking the perturbation theory to the second order in the framework of the kinematic diffraction approach, it follows that, for the case of spherical crystallites with diameter, $\mathbf{D}$, the forward-scattering amplitude, $\psi(0)$, nonaveraged over all orientations of vector τ for light of eigen polarization $(\mathbf{e}_\tau)$, which is diffracted by the helical structure, may be expressed as [16]

$$\psi(0) = \frac{\left[(\mathbf{Re}\delta)^2 - (\mathbf{Im}\delta)^2 + 2i\mathbf{Re}\delta\mathbf{Im}\delta\right]\left[\kappa_0^2\tau^2 + (\overrightarrow{\kappa_0\tau})^2\right](\overrightarrow{\kappa_0\tau})^2}{64\alpha^4\kappa_0^5\tau^4\varepsilon_0^2 i}$$

$$x\cdot[1+x^2/2-ix^3/3-(1+ix)\exp(-ix)]$$

(4.7)

where $\mathbf{x} = \alpha\,\kappa_0\mathbf{D}$, $\alpha = [\tau^2 + 2(\overrightarrow{\kappa_0\tau})]/2\kappa_0^2$, and $\mathbf{Re}\delta$ and $\mathbf{Im}\delta$ are real and imaginary parts of the dielectric anisotropy of cholesterics, respectively.

After averaging $\psi(0)$ over all possible orientations of the dispersed particles, the eigen polarizations can only be circular because, in a disordered, dispersed system, there is no preferential direction.

Carrying out the projecting polarization $(\mathbf{e}_\tau)$ over the circular ones and averaging $\psi(0)$ over all orientations of vectors (τ), one can obtain from the real and the imaginary parts of Equations 4.6 and 4.7 expressions that describe the optical rotation of the plane of polarization $(\alpha/\mathbf{L})$ and the coefficients of transmittance $(\mathbf{I}_+)$ of waves with circular polarizations:

$$\mathbf{I}_\pm = \exp\left[-\frac{3\kappa_0^2 D\mathbf{L}}{16\varepsilon_0^2}\int_{-1}^{1}\left\{\left[(\mathbf{Im}\delta)^2 - (\mathbf{Re}\delta)^2\right]\left(1+x^2/2-\cos x - x\sin x\right)+\right.\right.$$

$$\left.\left. + 2\mathbf{Re}\delta\mathbf{Im}\delta\left(\sin x - x\cos x - x^3/3\right)\right\}\left(1+y^2\right)(1+sy)^2 x^{-4}dy\right]$$

(4.8)

where $\mathbf{x} = \tau D(\mathbf{y} + \tau/2\kappa_0)$; τ is the reciprocal lattice vector, which is proportional to $4\pi/(\mathbf{P})$; κ_0 is the wave vector of incident radiation; $\mathbf{L}$ is the thickness of the "effective" DNA layer, determined from formula $\mathbf{L} = C\mathbf{l}/\rho$, where C and ρ are the concentration and density (g/cm^3) of DNA, respectively, $\mathbf{l}$ is the thickness of the optical cell, and $\delta = (\varepsilon_1 - \varepsilon_2)/2$ is the dielectric anisotropy; ε_1, $\varepsilon_2 = \varepsilon_3$ are the principal values of the tensor of dielectric permittivity; and $s = \pm 1$, depending on the sense of twist of the cholesteric helix.

Let us assume that an individual DNA dispersion particle has indeed a cholesteric structure with pitch $\mathbf{P}$ and a dielectric anisotropy δ, which contains resonant components for DNA as well as for the absorption band of an "external" chromophore molecule, that is, additional chemical or biologically active compounds, which can be incorporated into this system. For further consideration, it is important to note that, as the DNAs and the "external" chromophores have a local anisotropy of absorption, the anisotropy of the dielectric constant δ has an imaginary part for corresponding wavelengths. In the expression for dielectric anisotropy, it is convenient to specify the component originating from the absorption bands, that is, to present δ in the form

$$\delta = \bar{\delta} + \sum_i \delta_i \qquad (4.9)$$

To take into account the absorption of the nitrogen bases, as well as "external" chromophores such as, in particular, molecules of colored antibiotics, one can assume that one of the main values of ε_I and hence, of δ_i has the resonant form

$$\delta_i \approx \frac{\mathbf{r}_i}{\omega_{0_i}^2 - \Delta - \omega^2 + i\gamma_i\omega} \qquad (4.10)$$

where the components $\mathbf{r}_i$ are proportional to effects of the concentration of chromophores and the strength of the oscillator, and Δ is a factor conditioned by polarization of the medium.

Equations 4.8 and 4.9 make it possible to analyze theoretically the dependence of the shapes of CD spectra on such parameters as the dispersion particles' diameter, **D**, and the pitch, **P**, of the cholesteric twist, and the presence of additional chromophores with anisotropy δ_i within the DNA molecule.

A. There is an assumption according to which the low-molecular-mass double-stranded DNA molecules are as ordered in particles of dispersions as they are in the continuous cholesteric liquid-crystalline phase (moreover, helically twisted [cholesteric] packing is typical of these molecules). If this is true, then the anisotropy of absorption of nitrogen bases (chromophores) for the linear polarizations of the light must show itself as an intense band in the CD spectrum. The sign of the band in the CD spectrum in the region of the absorption of nitrogen bases will depend on the orientation of the planes of these components in respect to the long axis of the DNA molecule. The theoretically calculated CD spectrum of a DNA dispersion formed in a PEG-containing solution is given in Figure 4.2 (where $C_{DNA} = 10\ \mu m/mL$;

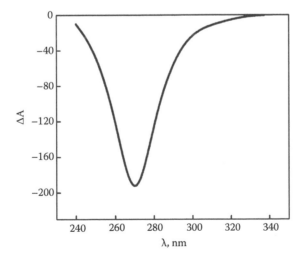

FIGURE 4.2 Theoretically calculated CD spectrum of DNA dispersion.

$\Delta A = (A_L - A_R)$ $(2.5 \cdot 10^5)$. One can see that, in accordance with the theoretical assumption given previously, the formation of dispersion, as well as a cholesteric phase (see Figure 4.1), is followed by a band in the CD spectrum located in the region of the nitrogen base absorption.

B. The shape of the band in the CD spectrum of the DNA dispersion is similar to the shape of its absorption band, while the maximum of the band in the CD spectrum is "red" shifted.

It can also be concluded from Figure 4.2 that the value of molecular circular dichroism ($\Delta\varepsilon$) calculated for the negative band in DNA dispersion CD spectra is about 100 units, and it is much larger than that of the value characterizing the molecular circular dichroism of nitrogen bases within the initial linear DNA molecules ($\Delta\varepsilon \sim 2.0$ units).

Intense bands in the CD spectra of both the cholesteric liquid-crystalline phase and DNA dispersion have the same sign (Figure 4.1). This points to the same twist of the cholesteric helix formed by DNA molecules ordered in the dispersion particles and in thin layers of the liquid-crystalline phase.

Double-stranded DNA molecules are ordered in the particle at distances of 3.0–5.0 nm; that is, they acquire the properties of a crystal, but molecules in the neighboring layers are mobile; that is, they retain the properties of a liquid. To stress this fact, the term "liquid-crystalline dispersions" (LCDs) was used to signify these dispersions.

According to the foregoing theoretical considerations, the intense band in the CD spectrum located in the absorption region of chromophores incorporated in the content of the bulk liquid-crystalline phase is direct evidence of the formation of helically twisted structure. Besides, the sense of helicity is reflected in the sign of the intense band in the CD spectrum (i.e., a positive band reflects a right-handed helical arrangement, and a negative band reflects a left-handed twist of neighboring quasi-nematic layers of DNA molecules). Random packing of nucleic acids was unaccompanied by any considerable change in the amplitude of the CD band, except those produced in the secondary structure.

An appearance of intense bands both in the CD spectra of the DNA liquid-crystalline phase and the dispersion means that the purine and pyrimidine nitrogen bases do play the role of "chromophores," providing information about the spatial packing of DNA molecules both in the bulk phase and in the particles of liquid-crystalline dispersions.

The coincidence of similar shapes in the theoretically calculated and experimentally measured CD spectra for the DNA dispersions shows that the method used to calculate the CD spectra of these objects, although phenomenological in its background, describes the optical properties of the DNA dispersions properly. This allows one to draw the conclusion that the appearance of an intense band in the CD spectra unequivocally reflects the left-handed twist of neighboring quasi-nematic layers of DNA molecules packed not only in the bulk phase but also in the particles of dispersion. To stress the twisting of quasi-nematic layers, the term "cholesteric liquid-crystalline dispersions" (CLCDs) was used to signify these dispersions [15]. This deduction corresponds to a common statement that the cholesteric packing is a specific property of any molecule having geometrical and optical anisotropy. Hence, rigid, anisotropic, double-stranded nucleic acid molecules (DNA, RNA, etc.) tend to realize their potential tendency to the cholesteric mode of packing in the particles of LCDs.

Figure 4.1 shows that the bands in the CD spectra both for DNA CLC and DNA CLCD have shapes similar to the band of DNA absorption (Figure 3.2), but the maxima of the bands in the CD spectra are "red" shifted ($\lambda \sim 270$–300 nm). The experimental displacement of the maximum of the band in the CD spectra for DNA CLC and DNA CLCD compared to the DNA absorption band is connected with diffraction effects (together with the known polarization displacement due to the "effective field" effects acting on molecules in the cholesteric phase). This displacement should, in principle, depend on the size of the particles of the DNA CLCDs.

Finally, one can conclude that the negative sign of the band in the CD spectrum (Figure 4.1) proves the left-handed cholesteric twist of quasi-nematic layers formed by the right-handed DNA molecules (B-form) in particles of LCDs.

Here, additional remarks are necessary. The value of $\Delta\varepsilon_{270}$ ($\sim$–100 units), which can be used to "reflect" the value of circular dichroism of the nitrogen bases in the structures of DNA CLC and DNA CLCD, is far larger than the molecular circular dichroism, $\Delta\varepsilon$ ($\sim$2.5 units, Figure 4.1), that is, the physical constant usually used for description of the peculiarities of isolated nitrogen bases and individual molecules of nucleic acids. The value of $\Delta\varepsilon$, as the constant in the case of isolated nitrogen bases as well as of individual, initial, linear DNA molecules, can be calculated theoretically. To evaluate the CD of nitrogen bases, and hence, to estimate the value of $\Delta\varepsilon$ it is necessary to calculate dipole strengths, rotational strengths, and transition frequencies. The key assumption in the theoretical calculations is that there is no electron exchange between nitrogen bases in solutions. There are two different strategies used in calculating their optical properties, depending on whether the corresponding transition in the nitrogen bases is electrically allowed (such as $\pi\rightarrow\pi^*$) or magnetically allowed (such as n$\rightarrow\pi^*$). Under the standard conditions (in solution), a density of "chromophores" (nitrogen bases) is much less that 1 chromophore/nm^2. In these circumstances, the delocalization of the excitation via the various coupling mechanisms is not significant and will involve only nearest-neighbor delocalization. Such a system will show a small, conservative CD spectrum (Figure 4.1). The calculations also show that the alterations of the double-stranded DNA secondary structure (denaturation, transition between the conformations such as A-, C-, or Z-forms) are accompanied by a change in the $\Delta\varepsilon$ value within 1–5 units. On the other hand, one can see that, in the case of the DNA CLCDs or the cholesteric liquid-crystalline phase, the CD spectra do not resemble the spectra of any known DNA conformations.

Despite the intuitively reasonable "feeling" that twisting of quasi-nematic layers should give rise to n intense band in the CD spectra, an examination of known theories of optical activity shows that a chiral arrangement by itself is not enough for amplification of the band in the CD spectrum. The chromophores (nitrogen bases, in our case) must also be coupled in some way, that is, the absorption of one chromophore must be affected by the presence of other chromophores and by the chiral relationship they bear to each other. A coupling between the excitation of neighboring choromophors (nitrogen bases) results in transitions with new polarizations and energies. This means that the individual nitrogen bases in quasi-nematic layers formed by DNA molecules must not respond independently to the incident light. Hence, the nitrogen bases in the content of neighboring DNA molecules in quasi-nematic layers could be significantly coupled to each other. This effect will occur in large molecular spatial structures due to their ability to delocalize their excitations and respond collectively to the incident radiation. One can stress that the relative importance of various mechanisms of coupling between any pairs of dipoles (for instance, exciton coupling, vibronic coupling, crystal field mixing) will depend on the dimensionality of the formed structure. The long-range couplings are possible if the liquid-crystalline system is large and dense enough. These effects are never observed in isotropic solutions of nitrogen bases or isolated DNA molecules.

However, the condensation of DNA molecules, by itself, is not a sufficient condition for the appearance of an intense band in the CD spectrum since many aggregated forms of DNA (for instance, the aggregates formed by single-stranded DNA molecules) fail to show the intense band in the CD spectra. Therefore, on the one hand, an appearance of an intense band in the CD spectrum is connected with the long-range coupling of the dipoles of nitrogen bases (i.e., the nitrogen bases are the main contributors to the optical behavior of the CLCD), in the amplitude of an intense band in the CD spectrum. On the other hand, taking into account that nitrogen bases are fixed rigidly enough in the secondary DNA structure, an appearance of this band reflects the specific type condensation of the DNA molecules, that is, the helical array of DNA molecules. By a helical array, one can consider again a parallel organization of DNA molecules that is then twisted slightly so that each quasi-nematic layer of DNA molecules is at a slight angle of twist with respect to the two neighboring ones. As the DNA molecules condense into particles of dispersion, the local density of nitrogen bases increases to the point at which significant delocalization can occur. This means that the amplitude of an intense band in the CD spectra of the DNA CLCDs, formed by ds DNA molecules that possess native secondary structure and fixed properties of nitrogen bases, is connected with both the long-range collective behavior of the nitrogen bases and a local density of DNA molecules in the particles of CLCDs. This, in turn, depends on the secondary structure of these molecules and on the properties of the solvent. This means that the amplitude of the intense band in the CD spectra can vary in a broad range of values. Hence, the value of $\Delta\varepsilon$ is not a constant. Therefore, the use of the $\Delta\varepsilon$ value may have only an illustrative character, and it seems to be illogical to apply this parameter for the comparison of the peculiarities of the CD spectra of various DNA CLCDs. To stress the difference between "molecular" and a so-called structural circular dichroism, the term ***abnormal band*** was used to signify an intense band in the CD spectrum [15]. With an allowance for this, the use of the amplitude of this band in CD spectra, expressed simply as experimentally measured $\Delta\mathbf{A}$ value, is more reasonable. This value will be used in this book very often.

C. The value of the abnormal band amplitude in the CD spectra of DNA CLCDs depends on both the size of dispersion particles and the pitch, **P,** value of the cholesteric structure.

The theoretical CD spectra of DNA CLCDs whose particles' size varies while the cholesteric helical pitch is constant are given in Figure 4.3, where the particles'

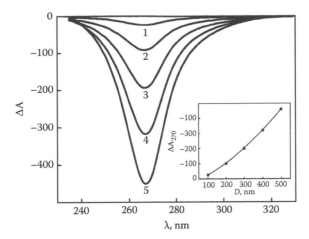

FIGURE 4.3 Theoretically calculated CD spectra of DNA CLCDs with different size of particles.

diameter is equal to 100, 200, 300, 400, and 500 nm, respectively, (curves 1–5); P = 2,000 nm; $\Delta A = (A_L - A_R) \cdot (2.5 \cdot 10^{-5})$. The amplitude of the abnormal band in the spectra depends on the CLCD particles' diameter and increases as the diameter increases (see Figure 4.3, where the dependence of ΔA on the particles' diameter is represented).

The important conclusion following from these results is that the "size effect" takes place in the case of DNA CLCD. The calculation shows that, if the diameter of a DNA CLCD particle reaches the minimal value about of 50 nm, the amplitude of the intense band in the CD spectra decreases so sharply that it no longer can be distinguished from that in the CD spectrum typical of the initial linear DNA [19]. This result means that, in the case of the formation of CLCDs with a diameter of particles ~50 nm, their presence cannot be registered by CD spectroscopy.

The effect of a cholesteric helical twist on the abnormal band amplitude in the CD spectra can be evaluated by fixing the DNA CLCD particles' size (assuming that the size, **D**, equals to, for instance, 5,000 Å). The calculated CD spectra of DNA CLCD whose particles are characterized by a cholesteric twist with a different pitch, **P**, value are represented in Figure 4.4 (where **P** = 2,000, 4,000, 6,000, 8,000, and 10,000 nm [curves 1–5, respectively]; **D** = 500 nm; $\Delta A = (A_L - A_R) \cdot (2.5 \cdot 10^{-5})$). It can be noticed that the smaller the **P** value of the DNA cholesteric structure (i.e., the greater the twist—the angle between the adjacent quasi-nematic layers of DNA in the helical structure of the cholesteric), the more intense the band in the CD spectrum (see Figure 4.4). Conversely, the more untwisted the cholesteric structure, the lower the amplitude of the band in the CD spectrum of the CLCD.

The theoretical treatment has also shown that at a **P** value of about 30 µm and with the constant structural properties of DNA molecules, the amplitude of the negative band in the CD spectrum is quite close to the amplitude of the band characteristic of isolated linear DNA molecules (Figure 4.1).

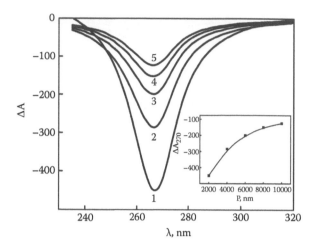

FIGURE 4.4 Theoretically calculated CD spectra of DNA CLCDs whose particles have different values of cholesteric pitch.

D. The sign of the band in CD spectra of DNA CLCDs depends on the sense of twist of the spatial cholesteric structure formed by the DNA molecules packed into CLCD particles (Figure 4.5, where $C_{DNA} = 10$ μg/mL; $\Delta A = (A_L - A_R) \cdot (2,5 \cdot 10^{-5})$).

This result of theoretical calculations means that the change in sense of the DNA molecules' packing causes the change of the negative sign of the intense band to positive, while the CD spectra shape does not change. In the framework of the foregoing theoretical considerations, there is a very important question regarding the role of the secondary structure of nucleic acids in the appearance of the intense band in the CD spectrum.

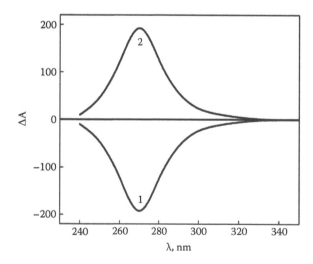

FIGURE 4.5 Theoretically calculated CD spectra of DNA CLCDs with left- (curve 1) and right-handed (curve 2) twist of spatial structure.

Indeed, experiments showed that CLCDs formed in a PEG-containing solution by right-handed, synthetic, double-stranded polynucleotides of the B-family (poly[dA] × poly[dT] and poly[dA-dT] × poly[dA-dT] with identical molecular mass but different nitrogen base sequences in the chains) have the same light scattering but differ in the signs of the intense band in the CD spectrum. The CD spectra are mirror images of each other. In view of the theoretical calculations, the change in the sign of the band in the CD spectrum shows that, in contrast to the left-handed cholesteric dispersions formed by molecules of poly(dA-dT) × poly(dA-dT), molecules of poly(dA) × poly(dT) form CLCDs with a right-handed twist.

Hence, very small alterations in the base pair sequence or in parameters of the secondary structure of molecules can be sufficient to cause the change from the left-handed to the right-handed twist of the quasi-nematic layers of the particles of CLCD. This means that these are two types (left-handed and right-handed) of cholesteric structures formed by double-stranded nucleic acids.

E. The theoretical analysis based on experimental data describing the process of double-stranded nucleic acids' phase exclusion (DNA and RNA) makes it possible to represent the formation of CLCD particles in these molecules as a scheme in Figure 4.6. Here, the CLCD particle is shown as an oval (2); the structure of the quasi-nematic layer with distance, **d**, between the adjacent DNA molecules is also shown. The formation of CLCD is followed by an appearance of the abnormal band in the CD spectra (3).

Judging by the picture, the CLCD particles' formation from native double-stranded nucleic acid molecules is followed by the appearance of an abnormal band in the CD spectrum that, as a rule, has a negative sign in the case of DNA molecules.

Figure 4.6 also shows that an extra chromophore whose optical properties are significantly different from the optical properties of the DNA nitrogen bases can be chemically incorporated into the secondary structure of neighboring DNA molecules, and, consequently, into the structure of the CLCD particle quasi-nematic layer.

The theory just considered predicts the appearance of an abnormal band in CD spectra if CLCD particles formed by double-stranded DNA molecules bind with compounds rigidly fixed in respect to the long axis of the DNA molecules (so-called external chromophores with an absorption band that does not coincide with the DNA

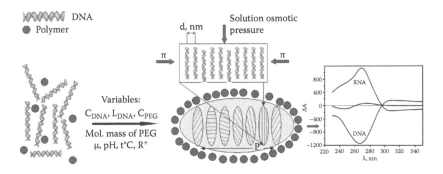

FIGURE 4.6 (*See color insert.*) Scheme illustrating the formation of CLCD particles from DNA molecules at their phase exclusion from water–salt polymer-containing solution.

nitrogen base absorption band). Meanwhile, at the low concentration of "external" chromophores introduced into the structure of particles of DNA CLCD, the mode of packing of these macromolecules in the CLCD particles will not change. Such a situation is possible, in particular, at the insertion (intercalation) of colored biologically active compounds, for instance, antibiotics of the anthracycline group, between the DNA base pairs. As all of the theory's propositions are applicable both to DNA chromophores (nitrogen bases) and "external" chromophores introduced into the structure of CLCD particles, for this case the theory predicts the appearance of two bands located in different regions of the CD spectrum.

The theoretically calculated CD spectra for DNA CLCD particles treated with colored compound—an antibiotic of the anthraquinon group, such as mitoxantrone (MX) and an antibiotic of anthracycline group, that is, daunomycin (DAU)—are given in Figure 4.7 and Figure 4.8. In both cases, the incorporation of colored compounds ("external" chromophores) into the DNA secondary structure and, consequently, into the CLCD particle structure, is accompanied by the appearance of two bands in different regions of the CD spectra.

One of the bands is located in the region of DNA chromophore (nitrogen base) absorption ($\lambda \sim 270$ nm); the other one is situated in the region of the antibiotic chromophores' absorption ($\lambda \sim 700$ nm for MX and $\lambda \sim 500$ nm for DAU). The appearance of the two bands in different regions of the CD spectra is definite evidence in favor of the cholesteric mode of packing of the DNA molecules in the CLCD particles. Both of the bands have negative signs, and their amplitudes are much larger in comparison to that typical of the molecular circular dichroism of nitrogen bases and antibiotics' chromophores. The coincidence of the signs of these bands shows that the antibiotics' molecules are located in respect to the DNA molecules' long axis the same way as the angle between the "external" chromophore molecules' plane and the DNA molecules' long axis in the nitrogen base pairs is close to 90°. Such coincidence of the signs of two abnormal bands in different regions of the CD spectrum is

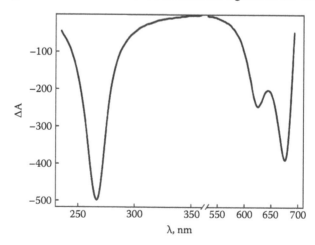

FIGURE 4.7 Theoretically calculated CD spectrum of DNA CLCD treated with mitoxantrone.

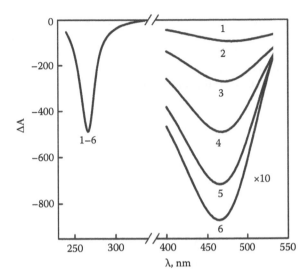

FIGURE 4.8   Theoretically calculated CD spectra of DNA CLCD treated with daunomycin.

only possible if the "external" chromophore molecules intercalate between the DNA nitrogen base pairs. As Figure 4.8 shows, the amplitude of the abnormal band in the region of colored compound absorption, in this case DAU, increases as the extent of DAU molecules' intercalation between the DNA nitrogen base pairs increases in reaching the equilibrium state. It is interesting that, in such cases, the amplitude of the abnormal band in the DNA absorption area does not change practically, which points to a very slight (if any) perturbation of the CLCD structure.

However, if the DNA secondary structure is noticeably altered because of the action of physical or chemical factors, the abnormal band in the CD spectra of DNA CLCD particles begins to decrease up to full disappearance.

4.2 EFFECT OF DIFFERENT FACTORS ON FORMATION AND PROPERTIES OF CLCD PARTICLES

4.2.1 Factors Determining CLCD Particle's Structure at "Moment of Formation"

4.2.1.1 DNA Molecule's Length

Unlike the common DNA molecules' aggregation, which can occur at any molecules' length and is not followed by the appearance of abnormal optical activity, the DNA CLCD is formed only in such cases when the molecules' length is above 150 Å (~50 base pairs). This result shows that there is a lower limit in the DNA molecules' length at which the optically active dispersion is formed. The upper DNA molecular mass limit at which the molecules maintain the ability to form CLCD is close to $3 \cdot 10^6$ Da [20,21]. Further increase in the DNA molecules' length causes decrease in the abnormal band amplitude in the CD spectrum of CLCDs, despite the formation detected by the appearance of the "apparent" optical density in the absorption spectrum. At a DNA molecular mass of more than $10 \cdot 10^6$ Da, the CD spectra of the formed dispersions differ only slightly from the CD spectra of the linear DNA. The absence of abnormal optical activity at the formation of dispersion by high-molecular-mass DNA molecules shows that the role of kinetic factors in the DNA packing process is very important.

4.2.1.2 Structure of Rigid, Double-Stranded, Nucleic Acid Molecules

Only DNA (RNA) rigid, linear, double-stranded molecules can form dispersions with cholesteric spatial packing of these molecules in their particles by phase exclusion. The detailed study of dispersions formed by a different synthetic (polyribonucleotides of A-family) showed that rigid, double-stranded RNA or poly(I) × poly(C) molecules create two types of CLCDs, which differ in the sign of the intense bands in the CD spectra. The transition between dispersions, which differ in the sign of this intense band, can be caused by changing either the PEG concentration in the solution (for RNA) or the ionic strength (for poly(I) × poly(C)). Additionally, rigid poly(dG-dC) × poly(dG-dC) molecules at high ionic strength ($\mu > 2.6$), belonging to the left-handed, helical Z-form, can generate two families of dispersions that differ only in the signs of the intense bands in the CD spectra.

These results mean that very small alterations in the structures of nucleic acids can be sufficient to cause the change from a left-handed to a right-handed spatial twist in the structure of the CLCD particles, and the "classical" cholesteric with the left-handed twist of quasi-nematic layers, formed by double-stranded DNA molecules, can be easily transformed into a right-handed twist in its spatial structure as a result of very fine changes in the properties of these molecules. This means that the minor peculiarities of the secondary structure of nucleic acids have a profound influence on the character of interaction between nucleic acids "at the moment of their recognition at close approach" and mutual twisting of these molecules at the formation of the CLCD particles. Hence, rigid, double-stranded nucleic acid molecules belonging to different families (right-handed B- and A- or left-handed Z-forms) can make CLCDs, which are characterized by the CD spectra with intense bands of different signs in the absorption region of the nitrogen bases.

The formation of dispersions from flexible single-stranded molecules of nucleic acids characterized by flexible structures and flexible orientation of their nitrogen bases does not result in an appearance of an abnormal band in the CD spectrum.

The most important event at the formation of CLCD particles by the rigid double-stranded DNA (RNA) is the "recognition" of neighboring nucleic acid molecules as they approach each other, caused by phase exclusion. The peculiarities of the "recognition" process depend on the peculiarities of the surfaces of these molecules and the solvent's properties. That is why it is necessary to determine the factors that can affect the mode by which nucleic acid molecules pack in CLCD particles. These factors can be divided into two groups: those acting at the moment of the CLCD particles' formation and those acting after the particles are formed.

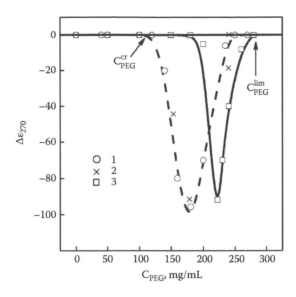

FIGURE 4.9   The dependence of the abnormal negative band amplitude in the CD spectrum of DNA ($\lambda = 270$ nm) on C_{PEG} for solutions of different salts.

4.2.1.3   Osmotic Pressure of PEG-Containing Solutions

The process of DNA (RNA) CLCD particle formation depends on the osmotic pressure of the solution determined by both the concentration of PEG in the solution and its molecular mass.

The increase in PEG concentration in a water–salt solution with a moderate ($\mu \sim$ 0.15...0.3) ionic strength where the DNA CLCD form is followed by not only the appearance of an abnormal band in the CD spectrum when the "critical" PEG concentration is reached but also its disappearance as the limit PEG (C^{lim}_{PEG}) concentration ~ 260–300 mg/mL in the solution is reached (Figure 4.9, where (1) = NaCl; (2) = $NaClO_4$; (3) = CsCl; PEG molecular mass = 4,000; and concentration of salts = 0.3 M).

The figure shows that, at a low concentration ($C_{PEG} < 150$ mg/mL), DNA molecules are in an isotropic state, and there is only a low-intensity band in their CD spectra typical of the DNA B-form (see Figure 4.1). As the concentration increases, particles with abnormal negative bands corresponding to CLCD in their spectra are formed. At $C_{PEG} > 300$ mg/mL, dispersion without any abnormal optical activity is formed.

The process of DNA (RNA) CLCD formation depends on PEG molecular mass. The curves given in Figure 4.10 describe the dependence of C^{cr}_{PEG} on PEG molecular mass at DNA CLCD formation in NaCl (curve 1) and CsCl (curve 2) solutions. It is noticeable that, regardless of the equal PEG concentrations in the solutions, the dispersion is formed more efficiently in an NaCl solution.

At the concentration corresponding to the area below curve 1 in the case of NaCl (curve 2 in the case of CsCl), the nucleic acid molecules (in particular, DNA) are

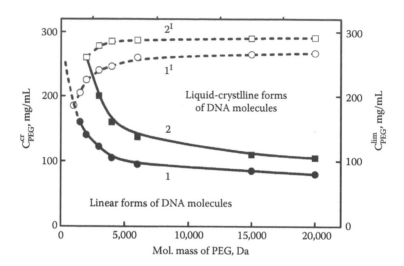

FIGURE 4.10 The dependence of the C^{cr}_{PEG}-value and the C^{lim}_{PEG}-value on molecular mass of PEG.

in a linear, isotropic state. Meanwhile, the dependence of C^{cr}_{PEG} necessary for the formation of such DNA CLCD particles with an abnormal band in their CD spectra (given in Figure 4.10) practically coincides with the dependence of C^{cr}_{PEG} on the PEG molecular mass determined earlier by the appearance of an "apparent" optical density in the DNA absorption spectra. Considering the fact that the PEG concentration at which the specific molecular structure of the PEG-containing solvent (determined by the dependence of viscosity of PEG solutions on PEG molecular mass) is formed is quite close to C^{cr}_{PEG}, the data shows that the efficiency of the DNA CLCD formation process is determined not only by the solvent's molecular structure but also by the efficiency of the DNA phosphate groups' screening with counterions. Consequently, regarding the properties of the PEG solution corresponding to the concentration above curves 1 and 2 (in the case of NaCl and CsCl, respectively), the nucleic acid molecules form liquid-crystalline dispersions with a cholesteric packing mode of adjacent molecules. Meanwhile, the analysis of the CD spectra shows that a set of properties limited by the C^{lim}_{PEG} value (i.e., such concentration above which DNA LCD particles are still formed but do not have an abnormal band in their CD spectra) is typical of PEG solutions in any PEG molecular mass.

Therefore, curves 1 and 2 form a "boundary," that is, the area below these curves corresponds to PEG solutions where DNA molecules exist in linear (isotropic) form, and the area above the curves corresponds to the solutions where DNA CLCDs are formed.

4.2.1.4 Effect of Ionic Composition of the Solvent on DNA CLCD Formation

The screening of negatively charged phosphate groups with cations plays a significant role in nucleic acid molecule condensation both *in vivo* and *in vitro*. In this connection, the effect of alkali and alkali–earth metal salts (whose cations screen DNA phosphate groups but do not form specific complexes with nitrogen bases) on CLCD formation in water–salt PEG solutions was investigated.

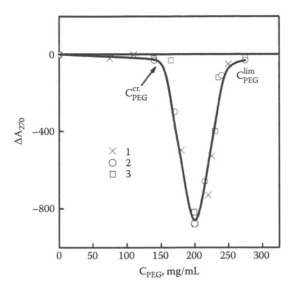

FIGURE 4.11   The dependence of abnormal negative band amplitude in DNA CD spectrum ($\lambda = 270$ nm) on C_{PEG} for solutions of different salts: 1 – KCL; 2 – KBr; 3 – KI.

The curves given in Figure 4.11 describe the dependence of abnormal negative band amplitude (expressed as $\Delta\varepsilon_{270}$ value) on the CD spectra of DNA CLCDs formed in 0.3 M solutions of different salts containing the same cation, while having different anions on C_{PEG}. It is noticeable that the type of $\Delta\varepsilon_{270}$ changes that are depending on C_{PEG} for all the salts is the same and similar to the dependence for NaCl (see Figure 4.9). Consequently, the efficiency of DNA CLCD formation is determined by the nature of the cation, while the anion nature does not exert any impact on this process (at least in solutions with a salt concentration under 0.3 M). A similar situation takes place in the case of DNA and PEG samples with different molecular mass.

Therefore, the received data show that the formation of DNA CLCD in solutions of different salts takes place only under the condition of DNA phosphate groups' screening, as determined by the nature of the cation.

A comparison of DNA CLCD particles' formation in PEG-containing solutions of NaCl, KCl, CsCl, and LiCl with different ionic strength has shown that the process is realized most efficiently in NaCl solutions. The data describing the dependence of C^{cr}_{PEG} on cation radius are given in the Figure 4.12. The minimal value of C^{cr}_{PEG} at any ionic strength is achieved when using Na$^+$ ions. For example, at the formation of DNA CLCD in CsCl solutions, C^{cr}_{PEG} is greater than the value corresponding to NaCl solutions. The fact that the formation of DNA CLCD in NaCl solutions occurs at the lowest PEG concentration can be understood considering the nature of DNA polyphosphate ([22–24]; also see Chapter 1). The more efficient the DNA phosphate groups' screening with cations, the more the DNA molecules aspire to the phase exclusion and the formation of CLCD.

It should be added that, in the absence of PEG, DNA sodium salts remain soluble up to an NaCl concentration corresponding to the saturated solution (~ 5.3 M), but

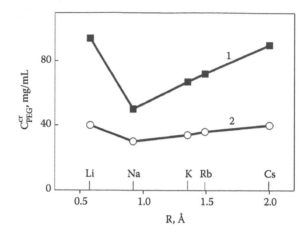

FIGURE 4.12   The dependence of C^{cr}_{PEG} on the radius of alkali metal cations in a solution; 1 and 2: the solution ionic strength 0.3 and 0.7, respectively; PEG molecular mass = 20,000.

only LiCl 10 M solution and 6.8 M CsCl solution provide the conditions for incompatibility of DNA salts with the solvent [25]. In other words, though in PEG-containing solutions a certain extent of DNA phosphate groups' screening is required, the conditions of DNA incompatibility are created in these solutions at a lower extent of the screening of negative charges of phosphate groups, in comparison to the screening in water–salt solutions without PEG.

The dependence of PEG-critical concentration on NaCl concentration in the solution is given in Figure 4.13 (PEG molecular mass = 4000; the area of DNA dispersion particles with abnormal optical activity is shaded). According to this, a lower C^{cr}_{PEG} value corresponds to higher solution ionic strength, that is, the higher the efficiency

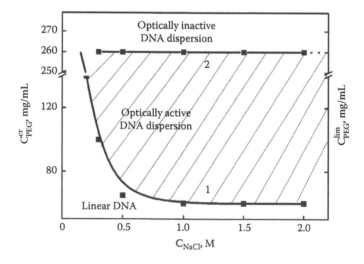

FIGURE 4.13   The dependence of C^{cr}_{PEG} (1) and C^{lim}_{PEG} (2) on NaCl concentration in PEG solutions.

of the phosphate groups' screening, the lower the concentration of PEG required for the formation of DNA CLCD. It is significant that the maximum amplitude of the negative band in the CD spectra of DNA CLCDs formed at different ionic strengths remains constant, regardless of the ionic strength of the solution. The straight line 2 in Figure 4.13 is drawn over the points corresponding to the "limiting" PEG concentration, that is, the concentration at which DNA LCD particles are formed but have no abnormal band in the CD spectra. Regardless of the existence of the dependence of the "critical" PEG concentration on the solution's ionic strength (curve 1), the disappearance of the abnormal band in the DNA CLCD CD spectrum occurs at C_{PEG} ~ 260 mg/mL, regardless of the solution's ionic strength.

The comparison of Figure 4.13 to Figure 4.10 shows that the area of DNA CLCD with abnormal optical activity is limited not only by the PEG concentration but also by the ionic strength of the solution. The area of PEG solutions with different ionic strengths (Figure 4.13) under curve 1 can be called the area of DNA isotropic phase; the area between curves 1 and 2 corresponds to the DNA dispersion with abnormal optical activity, while the area above curve 2 corresponds to DNA dispersions whose particles have no abnormal optical activity. This means that CLCD particles exist only within the limits of a certain solvent osmotic pressure interval determined by the PEG concentration in the solution. Crossing the lower limit causes the transition to the isotropic state, while crossing the upper limit leads to a transition to the area of existence of particles with no abnormal optical activity.

Consequently, the spatial structure of DNA cholesterics is quite flexible. The fact that DNA CLCDs are formed in water–polymer solutions with a certain osmotic pressure and dielectric constant proves the important role of the solvent in the determination of the sense of spatial twists of the cholesteric structure.

It can be concluded from Figure 4.12 that the formation of DNA CLCD at a fixed PEG concentration can be induced by the change in cation content without changing the total cation concentration in the solution. This prompts the assumption that the CLCD formation process can be initiated by the change of Ka^+ ions to Na^+ ions ($K^+ \rightarrow Na^+$), which are bound to DNA more efficiently, while the process of DNA molecules' transition from the condensed state to the isotropic state can be caused by the reverse cation change ($Na^+ \rightarrow K^+$).

Therefore, the efficiency of DNA CLCD formation in PEG-containing water–salt solutions can be regulated not only by changing the molecular structure of the solvent but also by changing one cation to another as a result of the different efficiencies of cations' binding with DNA conducted and strengthened under the conditions of a structured solvent. There are reasons to assume that these peculiarities of the DNA CLCD particle formation and its destruction processes at certain properties of water–salt PEG solutions reflect the main features and factors of the corresponding processes in viruses and chromosomes.

The fact that the abnormal band in a CD spectrum typical of DNA LC thin layer also exists in the case of DNA CLCD made it possible to determine the "local" parameters of phases formed from CLCD particles as a result of their low-speed centrifugation. The study of small-angle x-ray scattering from such phases formed at a certain PEG concentration (i.e., in solutions with a certain osmotic pressure) makes it

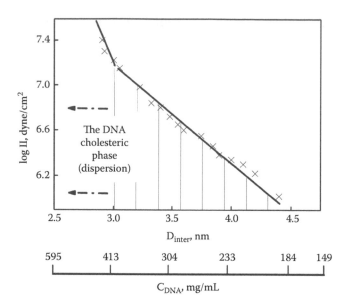

FIGURE 4.14 The dependence of the average distance (D) between DNA molecules in the phases formed in PEG-containing water–salt solutions on the osmotic pressure of the solutions.

possible not only to verify the assumption about the ordered state of DNA molecules in CLCD particles but also to evaluate DNA concentration in CLCD particles [26].

The x-ray parameters of the DNA cholesteric liquid-crystalline phases and their CD spectra, pointing to the existence of dispersions with different optical properties, permit one to compare the borders for existence of main packing types of DNA molecules in phases and dispersions. (Figure 4.14, where the area of DNA cholesteric liquid-crystal dispersion is shaded. The second horizontal axis corresponds to DNA concentration in phases formed as a result of low-speed centrifugation of particles of DNA liquid-crystalline dispersions.) It is noticeable that CLCD particles are formed at the solution osmotic pressure that provides the distance between the molecules within 30–50 Å limit. At the PEG solution osmotic pressure that reduces the distance between DNA (RNA) molecules to 29 Å or less, these molecules form a hexagonal phase that has no abnormal optical activity, while the right-handed helical twist and initial secondary structure of individual nucleic acid molecules is still maintained.

The phase diagram represented in Figure 4.14 drawn on the basis of properties of DNA dispersions in water–polymeric solutions corresponds quite exactly to the diagram of DNA phases received earlier for DNA water–salt solutions (see the Introduction).

4.2.1.5 Ionic Content of Solutions and Efficiency of DNA CLCD Formation

The results compared in Figure 4.15 demonstrate the efficiency of DNA CLCD formation in PEG-containing solutions of NaCl and $MgCl_2$ [27]. This comparison shows that there are significant differences in the details of the CLCD formation in these solutions. First, the DNA CLCD formed in $MgCl_2$-containing PEG solutions

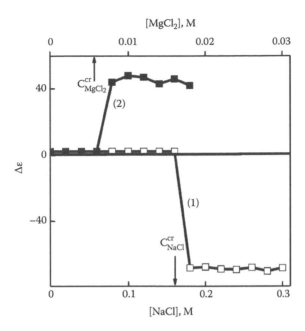

FIGURE 4.15 The dependence of the band amplitude in the CD spectra (λ = 270 nm) of PEG-containing DNA solutions on the concentrations of NaCl (1) and MgCl$_2$ (2). (C_{PEG} = 170 mg/mL; PEG molecular mass = 4000).

are characterized by an abnormal band in the CD spectrum that has a positive—not negative sign. The different signs of the bands in the CD spectra of the DNA CLCD formed in NaCl and MgCl$_2$ solutions show that the character of interaction between DNA molecules neutralized by Na$^+$ and Mg^{2+} cations is different. This difference is a sufficient condition for the formation of two DNA CLCD "families" that differ by the signs of the abnormal optical activity. As the sign of the abnormal band in the CD spectrum of CLCD is determined by the direction of DNA molecules' helical twist in CLCD particles, according to the results of theoretical calculations, Figure 4.15 shows that the DNA CLCD formed in NaCl and MgCl$_2$ PEG-containing solutions have different directions of the cholesteric helical twist. Second, there is a significant difference between the "critical" concentrations of Na$^+$ and Mg^{2+} cations necessary for the formation of DNA CLCD particles. At fixed PEG concentration (for instance, at C_{PEG} = 170 mg/mL and PEG molecular mass = 4000), the "critical" concentrations of NaCl and MgCl$_2$ are equal to 0.15 and 0.003 M, respectively. This result corresponds to the data according to which the screening of negatively charged DNA phosphate groups with Mg^{2+} ions is more efficient than screening with Na$^+$ ions.

The data in Figure 4.16 show that the direction of the cholesteric spatial twist in DNA CLCD particles is determined by the relation between Na$^+$ and Mg^{2+} cations in the PEG-containing solution. At the relation of Na$^+$/Mg^{2+} > 4, the DNA CLCD have left-handed spatial twist; at the relation of Na$^+$/Mg^{2+} < 4, the spatial twist is right-handed [27]. This means that, by changing the Na$^+$/Mg^{2+} relation in the solution, the mode of DNA molecules' spatial ordering in CLCD particles can be regulated.

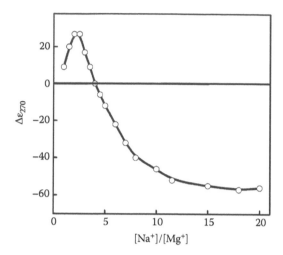

FIGURE 4.16   The dependence of $\Delta\varepsilon_{270}$ value in the CD spectra of DNA CLCDs on the relation of molar concentration of (Na^+/Mg^{2+}) cations in a PEG-containing water–salt solution.

Meanwhile, the higher the Na^+ ions concentration determining the formation of DNA CLCD with a negative band in the CD spectra, the higher concentration of Mg^{2+} ions necessary for the formation of CLCD with the positive band in the CD spectra under such conditions. This means that, under certain concentrations, there appears to be competition between Na^+ and Mg^{2+} cations for the DNA-binding sites. As there are certain conditions at which Na^+ and Mg^{2+} cations compete for the DNA-binding sites in a PEG-containing solution, the fraction of DNA phosphate groups' neutralization necessary for the formation of a dispersion can be evaluated with the help of the Manning theory [28]. According to the theory, there is a relationship between the concentrations of counterions competing for the phosphate groups and causing tDNA molecules' condensation, which can be described with the equation

$$\ln C_2 = const + Z_2\zeta(1 - r)\ln C_1 \qquad (4.11)$$

where C_1 and C_2 are the molar concentrations of Na^+ and Mg^{2+} cations, ζ is the linear charge density of the polyelectrolyte ($\zeta = 4.2$ in the case of DNA B-form); r is the fraction of DNA phosphate groups' neutralization with counterions: $r = Z_1\theta_1 + Z_2\theta_2$; Z_1 and Z_2 are the charges of Na^+ and Mg^{2+} ions; and θ_1 and θ_2 are the fractions of bounded Na^+ and Mg^{2+} ions.

It follows from Equation 4.11 that the dependence between the logarithms of C_1 and C_2 concentrations can be described as a straight line whose slope angle tangent (**tg** α) is equal to **tg** $\alpha = Z_2\zeta(1 - r)$. The values of **tg**α in the case of DNA dispersions' formation initiated by different combinations of competing counterions are given in Table 4.1. The theoretical evaluations of **tg**α values [29–32] show that **tg**α varies from 1.51 to 1.85 and corresponds to 88–90% neutralization of DNA phosphate groups.

TABLE 4.1

Some Characteristics of DNA Condensation in Solutions Containing Various Counterions

DNA Sources	DNA Molecular Mass	Solvent	Counterions	tgα
Calf thymus	~15·10^6 Da	H$_2$O	Spermidin/Mg^{2+}	2.6 ± 0.1
			Spermidin/Na$^+$	2.6 ± 0.1
Calf thymus	~(10–20)·10^6 Da	H$_2$O	Spermidin/Na$^+$	~1.1
Phage λ	~32·10^6 Da	H$_2$O	Co^{3+}(NH$_3$)$_6$/Na$^+$	~1
			Co^{3+}(NH$_3$)$_6$/Mg^{2+}	~1
Phage T7	~(25–30)·10^6 Da	H$_2$O	Spermidin/Na$^+$	~1.46
			Na$^+$/Mg^{2+}	~0.75

The appearance of such a state of DNA LCD in which the intense bands in the CD spectrum are absent, and which makes it possible to form any of the two CLCD families with abnormal bands with different signs in the CD spectra, can be used as a criterion of the Manning theory applicable to the description of the DNA CLCD formation process.

The data given in Figure 4.17 show that there is a direct relationship between $\log[Mg^{2+}]$ and $\log[Na^+]$ calculated for the case of $\Delta\varepsilon_{270} = 0$ only; that is, the "behavior" of a certain fraction of counterions can be described with the help of the Manning theory. Replacing the experimental value of **tgα** and other parameters to the Equations 4.5–4.11 permits one to calculate that the value of **r**, that is, the extent of DNA molecules' neutralization during CLCD particles' formation, is equal to 82%. The difference from the **r** value predicted theoretically (88–90%) can reflect the difference of dielectric constants typical of water–salt and PEG-containing solutions.

Several conclusions can be made based on the received results. First, the formation of DNA CLCD in PEG-containing solutions is only possible if ~82% of the negative charges of the DNA phosphate groups are neutralized with counterions. Second, the cation composition of the solvent determines the efficiency of the transition of DNA molecules from the isotropic state to the liquid-crystalline state. Finally, the interaction between DNA molecules whose phosphate groups are neutralized with different counterions acts as the factor regulating the DNA spatial packing in LCD particles.

4.2.1.6 Dielectric Constants of PEG Water–Salt Solution

The change in total dielectric constant of the PEG solution as a result of adding different organic solvents to the solution, such as methanol, ethanol, isopropanol, morpholine, dioxane, and so forth [33], is accompanied by the formation of CLCDs from right-handed DNA molecules that possess both the negative and positive abnormal bands in the CD spectra.

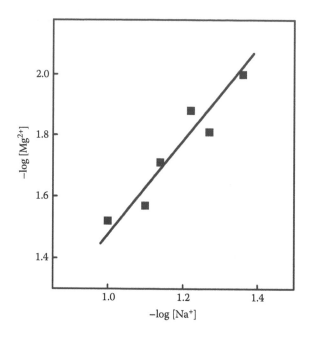

FIGURE 4.17   The dependence of log[Mg^{2+}] on log[Na^+].

4.2.2   Factors Influencing Type of Nucleic Acid Molecules' Spatial Packing in CLCD Particles after Their Formation

4.2.2.1   Role of Solution Temperature

Temperature increases the diffusion mobility of DNA molecules in the particles of the CLCDs and initiates the transition of the spatial structure of LCD particles into optically inactive state. As the temperature of the solution increases, the amplitudes of the abnormal band in the CD spectra of various nucleic acid CLCDs decrease, regardless of their signs [34]. The process that reflects this decrease is called "CD-melting" and is similar to the melting of the secondary structure of double-stranded nucleic acid molecules (denaturation). CD-melting is characterized by the temperature of the transition, varying from 30° to 80°C (depending on the PEG concentration and nucleotide composition of the nucleic acid molecules), which is marked as τ_m, similarly to the melting temperature (T_m) of nucleic acids' secondary structure.

Figure 4.18 shows the melting curves of the CLCD particles formed by poly(dA) × poly(dT) molecules under different conditions and registered by different methods (PEG molecular mass = 4,000; curve 1—80 mg/mL of PEG; curve 2—100 mg/mL of PEG; curve 3—the curve of melting of the poly(dA) × poly(dT) secondary structure registered by the change in optical density of this dispersion at 257 nm). It can be noticed that, in the case of dispersions of polyribonucleotides, the amplitude of the positive band in the CD spectra (curves 1–2) decreases and, at a certain temperature, the band disappears; meanwhile, the CD spectrum becomes similar to the polynucleotide spectrum under the absence of PEG. At relatively low values of C_{PEG},

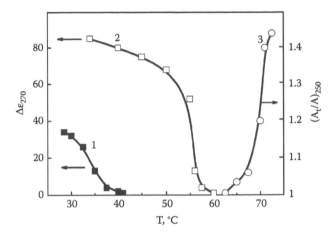

FIGURE 4.18 The dependence of the amplitude of the positive band in the CD spectra of CLCDs formed in PEG-containing solutions by poly(dA) × poly(dT) (1, 2) molecules on the temperature; (3) the melting curve of poly(dA) × poly(dT) molecules.

the abnormal band in the CD spectrum disappears at a temperature lower than the temperature of polynucleotide denaturation (curve 3).

The cooling of the PEG solution after CD-melting always results in the full restoration of, and even an in increase in, the abnormal band amplitude in the CD spectrum of the CLCD. Such "thermal training" is often used to achieve the maximum optical activity of CLCDs of different nucleic acids.

As for the denaturation of double-stranded polynucleotides, as well as DNA and RNA, the following fact should be noted: The destruction of the secondary structure of these polynucleotides or nucleic acids (melting) within the CLCD particles is followed by the hyperchromic effect, while the optical parameters of the melting of these molecules with or without PEG almost coincide. The hyperchromic effect corresponding to the denaturation of these polynucleotides within CLCD (curve 3 in Figure 4.19) is observed at the temperature ~85°C. The precise value of this temperature depends on the nucleotide composition of the polynucleotides and only slightly (by 2°–7°C) increases under the increase in PEG concentration used for the preparation of the dispersions. The relatively high percentage of denaturation (80–90%) registered by the change in CLCD absorption spectra should also be marked. Microcalorimetric measurements of the thermal denaturation of linear DNA molecules and DNA within CLCD particles made it possible to evaluate [35,36] and compare the thermodynamic parameters of this process (Table 4.2) calculated by nucleotide pairs.

According to calorimetric measurements, the percentage of DNA renaturation (~20%) remains almost constant in the area of PEG concentrations under the "critical" concentration, while at "critical" PEG concentration, the DNA renaturation extent sharply increases to 80%, and continues to increase under the increase in PEG concentration. It is significant that the calculation of ΔF value points at higher DNA stability within CLCD particles in comparison to an initial, linear DNA. The high

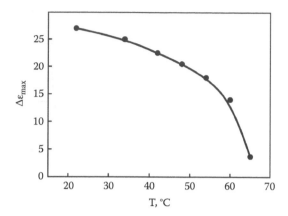

FIGURE 4.19 The dependence of the abnormal band amplitude in the CD spectrum of CLCD formed by poly(I) × poly(C) molecules on temperature.

TABLE 4.2

Thermodynamic Parameters of the Thermal Denaturation of DNA in PEG Solutions at Different Concentrations

C_{PEG}, mg/mL	ΔH, kJ/M_{bp}	ΔS, J/M·K^{-1} bp	T_{m}, °C	ΔF, kJ/M_{bp}	Renaturation Degree, %
0	32.8	78.1	87.9	4.6	20
50	35.1	83.2	88.6	5.0	21
90	37.2	87.7	89.4	5.4	22
130	41.1	95.5	93.0	6.3	81
150	43.7	100.0	94.6	6.9	85
170	42.3	96.7	95,0	6.7	89

percentage of DNA renaturation within the CLCD particles is caused by the fact that, in the case of dense packing DNA molecules, the separation of DNA strands in the neighboring molecules in quasi-nematic layers is not realized due to the sterical limit (between these molecules enough "free" space does not exist for locating the coils formed by denatured DNA molecules).

The comparison of different methods shows that the increase in the temperature of the solution containing DNA CLCD is followed by the "defrosting" of the diffusion degrees of freedom of these molecules (or the nitrogen bases within their content), which results in the disappearance of the cholesteric mode of packing of the adjacent DNA molecules without the destruction of their secondary structure. Thus, it is arguable that at an increase in temperature, the transition of DNA molecules from the cholesteric state to a similar state, probably a nematic state, takes place because the osmotic pressure of the solution prevents the DNA molecules from arranging randomly (i.e., transition into an isotropic state under these conditions is not possible). Hence, the temperature induces untwisting of the cholesteric helical structure of the DNA CLCDs. Under these conditions, the abnormal band in the CD spectrum disappears, and the amplitude of the band in the CD spectrum becomes close to its value for the initial DNA.

One can add that the high extent of renaturation of DNA molecules fixed within CLCD particles following from optical and calorimetric measurements may be biologically significant because this result means that CLCD particles provide conditions for fast and accurate restoration of native structure DNA molecules.

4.2.2.2 Change in Mode of Interaction between Adjacent DNA Molecules Fixed in Structure of CLCD Particles

In contrast to these temperature effects, one can assume that the neutralization of the electrostatic interaction between the adjacent DNA molecules fixed in the structure of CLCD particles, and the strengthening of the dipole–dipole interaction between these molecules, can result in such a change in the mode of packing of DNA molecules as can be accompanied by quite a different transition, for instance, transition from a "liquid" to a "solid" state. This process can be followed not by the decrease but by the increase (amplification) of the abnormal band in the CD spectrum of CLCD [37]. It is possible that this is exactly what happens when DNA CLCD particles are treated with gadolinium salts. During such processing, the gadolinium ions not only displace the sodium ions that neutralize the negative charges of DNA phosphate groups and decrease the solubility of the DNA molecules fixed in the spatial structure of CLCDs, they also induce modification of the DNA secondary structure. Under these conditions, CLCD particles lose their solubility and exist even in a solution that contains no PEG. These DNA CLCD particles, unlike the initial DNA CLCD particles, can be immobilized with a nuclear membrane filter, and their size can be directly measured. The evaluation shows that the diameter of these particles varies from 450 to 500 nm, which coincides with the results of the theoretical evaluations of sizes of the initial CLCD particles. The DNA–gadolinium complexes' CLCD particles obtain the properties of a "solid" substance while maintaining their abnormal optical activity.

The results given here show that the formation of nucleic acid CLCDs in water–polymeric solutions is regulated by a complicated mechanism. This mechanism does not follow directly from the van der Waals interaction between the right-handed twisted double-stranded DNA molecules. It is also obvious that the comparison of the formation of DNA or synthetic polynucleotide CLCDs with an idealized model of the "screws" packing, when only the direction of the spiral twist of the initial screw molecules without considering that their dispersive interaction and the excluded volume effects are important, is an extreme simplification. The foregoing facts make it possible to assume that the DNA molecules' packing at the formation of CLCDs in water–polymeric solutions are somehow connected to the properties of water molecules situated between the DNA molecules that "recognize" each other during the phase exclusion.

4.3 ORDER PARAMETER OF NUCLEIC ACID MOLECULES IN CLCD PARTICLES

As the biological objects are characterized by a high stability within a wide range of conditions, the stability of the DNA CLCDs under all kinds of external factors is an issue of interest.

In the physics of liquid crystals, as a parameter that is connected with the stability of the system, the parameter used is of long-range orientational order (S), which is described by the following analytical expression [38]:

$$S = (3\cos^2\theta_n - 1)/2 \tag{4.12}$$

where θ_n is the angle between the effective long axis of the molecule and the "director" of the quasi-nematic layer (it must be remembered that, for molecules as complex as DNA, the long axis is not defined by molecular symmetry but is an operational definition). The ideal parallel orientation of molecules **S** is equal to 1, and, at the chaotic orientation, it is equal to **0**.

Assuming that the double-stranded DNA or synthetic polynucleotide molecules in CLCD particles are packed similarly to their packing in liquid-crystalline phases, to analyze the transition from the cholesteric to the optically inactive state under the heating of DNA CLCD, followed by the change in the abnormal optical activity ΔA, the following equation can be used:

$$\Delta A = (A_L - A_R)_i = Pv_i^3 \Delta n(A_\| - A_\perp)i/2(v_i^2 - v_o^2) \tag{4.13}$$

where $(A_L - A_R)_i$ is the circular dichroism of the colored cholesterics at the v_i frequency; v_0 is the frequency corresponding to the cholesteric selective reflection band connected to the pitch, **P**, of its spatial structure: $v_0 = \mathbf{nP}$ (**n** is the average refractive index of the quasi-nematic layer; because the **P** value of the cholesteric structure formed by DNA molecules is ~ 2.5 μm, the v_0 value is situated in the infrared area); $\Delta \mathbf{n}$ is the optical anisotropy of the quasi-nematic layers, which is negative in the case of DNA [39]; $(A_\| - A_\perp)_i$ is the linear dichroism of the quasi-nematic layer at the frequency v_i, which is negative in the case of DNA [40]; **P** is the pitch of the cholesteric helical twist; depending on the direction of the cholesteric twist, **P** is either positive (right-handed twist) or negative (left-handed twist) [41].

The two actors in Equation 4.13, ($\Delta \mathbf{n}$) and $(A_\| - A_\perp)_I$, depend, respectively, on the value of order parameter, **S**, of the molecules forming the liquid crystal, that is, the molecules of nucleic acids, and on the order parameter of the molecules of chromophores (S_{chr}) added to the liquid-crystalline phase, that is, nitrogen bases within the structure of nucleic acid molecules.

As follows from the Equation 4.13, the amplitude of the band in the CD spectrum situated in the region of the chromophores' absorption makes it possible to evaluate the ordering of the liquid crystal if the values of all the actors of this equation are known.

The results of research [42] show that the dependence of the abnormal band amplitude in the CD spectrum in the region of the chromophores' absorption is a function to the temperature:

$$\Delta A = (A_L - A_R) = const \ [(T_0 - T)/T_0]^{2\gamma}, \qquad (4.14)$$

where $T_0 = T_m + \Delta T$ (the values of ΔT are chosen so that the dependence of $\log \Delta A$ on $\log[(T_0 - T)/T_0]$ is linear on the whole temperature interval); T means the temperature at which the continuous phase transition would take place.

The decrease in the band amplitude in the CD spectra in the region of the chromophores' absorption observed at the temperature increase can be attributed to the diminishing the extent of the molecules' ordering (S) or to the decrease in the ordering of the chromophores' molecules (S_{chr}). It is obvious that, to evaluate the dependence of the order parameter on the temperature, the information about the connection between Δn and $(A_\parallel - A_\perp)$ to S and S_{chr}, respectively, is necessary. In the case of nematic liquid crystals, it has been determined [42] that the connection between Δn and S is linear ($\Delta n \sim S$). It has also been shown that $(A_\parallel - A_\perp) \sim S_{chr}$ [43]. Besides, locally, the order that coincides with the order of nematic liquid crystals can be attributed to cholesteric liquid crystals [2,3]. Hence, we can accept that, in the case of cholesteric liquid crystals, there is the same connection between Δn and S and between $(A_\parallel - A_\perp)$ and S_{chr} that was noted for the nematics. Even considering the fact that the chromophores (in this case, the nitrogen bases of double-stranded nucleic acid molecules) are bound with the sugar-phosphate chain, it can be considered that their order parameter S_{chr} will not differ significantly from S. In this case, $S_{chr} \approx S$; it follows from Equation 4.2 that, in the case of colored cholesteric liquid crystals, $\Delta A = (A_L - A_R) \sim S^2$. Combining the received expression ($\Delta A \sim S^2$) with Equation 4.14, it is easy to show that the order parameter is a function of the temperature [42–44]:

$$S(T) = S_0(1 - T/T_0)^\gamma \qquad (4.15)$$

According to Reference 7, we can accept that the constant S_0 is equal to 1 and if $S \to 1$ under the extrapolation $T \to 0$.

Consequently, the temperature dependence of the band amplitude in the CD spectra of the colored cholesterics (including the particles of double-stranded DNA or synthetic polynucleotides CLCDs) makes it possible to evaluate both their order parameter and its temperature dependence.

The following data [45] describe the change of the abnormal optical properties at the heating of the CLCD formed in a PEG-containing water–salt solution from molecules of a synthetic, double-stranded polyribonucleotide poly(I) × poly(C), whose CLCD particles can be considered as an example of the colored cholesterics.

The temperature dependence of the amplitude of the abnormal band in the CD spectrum of poly(I) × poly(C) CLCD is given in Figure 4.19. This dependence is described (Figure 4.20) by the function

$$\log(\Delta \varepsilon) = 2\gamma \log(1 - T/T_0) + K \qquad (4.16)$$

where $\gamma = 0.14$; the constant $K = 1,46$; $T_0 = 71°C$ is the temperature corresponding to the transition of the molecules from liquid-crystalline to optically nonactive state. (Here, $C_{PEG} = 120$ mg/mL; PEG molecular mass = 4,000; 0.3 M NaCl + 10^{-2} M Na—the phosphate buffer.)

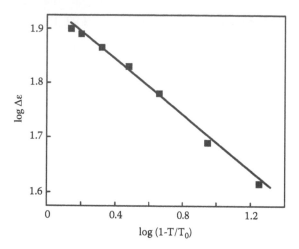

FIGURE 4.20   The dependence of log($\Delta\varepsilon$) on log(1 – T/T$_0$) for CLCD formed by poly(I) × poly(C) molecules.

The observed decrease in the band amplitude in the CD spectrum at heating poly(I) × poly(C) CLCD is caused by the diminishing of the molecules' ordering. The change in the value of the order parameter in the case of CLCD formed in PEG water–salt solutions for various nucleic acids and synthetic polynucleotides is shown in Figure 4.21 (here: $C_t = 19 \cdot 10^{-6}$ M; $C_{PEG} = 170$ mg/mL; PEG molecular mass = 4,000; 0.3 M NaCl + 10^{-2} M Na—the phosphate buffer [46]).

The received result is principally significant. First, the results obtained show that the order parameter for nucleic acids forming liquid-crystalline dispersions (or their complexes with drugs) is high (see Figure 4.21). The **S** value is about 0.8 and the temperature "jump" of **S** value at the "cholesteric-nematic" transition of double-stranded nucleic acids is ~0.3 unit. This coincides with changes of **S** values for known cases of different cholesterics. Second, the high value of the order parameter (**S**) means that the molecules of nucleic acids forming the dispersions tend to arrange along a common director at any time, even though this director may fluctuate from one configuration to another. A high value of the order parameter is the reason for the relatively stable cholesteric structure of the dispersions formed by the DNA or the (DNA-drug) molecules, that is, a mode of ordering of these molecules in the liquid-crystalline dispersions is determined at a "moment of formation" and then remains unchanged. Besides, the high value of **S** reflects a high "range of stability" of the DNA or (DNA–drug) dispersions in respect to the alteration of the DNA secondary structure (up to conditions when the secondary structure of DNA molecules can be destroyed). Finally, the ordered packing of molecules of nucleic acids in particles of liquid-crystalline dispersions remains despite the untwisting of neighboring quasi-nematic layers.

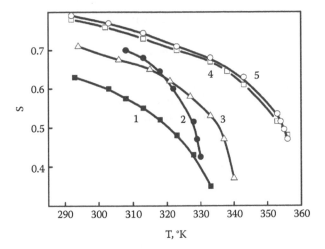

FIGURE 4.21 The temperature dependence of the order parameter (S) of molecules of natural double-stranded nucleic acids and synthetic polynucleotides in CLCD particles. Curve 1—RNA; curve 2—poly(dA) × poly(dT); curve 3—poly(I) × poly(C); curve 4—DNA; curve 5—(DNA-DAU) complex.

4.4 SUMMARY

Therefore, the results of theoretical calculations of the CD spectra of the particles of dispersions formed by double-stranded DNA molecules of low-molecular mass (or other double-stranded polynucleotides) in the water–polymeric solutions given in this chapter make it possible to draw several conclusions that have a significant practical meaning:

1. To define the dispersions that have the abnormal bands in their CD spectra, the term *cholesteric liquid–crystal dispersions (CLCD)* can be used.
2. The rigid double-stranded, anisotropic DNA molecules in CLCD particles realize their potential tendency to be packed in a cholesteric mode. The adjacent right-handed twisted DNA molecules are packed in the particles so that the spatial helical structure with the left-handed twist, characterized by an intense band in the CD spectrum, is apparent. The amplitude of this band is very high, so this band can be called the "abnormal band." The amplitude of this band depends on the CLCD particles' parameters if the DNA secondary structure and the properties of the nitrogen bases are constant. Hence, it is not reasonable to apply the molecular circular dichroism ($\Delta\varepsilon$) that reflects the properties of the isolated nitrogen bases of the initial linear DNA molecules and it can be calculated theoretically to describe the initial band in the linear DNA CD spectrum. Any modifications of the DNA secondary structure (denaturation, transition between the different

forms such as A, C, or Z) is followed by the change in $\Delta\varepsilon$ value from 1 to 5. In the description of further results, $\Delta\varepsilon$ will be used only as an illustration reflecting the abnormal nature of the band in the CD spectra. Considering what was stated above, it is more reasonable to use the experimentally measured amplitude of the abnormal band in the CD spectra, that is, ΔA, for description of the peculiarities of optical properties of the CLCDs.

The amplitude of the abnormal band the CD spectra of DNA CLCD is a simple and easily detectable criterion reflecting both the formation of CLCD particles from rigid, linear, double-stranded DNA molecules and the change in the CLCD properties under the influence of different factors on the initial DNA molecules and/or the formed CLCD particles. This means that, by registering the CD spectra, both the fine details of the CLCD particles formation process and the impact of different factors on this process can be determined.

3. The factors that regulate the properties of DNA CLCD are the properties of both the DNA molecules and water–salt–polymeric solution. The CLCDs, as well as liquid-crystalline phases, are characterized by the multiplicity of packing modes. Nevertheless, in the case of dispersions, the combination of factors that regulate the packing type is quite flexible, which makes the DNA CLCD particles' behavior similar to the behavior of biological objects.

REFERENCES

1. Yevdokimov, Yu.M., Skuridin, S.G., and Chernuha, B.A. The backgrounds for creating biosensors based on nucleic acid molecules. *Adv. Biosensors*, 1995, vol. 3, p. 143–164.
2. Belyakov, V.A. and Sonin, A.S. *Optics of Cholesteric Liquid Crystals* (Russian Edition). Moscow: Nauka, 1982. p. 360.
3. Sonin, A.S. *Introduction in Physics of Liquid Crystals* (Russian Edition). Moscow: Vysshaya shkola, 1983. p. 320.
4. Livolant, F. and Maestre, M.F. Circular dichroism microscopy of compact forms of DNA and chromatin in vivo and in vitro: Cholesteric liquid crystalline phases of DNA and single Dinoflagellate nuclei. *Biochemistry*, 1988, vol. 27, p. 3056–3068.
5. De Vries, H. Rotatory power and other optical properties of certain liquid crystals. *Acta Cryst.*, 1951, vol. 4, p. 219–226.
6. De Gennes, P.-G. *The Physics of Liquid Crystals*. London: Oxford University Press, 1974. p. 333.
7. Saeva, F.D., Sharpe, P.E., and Olin, G.R. Cholesteric liquid crystal induced circular dichroism (LCID) V. Mechanic aspects of LCID. *J. Am. Chem. Soc.*, 1973, vol. 95, p. 7656–7659.
8. Maestre, M.F. and Reich, C. Contribution of light scattering to the circular dichroism of deoxyribonucleic acid films, deoxyribonucleic acid-polylysine complexes, and deoxyribonucleic acid particles in ethanolic buffers. *Biochemistry*, 1980, vol. 19, p. 5214–5223.
9. Keller, D. and Bustamante, C. Theory of the interaction of light with large inhomo-geneous molecular aggregates. I. Absorption. *J. Chem. Phys.*, 1986, vol. 84, p. 2961–2971.

10. Keller, D. and Bustamante, C. Theory of the interaction of light with large inhomogeneous molecular aggregates. II Psi-type dichroism. *J. Chem. Phys.*, 1986, vol. 84, p. 2972–2980.

11. Kim, M.-H., Ulibarri, L., Keller, D., and Bustamante, C. The psi-type dichroism of large molecular aggregates. III. Calculations. *J. Chem. Phys.*, 1986, vol. 84, p. 2981–2989.

12. Purcell, E.M. and Pennypacker, C.R. Scattering and absorption of light by nonspherical dielectric grains. *Astrophys. J.*, 1973, vol. 186, p. 705–714.

13. Drain, B.T. The discrete-dipole approximation and its application to interstellar graphite grains. *Astrophys. J.*, 1988, vol. 333, p. 848–872.

14. Belyakov, V.A. and Dmitrienko, V.E. Optics of chiral liquid crystals. *Sov. Sci. Rev., A.*, 1989, vol. 13, p. 212.

15. Belyakov, V.A., Orlov, V.P., Semenov, S.V., et al. Comparison of calculated and observed CD spectra of liquid-crystalline dispersions formed from double-stranded DNA and from DNA complexes with coloured compounds. *Liq. Crystals*, 1996, vol. 20, p. 777–784.

16. Belyakov, V.A., Demikhov, E.I., Dmitrienko, V.E., and Dolganov, V.K. Optical activity, transmission spectra, and structure of blue phases of liquid crystals. *JETP*, 1985, vol. 62, p. 1173–1182.

17. Holzwarth, G., Chabay, I., and Holzwarth, N.A.W. Infrared circular dichroism and linear dichroism of liquid crystals. *J. Chem. Phys.*, 1973, vol. 58, p. 4816–4819.

18. Holzwarth, G. and Holzwart, N.A.W. Circular dichroism and rotatory dispersion near absorption bands of liquid crystals. *J. Opt. Soc. Amer.*, 1973, vol. 63, p. 324–331.

19. Belyakov, V.A., Osadchii, S.M., and Korotkov, V.A. Optics of imperfect cholesteric liquid crystals. *Crystallography Rep.* (Russian Edition), 1986, vol. 31, p. 522–527.

20. Yevdokimov, Yu.M. Liquid-crystalline dispersions of nucleic acids. *Bull. USSR Acad. Sci., Phys.* (Russian Edition), 1991, vol. 55, p. 1804–1816.

21. Yevdokimov, Yu.M., Golo, V.L., Salyanov, V.I., et al. The "phantom" structure of solvent and the packing of double-stranded molecules of nucleic acids in particles of mesomorphic dispersions. *Biophysics* (Russian Edition), 2000, vol. 45, p. 1029–1038.

22. Strauss, U.P. and Ross, P.D. Counterion binding by polyelectrolytes. V. The effect of binding of univalent cations by polyphosphates on the intrinsic viscosity. *J. Amer. Chem. Soc.*, 1960, vol. 82, p. 1311–1314.

23. Crutchfield, M.M. and Irani, R.R. A P^{31} nuclear magnetic resonans study of complexing between Li^+, Ca^{2+}, Mg^{2+} ions and lower condensed phosphate polyanions. *J. Amer. Chem. Soc.*, 1965, vol. 87, p. 2815–2820.

24. Strauss, U.P., Woodside, D., and Wineman, P. Counterion binding by polyelectrolytes. I. Exploratory electrophoresis, solubility and viscosity studies of the interaction between polyphosphates and several univalent cations. *J. Phys. Chem.*, 1957, vol. 61, p. 1353–1356.

25. Emanuel, C.F. Some physical properties of deoxyribonucleic acids dissolved in a high-salt medium: Salt hyperchromity. *Biochim. Biophys. Acta*, 1960, vol. 42, p. 91–98.

26. Yevdokimov, Yu.M., Skuridin, S.G., and Salyanov, V.I. The liquid-crystalline phases of double-stranded nucleic acids in vitro and in vivo. *Liq. Crystals*, 1988, vol. 3, p. 1443–1459.

27. Skuridin, S.G., Dembo, A.T., and Yevdokimov, Yu.M. Space liquid-crystalline ordering of double-stranded DNA molecules at different cation content of the solvent. *Biophysics* (Russian Edition), 1985, vol. 30, p. 750–757.

28. Manning, G. The molecular theory of polyelectrolyte solutions with application to the electrostatic properties of polynucleotides. *Quart. Rev. Biophys.*, 1978, vol. 11, p. 179–246.

29. Subirana, I.A. and Vives, J.L. The precipitation of DNA by spermine. *Biopolymers*, 1981, vol. 20, p. 2281–2283.

30. Marx, V.A. and Rubin, G.A. Studies of DNA organization in hydrated spermine-condensed DNA torus and spermine-DNA fibres. *J. Biomol. Struct. Dynam.*, 1984, vol. 1, p. 1109–1132.
31. Widom, J. and Baldwin, R.L. Cation-induced toroidal condensation of DNA: Studies with Co $^{3+}$(NH3)$_6$. *J. Mol. Biol.*, 1980, vol. 144, p. 431–453.
32. Wilson, J. and Bloomfield, V.A. Counterion-induced condensation of deoxyribonucleic acid. *Biochemistry*, 1979, vol. 18, p. 2192–2196.
33. Yevdokimov, Yu.M., Salyanov, V.I., and Dembo, A.T. Optical properties of DNA liquid-crystalline micophases in water-organic solutions. *Crystallography Rep.* (Russian Edition), 1986, vol. 31, p. 738–741.
34. Evdokimov, Yu.M., Pyatigorskaya, T.L., Belozerskaya, N.A., et al. DNA compact form in solution. XI. Melting of the DNA compact state, formed in water-salt solutions, containing poly(ethylene glycol). *Mol. Biol.* (Russian Edition), 1977, vol. 11, p. 507–515.
35. Grasso, D., Fasone, S., La Rosa, C., et al. A calorimetric study of the different thermal behaviour of DNA in the isotropic and liquid-crystalline states. *Liq. Cryst.*, 1991, vol. 9, p. 299–305.
36. Grasso, D., Campisi, R.G., and La Rosa, C. Microcalorimetric measurements of thermal denaturation and renaturation processes of salmon sperm DNA in gel and liquid crystalline phases. *Thermochim. Acta*, 1992, vol. 199, p. 239–245.
37. Golo, V.L., Kats, E.I., and Kikot, I.P. Effect of dipole forces on the structure of the liquid crystalline phases of DNA. *JETP Lett.*, 2006, vol. 84, p. 275–279.
38. Chandrasekhar, S. *Liquid Crystals*. Ed. by A.A. Vedenov and I.G. *Chistyakov* (Russian Edition). Moscow: Mir, 1980. p. 344.
39. Livolant, F. Precholesteric liquid crystalline states of DNA. *J. Phys.*, 1987, vol. 48, p. 1051–1066.
40. Yevdokimov, Yu.M. Liquid-crystalline dispersions of nucleic acids. *Bull. USSR Acad. Sci., Phys.* (Russian Edition), 1991, vol. 55, p. 1804–1816.
41. Gottarelli, G. and Spada, G.P. Application of CD to the study of some cholesteric mesophases. In Circular Dichroism: Principles and Applications. Eds. K. Nakanishi, N. Berova, and R.W. Woody. New York: VCH, 1994, p. 105–119.
42. Abdulin, A.Z., Bezborodov, V.S., Min·ko, A.A., and Rachkovich, V.S. *Formation of Textures and Structural Ordering in Liquid Crystals* (Russian Edition). Minsk: Universitetskoye, 1987. p. 176.
43. Prokhorov, V.V. and Kizel,· V.A. Circular dichroism in the absorption region in blue and cholesteric phases of liquid crystals. *Crystallography Rep.* (Russian Edition), 1985, vol. 30, p. 958–960.
44. De Jeu, W.H. *Properties of Liquid Crystalline Materials* (Russian Edition). Moscow: Mir, 1982. p. 152.
45. Skuridin, S.G., Badaev, N.S., Dembo, A.T., et al. Two types of temperature induced transitions of poly(I)×poly(C) liquid crystals. *Liq. Crystals*, 1988, vol. 3, p. 51–62.
46. Yevdokimov, Yu.M., Salyanov, V.I., Skuridin, S.G., et al. Liquid-crystalline dispersions of the (DNA-drug) complexes as a background for creation of a multifunctional biosensing units: First step. In *Evolutionary Biochemistry and Related Areas of Physicochemical Biology*. Moscow: Bach Institute of Biochemistry and ANKO, 1995. p. 315–326.

5 Polymorphism of Liquid-Crystalline Structures Formed by (DNA-Polycation) Complexes

5.1 SOME PECULIARITIES OF INTERACTION OF DNA MOLECULES WITH POLYCATIONS

The second way for ordering rigid, linear, low-molecular-mass nucleic acid molecules is the phase exclusion of these molecules from water–salt solutions because of the "correlation interaction" (attraction) [1,2] between nucleic acid segments whose phosphate group negative charges are neutralized with positively charged counterions. Since the moving force of this process is the change in the enthalpy of the system, this way of phase exclusion is called an "enthalpy condensation." For realization of this process, the polycations that neutralize a large number (80–90%) of negative charges of nucleic acid phosphate groups are needed. Then the attraction between the nucleic acid (NA) molecules becomes strong enough to induce a spontaneous condensation. The attraction forces at condensation are mostly electrostatic (electrodynamic) by nature, including the London dispersion forces and the dipole-induced dipole interaction. These forces, which are relatively small at long distances between NA molecules, increase sharply as the NA molecules approach one another, while the attraction energy increases as $1/r^5$. The value of the dispersion force is proportional to the Hamaker empiric constant $\mathbf{A}$, which is $\sim 4 \times 10^{-14}$ erg for organic molecules interacting in water, that is, about 1 $\mathbf{kT}$ at room temperature. For the attraction at a distance of approximately 30Å, the value of $\mathbf{A}$ varies from 2 to 5 $\mathbf{kT}$.

It is obvious that, when the NA molecules' surface charge density decreases (in particular, because of the neutralization of NA phosphate group charges with positive charges of polycations added to the solution), the dispersion forces balance and then exceed the electrostatic repulsion of the adjacent molecules, forming (NA–polycation) complexes. In this case, the "correlation attraction" [2] between almost neutral molecules causes the exclusion of the formed (NA–polycation) complexes from the solution.

And again, depending on the NA molecular mass, the obtaining of two different structures as a result of intra- or intermolecular condensation, that is, an NA compact form that consists of only one molecule of (NA–polycation) complex, and an aggregate containing many of these molecules, is possible.

The intramolecular condensation of high-molecular-mass DNA takes place as a result of the interaction between DNA and such polycations as polyamines, polypeptides, and so forth, whose positively charged groups neutralize the negatively charged DNA phosphate groups [3,4]. Under such conditions, the high-molecular DNA lose their solubility even under the osmotic pressure of the water–salt solution and, for the foregoing reasons, take a compact toroidal shape. The toroidal particle formed by a single DNA molecule loses almost all negative charges [5]. In particular, the total charge of a toroid particle formed by T4 DNA (166,000 of base pairs) in the presence of spermidin reaches only 10% of the initial value. The phase diagram describing the transition of high-molecular-mass DNA to the condensed state at interaction with such polyamine as spermidin [6] is given in Figure 5.1. All the DNA molecules have a linear shape up to the spermidin concentration of 0.6 mM. At spermidin concentrations of about 0.8 mM, there are both linear molecules and toroidal particles in the solution. When the spermidin concentration exceeds 1 mM, all the DNA molecules accept a toroidal shape with a diameter of 1,000–2,000 Å observed by electron microscope (see Figure 5.2). The calculation shows that the effective volumes of linear DNA molecule and toroidal DNA particles formed by the interaction with spermidin differ by more than 10^4 times.

It has been noted that DNA toroidal molecules do not aggregate with one another for at least several hours. The factors that determine the formation of certain high-molecular-mass DNA spatial shapes as a result of enthalpy condensation of these molecules are subjects of intensive theoretical investigations at the moment [7,8].

Therefore, high-molecular-mass DNA molecules can form single particles (see Figure 5.3), most of which have the shape of toroids with an outside diameter of about

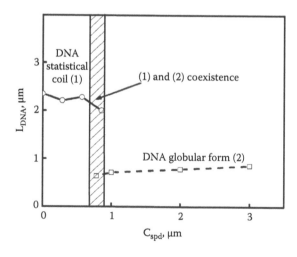

FIGURE 5.1 Dependence of maximal length of DNA molecules on spermidin concentration. (Here, molecular mass of PEG = 8,200 Da; 0.05 M NaCl; 0.01 M Tris-buffer.)

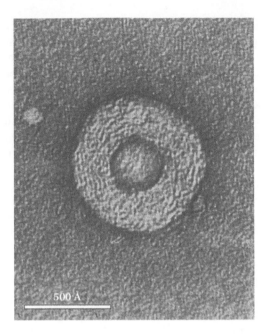

FIGURE 5.2 Electron-microscopic image of toroidal particle of bacteriophage T7 DNA obtained in the presence of spermidin.

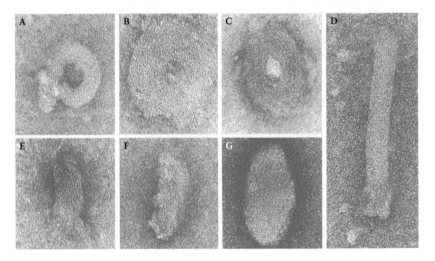

FIGURE 5.3 Electron-microscopic images of globular DNA forms of various bacteriophages. (A) T7 bacteriophage DNA condensed as a result of interaction with polylysine. (B) and (C) T4 bacteriophage DNA condensed as a result of interaction with polylysine. (D) "Baton-shaped" particle of T7 bacteriophage DNA. (E), (F), (G)—T4 bacteriophage DNA condensed in the presence of PEG (molecular mass of PEG = 6,000 Da).

1,000 Å, as a result of enthalpy condensation. One can say that as a result of phase exclusion the high-molecular-mass DNA molecules form a dispersed phase, and each particle of this phase consists of a single molecule of (DNA–polycation) complex.

The experimental research on the process of single high-molecular-mass DNA molecules' condensation as a result of interaction with polycations are quite difficult, which is caused both by the limited number of methods that make it possible to observe the change in DNA shape and by the use of conditions that eliminate, first of all, DNA intermolecular aggregation. That is why there are only a small number of studies that contain the evidence of the formation of toroidal particles from a single molecule of (DNA–polycation) complexes.

However, if rigid, linear, low-molecular-mass, double-stranded NA molecules are used for enthalpy condensation, their phase exclusion, as a result of intermolecular attraction, leads to the ordering of (NA–polycation) complex molecules, that is, to the formation of a dispersion whose particles may be characterized by the liquid-crystalline (LC) packing of (NA-polycation) complex molecules.

The most important event in the process of NA enthalpy condensation is the approaching of adjacent (NA-polycation) complex molecules. The peculiarities of the process depend on the NA molecules' surface properties, the polycation molecules' spatial structure, and the mode of charge distribution in these molecules, as well as on the solvent properties. This means that a small modification of the NA molecule surface, which can happen under the action of polycations, must lead to the change in properties of the forming particles of NA liquid-crystalline dispersions (LCDs). Polyamines, polyaminoacids, proteins (histones), dendrimers, and so forth, were used as polycations, causing the formation of LCDs [3,9].

The properties of particles of (NA–polycation) complex LCDs usually differ from the properties of NA LCD particles formed by entropy condensation. These differences consist of the following. First, polycation molecules fixed on the surface of NA molecules are always entered into the content of LCD particles formed in water–salt solutions. They can also play the role of a "dielectric medium" changing the character of interaction between the NA molecules. Second, the phase exclusion of molecules of the (double-stranded NA–polycation) complex takes place as the "critical" concentration of polycation in the solution, determined when the value of the interaction constant of polycation with NA is reached. The higher the interaction constant, the lower the "critical" concentration of polycation, which causes the phase exclusion. Meanwhile, the values of the interaction constants of polycations are a function of their molecular mass. Third, considering the fact that polycation molecules have a fixed spatial structure, the energy of the interaction between the (NA–polycation) molecules provides a constant (fixed) distance between the adjacent molecules. Fourth, the combination of the first and the third points shows that the polycations play a dual role: on the one hand, when interacting with the DNA, they change the homogeneity of the charge distribution on the surface of these molecules; on the other hand, they perform the function of a medium that modifies the efficiency of interaction between the adjacent NA molecules. Under these conditions, the type of "recognition" of molecules of the (NA–polycation) complex is different from the type of "recognition" of the initial anisotropic double-stranded NA molecules. The (NA–polycation) complex molecules are ordered so that, as a rule, a hexagonal (or very close to hexagonal), not cholesteric, packing is realized. Such packing of molecules corresponds to the Bragg distances within the limit of 26–29 Å, which are typical of the hexagonal packing of double-stranded NA molecules. Different models are used to describe the phase exclusion of (NA–polycation) complexes [2].

Fifth, depending on the empirical combination of several factors, notably the ionic strength of the solution, the spatial structure of polycation molecules, the content of positively charged groups, and so forth, only in some cases is it possible to form an LCD characterized by cholesteric (NA–polycation) molecules' packing in the formed LCD particles. In these cases, the abnormal optical activity typical of the (NA–polycation) complexes of CLCD particles makes it possible to observe the changes in the mode of the molecules' packing.

In the case of interaction of calf thymus DNA with poly-L-lysine and its analogs [10], the formation of dispersions with different signs of low-intense bands in the CD spectra, which is evidence of the formation of CLCD particles with different sense of spatial twist of the structure of the dispersion particles, was observed. The interaction of low-molecular-mass DNA with polycations, such as poly(amidoaminodendrimers) [11,12] or poly(ethyleneimines), is usually followed by the formation of an aggregated phase or a dispersion whose particles are characterized by hexagonal packing of adjacent (DNA–polycation) complex molecules.

The condensed phases of (DNA–polycation) complexes are interesting from the biological point of view because such complexes are used as carriers of genetic material to deliver it to cells. Therefore, the research on the enthalpy condensation of DNA under the influence of various polycations is important.

Among various polymeric compounds that can provoke DNA condensation, aminopolysaccharides, that is, naturally occurring polycations, which are widely distributed among living organisms, are the most interesting. These biodegradable polymers are involved in many cellular processes; they attract attention as the regulators of activity of numerous enzymatic systems and as the carriers for many biologically relevant compounds. One of the members of the aminopolysaccharide family, a biocompatible, biodegradable poly(aminosaccharide) named chitosan (a copolymer, consisting of β-(1→4)-2-amino-2-deoxy-D-glucopyranose and β-(1→4)-2-acetamido-2-deoxy-D-glucopyranose residues), attracts the attention of researchers in different domains of science and technology. Chitosan derivatives of different molecular masses are acceptable in reasonable amounts in laboratories; these compounds participate in intracellular processes as regulators of enzymatic activity and as carriers of biologically active substances and pharmaceuticals into cell. Also of interest are the features of the chemical and spatial structure of chitosans determining their ability to form specific complexes with various compounds. The chitosan (Chi) molecule has positively charged amino groups, and can form complexes with DNA molecules because their phosphate groups carry negative charges. Further interest in chitosan is created by the change in the content of amino groups in a molecule at a constant molecular mass of Chi samples, followed not only by a noticeable change in the properties of these molecules but also by a sharp change in the properties of the particles of LCDs formed by (DNA–Chi) complexes.

5.2 SPECIFICITY OF CHITOSAN BINDING TO DNA

Figure 5.4 shows diagrams characterizing the efficiency of the interaction of Chi with single-stranded oligonucleotides used as a model for the DNA molecule [13]. It is seen that Chi interacts with different fragments of single-stranded DNA in different cells. However, the fluorescent signal in all cells is the same order of magnitude. This testifies to a nonspecific character of Chi: DNA interaction. This conclusion is based on the fact that, in the case of the specific interaction of DNA with molecules of other biologically active polymers (polypeptides, proteins, etc.), the intensity of the fluorescent signal in certain biochip cells containing nitrogen base sequences in the content of the oligonucleotide specifically "recognizable" by these molecules differs from the signal intensity of neighboring cells 10–100 times. Consequently, the nucleotide composition of the DNA molecule does not affect the efficiency of Chi binding to DNA, that is, it can be considered that the complex of Chi and DNA is formed largely by interaction between positively charged Chi groups and negatively charged phosphate groups of DNA molecules [13]. This is significant as it allows one to simplify the explanation of certain effects described in the following text.

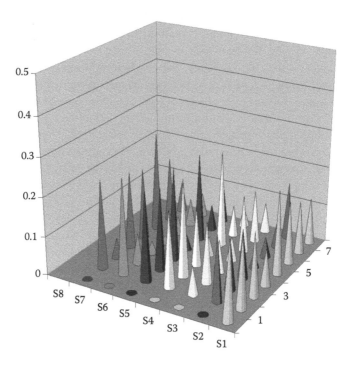

FIGURE 5.4 (*See color insert.*) Chitosan binding to biochip containing single-stranded oligonucleotides. (Cone height corresponds to the concentration of chitosan bound to oligonucleotides in the given cell. Only one cell in the first row contains oligonucleotides.)

5.3   FORMATION OF DISPERSIONS OF (DNA–CHITOSAN) COMPLEXES

The absorption spectra observed at the titration of a water–salt solution of double-stranded, low-molecular-mass DNA molecules with small portions of Chi solution (Figure 5.5) have many common features with the absorption spectra observed earlier in the entropy condensation of these molecules. When the "critical" Chi concentration is reached, the absorption appears in the spectrum in the region of the spectrum ($\lambda > 320$ nm) where neither DNA nor Chi are absorbed, and the absorption at $\lambda \sim 260$ nm increases [14]. The changes in the shape of the absorption spectrum, as well as in the case of entropy DNA condensation, are related to the formation of the dispersion of DNA–Chi complexes whose particles scatter UV radiation. The dependence of the "apparent" optical density (A_{app}) measured at $\lambda = 340$ nm (A_{340}) allows one to determine the C^{cr} value, that is, the concentration of Chi necessary for the formation of the dispersion of the DNA-Chi complex. The C^{cr} value depends on the DNA concentration; the higher the DNA concentration, the higher the C^{cr} [14].

Therefore, the interaction between Chi molecules carrying positively charged amino groups and double-stranded linear NA molecules carrying negatively charged phosphate groups in a water–salt solution is strong enough to form a dispersion whose particles consist of the molecules of these complexes.

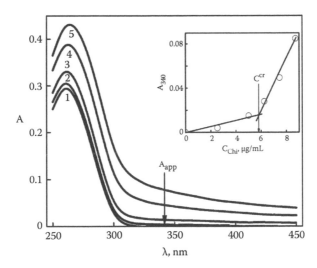

FIGURE 5.5   The DNA absorption spectra in the absence (curve1) and in the presence (curves 2–5) of chitosan in solution. Chitosan concentration in solution: 1—0; 2—2.5; 3—5.0; 4—7.5; 5—8.75 µg/mL (chitosan: 85% of amino groups; molecular mass 19 kDa); $C_{DNA} \sim$ 15.5 µg/mL, 0.15 M NaCl + 0.001M Na$^+$-phosphate buffer; pH 6.85. Insert: the dependence of the "apparent" optical density at $\lambda \sim 340$ nm upon chitosan concentration in solution.

5.4 CD SPECTRA OF DISPERSIONS FORMED BY (DNA–CHITOSAN) COMPLEXES

The CD spectra of the initial DNA (B-form, curve 1) and dispersions of DNA–Chi complexes (curves 2–6) are given in Figure 5.6. When the "critical" concentration of Chi is reached (C^{cr}, insert in Figure 5.6), where the dependence of the amplitude of the band ($\lambda = 270$ nm) in the CD spectrum of dispersions of DNA–Chi complexes on Chi concentration in the solution is shown—$\Delta A = AL - AR$ (1×10^{-3} opt. units)—an intense (abnormal) positive band appears in the CD spectrum in the region of DNA absorption ($\lambda \sim 270$ nm) [14].

The C^{cr} value calculated by the analysis of the CD spectra almost coincides with C^{cr} value calculated from the absorption spectra. This means that the appearance of an abnormal positive band in the CD spectrum takes place under the same conditions that apply when the DNA–Chi dispersion scattering UV radiation is formed.

It can be concluded from the CD spectra that, because of the interaction of DNA and Chi molecules, the LCD characterized by a cholesteric spatial structure is formed. It should be noted that the "classical" cholesteric formed by the phase exclusion of linear, double-stranded, right-handed twisted DNA molecules that belong to the B-family of polynucleotides has an abnormal negative band in the CD spectrum, which, according to the results of theoretical calculations, proves that the particles of LCD possess a cholesteric helical structure with a left-handed twist. In our case (Figure 5.6), an appearance of the abnormal band in the CD spectrum reflects, according to the theory, the obtaining of cholesteric liquid-crystalline dispersion from molecules of DNA–Chi complexes. However, unlike classical cholesterics,

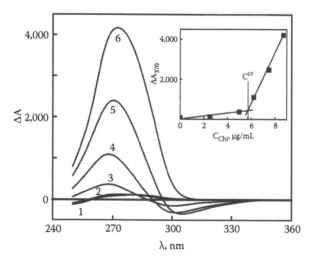

FIGURE 5.6 The CD spectra of DNA in the absence (curve 1) and in the presence (curves 2–6) of chitosan in solution. Chitosan concentration in solution: 1—0; 2—2.5; 3—5.0; 4—6.25; 5—7.5; 6—8.75 µg/mL; (chitosan: 85% of amino groups; molecular mass 19 kDa); $C_{DNA} = 15.5$ µg/m; 0.15M NaCl + 0.001M Na⁺-phosphate buffer; pH 6.85. Insert: the dependence of the amplitude of the band in the CD spectra of the DNA–chitosan dispersions at $\lambda \sim$ 270 nm in chitosan concentration in solution.

there is typically a positive band in the CD spectrum for the DNA–Chi cholesterics, and, consequently, a cholesteric spatial structure with a right-handed twist.

In the case of the DNA–Chi complex, the amplitude of the positive band expressed as $\Delta\varepsilon_{270}$ (about 85 units) is much higher than the value of $\Delta\varepsilon_{270}$ (~ 2), which corresponds to the molecular optical activity of nitrogen bases in isolated DNA molecules. It was interesting to determine the x-ray scattering parameters of phases obtained as a result of the precipitation of CLCD particles formed by DNA–Chi at low-speed centrifugation.

5.5   X-RAY PARAMETERS OF PHASES FORMED BY (DNA–CHITOSAN) COMPLEXES

Figure 5.7 shows the curves of small-angle x-ray scattering by a phase formed as a result of the concentration of the dispersion of the DNA–Chi complex at low-speed sedimentation.

There is one distinct Bragg reflection on the curves, corresponding to phases formed as a result of concentration of the CLCD DNA–Chi complex, which has a positive band in the CD spectrum. The magnitude of this reflection and the x-ray scattering parameters of the resulting phases are given in Table 5.1 [15].

The absence of highest orders of the small-angle reflection in the x-ray scattering patterns points to the absence of a regular three-dimensional order in the arrangement of the DNA–chitosan complexes, that is, the phase does not have the ideal crystalline structure. This means that only a "short-range" orientational order, typical of liquid-crystalline phases prepared from DNA molecules, can exist in the arrangement of neighboring molecules of the DNA–chitosan complexes.

Analysis of the data presented in Table 5.1 underlines a number of important facts. First, the distance $\mathbf{D}$ ($D = 2\mathbf{d}_{\text{Bragg}}/3^{1/2}$) between the adjacent DNA molecules

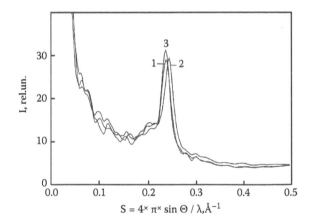

FIGURE 5.7   Small-angle x-ray scattering curves of the liquid-crystalline phase of the DNA–chitosan complex at different temperatures. Curve 1 was received at the temperature of 20°C; curve 2 at 80°C; curve 3 specimen 2 after its cooling to 20°C; C_{DNA} = 29.8 µg/mL; C_{Chi} = 20.4 µg/mL; chitosan: 85% of amino groups; molecular mass 19 kDa; 0.15 M NaCl + 0.001 M Na⁺-phosphate buffer; pH 6.86.

TABLE 5.1

X-ray Parameters for the Phase Formed by the (DNA–Chitosan) Complex

Sample	T, °C	d_{Bragg}, Å	β_s, rad	L, Å	r_m, Å	Δ/d_{Bragg}
Phase of (DNA–chitosan) complex	20	26.060	0.00584	264.115	416.891	0.100[a]
		26.579	0.00989	155.975	246.203	0.131[b]
Phase of (DNA–chitosan) complex	40	25.804	0.00525	293.738	463.645	0.094[a]
		25.323	0.01048	147.210	232.364	0.131[b]
Phase of (DNA–chitosan) complex	60	25.552	0.00523	295.119	465.821	0.093[a]
		26.060	0.0093	165.419	231.822	0.134[b]
Phase of (DNA–chitosan) complex	70	25.315	0.00523	295.122	465.821	0.093[a]
		26.060	0.01050	146.867	231.822	0.134[b]
Phase of (DNA–chitosan) complex	80	25.315	0.00582	265.236	418.650	0.098[a]
		25.804	0.00991	155.593	245.593	0.130[b]
Cooled sample	20	26.060	0.00525	293.735	463.645	0.095[a]
		26.579	0.00930	165.852	261.793	0.127[b]

Note: $d_{Bragg} = \lambda/2\sin\theta$—the average spacing between axis of adjusting (DNA–Chi) molecules in formed phase.

λ—the wavelength of x-ray irradiation (in our case, $\lambda = 1.54$ Å).

θ—the half of scattering angle.

β_s—the full width at a half-maximum intensity of a peak observed at a mean scattering angle of 2θ.

$L = \lambda/\beta_s\cos\theta$—the size of crystallites.

$r_m = (\pi/2.5)^2 \, (\lambda/\beta_s)$—the radius of interaction

$\Delta/d_{Bragg} = (1/\pi)\cdot(\beta_s d_{Bragg}/\lambda)^{-0.5}$—the disorder parameter.

Δ—mean-square deviation from d_{Bragg}-value.

[a]—the data were obtained taking into account with collimation of the x-ray beam.

[b]—the data were obtained without collimation of the x-ray beam.

---

in the formed phase in about 30 Å. Second, this distance decreases only slightly with the temperature change from 20° to 80°C. Third, the value of the small-angle reflection is completely restored upon cooling to 20°C.

Thereby, an ordered packing of neighboring molecules of DNA–Chi complexes is typical of formed phases.

The temperature behavior of the phase formed by the DNA–Chi complex is quite close to the behavior of the phases formed by other DNA complexes. This behavior has two differences from the behavior of "classical" cholesterics produced by phase exclusion of DNA molecules in PEG-containing solutions. First, the distance between DNA molecules in "classical" cholesteric phases is within 30 Å to 51 Å. Taking into account that the distance between neighboring DNA molecules in the phase of DNA–chitosan complexes is about 30 Å, one can conclude that the extent of ordering of adjacent DNA–chitosan complexes in the phase is higher than that of DNA molecules in the classical cholesteric phase. Second, in the case of "classical" cholesteric phases of initial double-stranded DNA, the change in temperature is accompanied by a marked (3 Å–6 Å depending on conditions) increase in the distance between DNA molecules [15,16]. A small change in the Bragg reflection at the increase of temperature shows that phases of DNA–Chi complexes are not as liquid as the initial DNA cholesteric phase. A very small change in the Bragg reflection (Table 5.1) on temperature increase allows one to suppose that the molecules of DNA–chitosan complexes in liquid-crystalline dispersions are not as "fluid" as pure DNA molecules in the cholesteric phase. Additional confirmation comes from the fact that the Bragg reflections coincide with the **d** values specific to the transition (intermediate) area between the cholesteric and hexagonal phases of DNA molecules (Figure 4.14).

The question of the magnitude of the abnormal optical activity observed in the case of CLCDs of DNA–Chi complexes is interesting. If the value of the abnormal optical activity is related only to the distance between the DNA molecules in the cholesteric dispersion, it could be expected that a small abnormal optical activity would correspond to a short distance between DNA molecules. Nevertheless, though the abnormal optical activity of DNA–Chi dispersions (Figure 5.6) is lower than the activity of the classical cholesterics, it is quite comparable. Considering that the abnormal optical activity is observed in the region of nitrogen bases absorption, it can be assumed that, when the CLCD of DNA–Chi complexes are formed, the diffusion degrees of freedom of both DNA molecules and nitrogen bases relative to the DNA molecules axis are more limited in comparison to the case of the CLCD of initial DNA molecules. Such "freezing" can be a result of a specific interaction of Chi with the DNA molecules (base pairs); it can lead to the increase in the average abnormal optical activity of the CLCD of DNA–Chi complexes, regardless of the short distance between DNA molecules in the dispersion particles.

Therefore, the mode of the spatial packing of adjacent molecules of DNA–Chi complexes in CLCDs depends not just on the distance between nucleic acid molecules. From this point of view it would be interesting to determine the effect of different conditions on the efficiency of CLCD formation from DNA–Chi complexes.

5.6   DEPENDENCE OF EFFICIENCY OF CLCD FORMATION BY (DNA–CHITOSAN) COMPLEXES ON VARIOUS FACTORS

Dependence of Efficiency of Formation of Nucleic Acid–Chitosan Complexes CLCD on Chitosan Molecular Mass. Figure 5.8 (where $C_{DNA} \sim 15.5\ \mu g/mL$; Cpoly(I) × poly(C) ~ 15.8 μg/m; 0.15M NaCl; 0.001M Na$^+$-phosphate buffer, pH6.85) shows the dependence of the C^{cr} value (C^{cr} is expressed as the molar concentration of chitosan-repeating units—sugar residues; the unit mean molecular mass is 220.5 Da) on the molecular mass of the chitosan samples used for preparing dispersions of poly(I) × poly(C) (curve 1) or DNA (curve 2). One can see that the efficiency of the NA–chitosan complex formation sharply increases with an increase in the chitosan molecular mass. This is accompanied by a decrease in the chitosan concentration necessary for NA condensation. However, starting from a molecular mass of chitosan about 4 kDa (i.e., at a certain chain length of the chitosan molecule), the efficiency of this process is practically independent of molecular mass. The dependence in this figure indicates the cooperative binding of chitosan to double-stranded NA molecules [17].

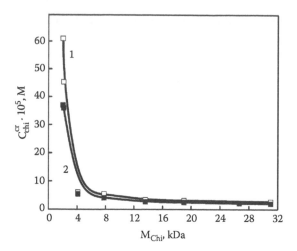

FIGURE 5.8 The dependence of chitosan critical concentration (C^{cr}_{Chi}) on molecular mass of chitosan (85% of amino groups) preparations for the poly(I) × poly(C)–chitosan (curve 1) and DNA–chitosan (curve 2) complexes.

Thus, there is a definite limit in the length of chitosan molecules that triggers the cooperative formation of NA–chitosan complexes.

The C^{cr} value necessary for the formation of dispersions of DNA–Chi complexes almost coincides with the corresponding concentration in the case of poly(I) × poly(C)–Chi complex if the molecular mass of the used Chi is equal. In particular, in the case of DNA C^{cr} is about 6 µg/mL, and in the case of poly(I) × poly(C), it is about 8 µg/mL if a Chi sample with the molar weight of 19 kDa is used.

The maximal abnormal optical activity of dispersions of DNA–chitosan and poly(I) × poly(C)–chitosan complexes show different dependences on chitosan molecular mass. An abnormal optical activity of the dispersions of DNA–chitosan complexes occurs within a wide range of chitosan molecular mass, whereas, for poly(I) × poly(C)–chitosan complexes, the region of occurrence of dispersions with abnormal optical activity is much narrower.

Thus, depending on chitosan molecular mass, one could produce dispersions of NA–chitosan complexes both possessing abnormal optical activity and having no such activity.

The dependence of formation efficiency of DNA–chitosan complexes' CLCD on the pH and ionic strength of solutions. The efficiency of DNA–Chi complexes CLCD formation depends both on pH and on the ionic strength of solutions. These dependences confirm the electrostatic nature of the interaction between Chi and DNA molecules.

The dependence of abnormal optical properties of DNA–chitosan complexes' CLCD on the distance between amino groups in chitosan molecules. The dependence of abnormal optical properties of DNA–Chi complexes on the distance between amino groups in Chi molecules is unique [17].

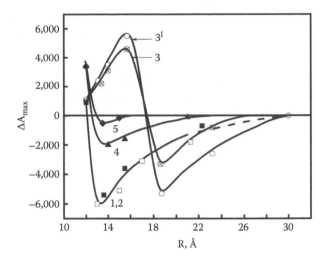

FIGURE 5.9 The dependence of the maximum amplitude of the band ($\lambda \sim 270$ nm) in the CD spectra of the DNA–chitosan LCDs upon the average distance **R** between amino groups in the chitosan molecules. Curves 1, 4, and 5—molecular mass of chitosan ~ 14.6 kDa; 0.05; 0.15, and 0.5 M NaCl, respectively; curve 2—molecular mass of chitosan ~ 8.4 kDa; 0.05 M NaCl; curve 3—molecular mass of chitosan ~ 5.0 kDa; 0.05 M NaCl.

The dependences of the maximum amplitude, ΔA_{max}, of the band in the CD spectra of DNA–chitosan CLCDs on the average distance, **R**, between amino groups in the chitosan molecules is given in Figure 5.9. The average distance between amino groups in Chi molecules was evaluated by the equation $R(\text{Å}) = 2(5.15 \times 100\%)/(\text{NH}_2)$. Here, 5.15 Å is the distance between two neighboring repeating chitosan units [18,19]. Coefficient 2 takes into account the fact that, due to steric restrictions, neighboring chitosan amino groups participate in the neutralization of the negative charges of phosphate groups of the DNA in an alternating mode; (NH_2) is the percentage of amino groups in the chitosan sample. The numbers above the curves represent the concentration of NaCl in the solution where the DNA–Chi complex was formed; curve 3 turns into curve 3′ when considering Chi that is not bound in the DNA complex ($C_{DNA} = 15.5$ µg/mL; 0.001 M Na phosphate buffer; pH 6.86; ΔA in optical units ($\times 10^{-3}$)).

The fact that abnormal optical activity is a complex function of the distance between amino groups in Chi molecules and the ionic strength of the solution attracts attention. The curves given in Figure 5.9 show that the combination of the two parameters, that is, the ionic strength of the solution and the percentage of amino groups, determines the possibility of the formation of two types of CLCDs by DNA–Chi complexes characterized by either negative or positive abnormal bands in the CD spectra. Furthermore, a definite combination of these parameters results in dispersions having no abnormal optical activity, while ΔA_{max} becomes zero at both long and short distances between amino groups in the Chi molecules.

In this connection, the dependence of the amplitude of the abnormal band in the CD spectrum on the concentration of salt in the solution is an issue of interest. It has been shown that the area of existence of CLCDs by DNA–Chi complexes that have only one positive band in the CD spectrum in the case of a Chi sample with 80% of amino groups is relatively wide. At the same time, in the case of Chi that contains 46% of amino groups, this area is quite narrow; moreover, in this case, CLCD that has only a negative band in the CD spectrum appears. Nevertheless, in the case of a Chi sample with 75% of amino groups, CLCD with both negative and positive stripe in the CD spectrum can be formed.

Therefore, depending on certain conditions for the formation of dispersions, a multiplicity of spatial forms is typical of CLCDs in DNA–Chi complexes. The packing of molecules of the DNA–Chi complex in CLCD particles depends on two fundamental parameters of a Chi molecule—the degree of its deacetylation and its molecular mass, while the conformation of the Chi main chain combined with the type and position of the ionized amino groups are important factors that determine the direction of the spatial twist of neighboring DNA–Chi molecules in CLCD particles.

5.7 PECULIARITIES OF INTERACTION OF CHITOSAN MOLECULES WITH NUCLEIC ACIDS

Based on the results and data from the literature, some molecular peculiarities of the interaction between chitosan and nucleic acid molecules can be considered [13–15,17].

The formation of complexes between macromolecules is significantly different by its parameters from the formation of complexes between low-molecular-mass compounds. This difference is caused by *polymer effects,* which form relatively stable complexes even when the energy of interaction between molecules is low. That is why the apparent equilibrium constant is extremely high, and the reaction between macromolecules is essentially irreversible.

A complex between macromolecules is always more stable in comparison to a monomer–monomer or monomer–macromolecule complex. The change in the free energy for macromolecular complexes is a function of the degree of polymerization, while the equilibrium constant sharply increases at a certain critical length of the polymer chain. The formation of this particular complex between Chi and NA molecules is confirmed, first, by the fact that D-glycosamine (the main Chi component) does not form an insoluble complex with NA and, second (Figure 5.8), the efficiency of NA LCD formation sharply depends on Chi molecular mass. Assuming the "average" molecular mass of a chitosan monomeric unit to be about 200, one may say that, starting from a length of chitosan equivalent to (4000/200) ~ 20 monomeric units (i.e., a length of about 100A° [18,19]), chitosan begins to interact cooperatively with NA molecules. This evaluation ignores the fact that, under the selected conditions, only 50% of chitosan amino groups can interact electrostatically with NA phosphate groups and, furthermore, that due to features of the spatial structure of chitosan molecules [18,19], amino groups could interact with negatively charged phosphate groups of NA in an alternate manner. The foregoing result means, for instance, that in the case of double-stranded DNA the minimal fragment capable of forming an "insoluble" complex with chitosan should involve 100 Å/3.4 Å (i.e., about 30 base pairs [17]); this agrees well with the assessment (15–20 base pairs) of other authors based on the precipitation of DNA molecules under the action of chitosan [20,21].

At the initial Chi addition, when there is an excess of negatively charged phosphate groups available to be bound with positively charged Chi amino groups on the surface of NA molecules, the interaction between NA and Chi is followed by the release of heat, which corresponds to an *exothermal* process [21]. An exothermal process proves the formation of a complex by the electrostatic interaction between DNA and Chi molecules. The presence of an electrostatic contribution to the mechanism of binding of DNA and Chi molecules can be illustrated by the change of binding constant in the presence of univalent salts.

In this connection the data given in Figure 5.10 (where the molecular mass of Chi samples is 14.6 kDa; the mean molecular mass of Chi monomer units with different percentage of amino groups was accepted to be 212.66, 216.86, 218.75, and 220.85 Da, respectively; $C_{DNA} = 15.5$ µg/mL; 0.001M Na phosphate buffer; pH 6.86), related to the formation of a DNA–Chi complex in solutions with different ionic strength, are rather interesting. Accepting that C^{cr} is connected to the constant of DNA–Chi complex formation, it can be expected that the dependence of **$logC^{cr}_{Chi}$** on **$log(NaCl)$** will be linear (Chapter 4).

Indeed, such a dependence (Figure 5.11) is observed for chitosan preparations containing 66%, 75%, and 85% of amino groups (at ionic strength below 0.5). However, the straight line obtained does not include points related to a chitosan sample containing 46% of amino groups (see curve 1 in the Figure 5.10). This indicates that the conformation of Chi main chain in combination with the distribution (position)

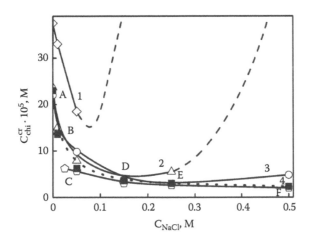

FIGURE 5.10 The dependence of C^{cr}_{Chi}-value on NaCl concentration in solution. 1—46; 2—66; 3—75; and 4—85% NH_2 groups in the chitosan molecules, respectively. Points A, B, C, D, E, and F are average values of C^{cr}_{Chi} at appropriate NaCl concentration in solution.

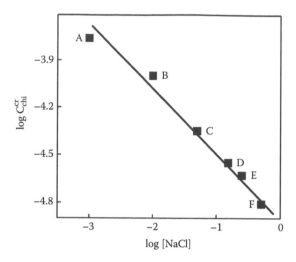

FIGURE 5.11 The dependence of $\log C^{cr}_{Chi}$ on $\log(NaCl)$ in solution (to draw this dependence, points A, B, C, D, E, and F from Figure 5.10 were used).

of ionized amino groups are important factors for the formation of polyelectrolyte complexes of this aminosaccharide with NA.

The exothermal process makes a significant contribution to the free energy of the interaction between Chi and NA. As the binding sites are filled, the efficiency of heat release decreases, and when all the sites are bound, the complex becomes electrically neutral and Chi does not bind any more. A small *endothermal* effect observed at the fraction of charges 1:1.2 (–/+) can be considered as evidence of entropy becoming the primary component of the free energy. The restructuring of chitosan and possibly NA molecules with the subsequent phase exclusion and formation of dispersion by the NA–Chi complex provides the entropy component of the free energy of the NA and Chi interaction process.

Therefore, certain conditions can be provided in a water–salt solution so that the interaction between Chi molecules carrying positively charged amino groups, and double-stranded linear NA molecules carrying negatively charged phosphate groups, is strong enough to cause the phase exclusion of molecules of NA–Chi complexes and the formation of CLCDs of these complexes.

It can be assumed that a Chi molecule interacts with NA so that the Chi amino groups not only neutralize the negative charges of phosphate groups but also form a specific distribution of positively charged amino groups near the NA surface. The Chi molecules bound to NA molecules perform the function of a "dielectric medium" that influences the mode and efficiency of the interaction between the adjacent NA molecules as they approach one another, which is necessary for the formation of the particles of dispersion. This means that a significant contribution to the efficiency of the interaction between Chi and NA must be made by the conformation of the carbohydrate chain of a Chi molecule, and its peculiarities should be considered when creating a rational theory describing the interaction between Chi and NA molecules.

The fact that the conformation of Chi molecules affects the interactions between the adjacent NA molecules as they approach each other and consequently the direction of the spatial twist of complex molecules in the formation of LCD particles is confirmed by the fact that the LCD of a complex (poly(I) × poly(C)–Chi) with a molecular mass of 19 kDa has no abnormal optical activity, while the LCD formed by the complex of this polynucleotide with Chi (molecular mass 2 kDa) has such optical activity.

Therefore, the conformation of the Chi main chain combined with the type and position of the ionized groups is an important factor that determines the abnormal optical activity of CLCD particles formed by NA–Chi complexes.

The time stability of optical properties of the CLCDs of NA and Chi complexes is an issue of interest. It can be assumed that, because of the peculiarities of the spatial structure of the NA–Chi complex, the particles of CLCD of such complexes have a certain electrostatic potential that is higher in comparison to the potential energy of their van der Waals interaction, and this circumstance provides the temporal stability of the particles relative to their subsequent aggregation.

5.8 ATTEMPT AT A THEORETICAL DESCRIPTION OF INTERACTIONS OCCURRING IN THE (DNA–CHITOSAN) COMPLEXES AND RESULTING IN THE FORMATION OF LIQUID-CRYSTALLINE DISPERSIONS WITH DIFFERENT OPTICAL PROPERTIES

According to current theoretical concepts [22], the ordered arrangement of electric dipoles along the long axes of helical polymeric molecules can affect the properties of the liquid-crystal phase produced by the phase exclusion of these polymers. It is implied that the dipoles are perpendicular to the polymer molecular axis and smoothly rotate around it to form their own helical structure. Polymeric molecules are assumed to involve short-range steric repulsions that favor the nematic ordering of neighboring molecules in the nascent phase, and dipole–dipole interactions that produce perturbations resulting in the helical twisting of the phase, (i.e., the formation of the cholesteric phase). Depending on the direction of the spatial twist of the helical structure in the nascent phase, there results an abnormal optical activity of a different sign.

Considering the fact that the type of DNA molecules' packing in CLCD particles is quite close to the type of molecules' packing in the cholesteric liquid-crystalline phase, this theory can be used for the qualitative evaluation of properties of CLCD particles formed from DNA–Chi complexes.

The properties of cholesteric liquid crystals are characterized by such physical quantities as the ordering parameter, density of free deformation energy, and director field components. The distances at which these quantities significantly change are about 1 μm, which is far more than the size of a molecule (~20 Å). This means that the deformation can be described by means of a theory [23] that ignores the details of structure comparable with the size of molecules. The theory makes it possible to find the shape of a cholesteric helix and study the dependence of the pitch of the cholesteric on such factors as temperature, solvent properties, and the presence of compounds that perturb the structure of the investigated molecules. Such research is based on the consideration of the expression for the additive to the free energy of the crystal, which evolves because of the distortion—the deviation of the director field $\vec{n}(\vec{r})$ from the homogeneous direction $\vec{n}_0(\vec{r})$ typical of a nematic. In the case of turning followed by the formation of a helical twist of a nematic liquid crystal, a phenomenological expression for the free energy of deformation of the system is the following [23,24]:

$$F_d = \frac{1}{2}K_{22}(\vec{n}(\vec{r}) \cdot \mathbf{rot}\,\vec{n}(\vec{r}) + q_0)^2 \tag{5.1}$$

where the unit vector $\vec{n}(\vec{r})$ is a director that defines the average orientation of the long axes of molecules producing the liquid crystal; K_{22} is the elastic twist constant; **and** q_0 is the relation of the two constants: $q_0 = -K_2/K_{22}$ (the constant K_2 also describes the elastic properties of a liquid crystal).

The measurements for a number of liquid crystals show that the values of K_{22} vary from 10^{-7} to 10^{-6} dynes. For example, for *p*-azoxyanisole (NAA) at 120°C, $K_{22} = 0,43 \times 10^{-6}$ dyne [25–27].

For a pure twist around **z** axis, the director field $\vec{n}(\vec{r})$ is described with a number of equations:

$$n_x = \cos\theta(z); \; n_y = \sin\theta(z); \; n_z = 0 \tag{5.2}$$

Then, the expression for the deformation free energy F_d (5.1) looks as follows:

$$F_d = \frac{1}{2}K_{22}\left(\frac{\partial\theta}{\partial z} - q_0\right)^2 \tag{5.3}$$

Consequently, the equilibrium distortion corresponding to the minimum F_d ($F_d = 0$) is a cholesteric helix with a constant wave number $\partial\theta/\partial z = q_0$. The director is spatially twisted around the axis **z**, and its direction changes by the law:

$$n_x = \cos q_0 z, \; n_y = \sin q_0 z, \; n_z = 0$$

The sign of the wave number q_0 determines the direction of twist of the cholesteric, that is, if q_0 is positive, the cholesteric has a right-handed twist; if q_0 is negative, it has a left-handed twist.

The interval of the cholesteric spiral is constant and determined by the equation

$$P = 2\pi/q_0 \tag{5.4}$$

As (5.1) shows, the free energy of deformation is equal to (accurate within the additive constant)

$$\mathbf{F_d} = \frac{1}{2}\mathbf{K_{22}}\left(\vec{\mathbf{n}}(\vec{\mathbf{r}})\mathbf{rot}\vec{\mathbf{n}}(\vec{\mathbf{r}})\right)^2 + \lambda\vec{\mathbf{n}}(\vec{\mathbf{r}})\mathbf{rot}\vec{\mathbf{n}}(\vec{\mathbf{r}}) \tag{5.5}$$

$$\lambda = \mathbf{K_{22}q_0} \tag{5.6}$$

That is why, if the coefficient, λ, is theoretically calculated with the help of the microscopic approach, considering the energy of interaction between two molecules of a liquid crystal when calculating the free energy at a contribution to the free deformation energy proportional to $\vec{\mathbf{n}}(\vec{\mathbf{r}})\mathbf{rot}\vec{\mathbf{n}}(\vec{\mathbf{r}})$, the pitch of the cholesteric helix is calculated from the equation

$$2\pi/\mathbf{P} = \lambda/\mathbf{K_{22}} \tag{5.7}$$

Let us note that, for the cholesteric helix considered above (i.e., for the case of $\mathbf{n_x} = \mathbf{cosqz}$, $\mathbf{n_y} = \mathbf{sinqz}$, $\mathbf{n_z} = \mathbf{0}$, for the value of the scalar product of $\vec{\mathbf{n}}(\vec{\mathbf{r}})\mathbf{rot}\vec{\mathbf{n}}(\vec{\mathbf{r}})$), the following equation exists:

$$\vec{\mathbf{n}}(\vec{\mathbf{r}})\mathbf{rot}\vec{\mathbf{n}}(\vec{\mathbf{r}}) = -\mathbf{q}$$

The microscopic model of a cholesteric liquid crystal considering the peculiarities of the structure of chiral molecules forming this crystal has been drawn in Reference 28. The analysis was conducted for the case of polypeptide molecules in an α-spiral conformation. It was supposed that the molecules are characterized by a uniform distribution of dipoles along the long axes. The direction of the dipoles is perpendicular to the long axis of the polypeptide; the dipoles form a spiral structure relative to the molecule's axis. The pitch and direction of the helical twist of the neighboring polypeptide molecules are determined by the wave number q. If $\mathbf{q} > \mathbf{0}$, the polypeptide molecules form a right-handed, helically twisted structure; if $\mathbf{q} < \mathbf{0}$, the structure is left-handed twisted. There is a short-range steric interaction between the polypeptide molecules that leads to the formation of a nematic, and a long-range dipole–dipole interaction that can be considered under the perturbation theory.

With the help of the thermodynamic theory of perturbations, the free energy of a single dipole $\mathbf{f_A}$ fixed in the point of coordinate origin and belonging to the molecule located in the plane $\mathbf{z} = \mathbf{0}$ with all of the liquid-crystal molecules situated in the plane $\mathbf{z} = \mathbf{const} > 0$ was calculated. This energy is described with the following equation [28]:

$$\mathbf{f_A} = \mathbf{f_{A0}} + \mathbf{Aq^3z}\,\mathbf{sin(2\varphi)} \tag{5.8}$$

where $\mathbf{A}$ is a positive constant determined by the parameters of the liquid-crystal molecules, and angle φ is the rotation angle of the polypeptide molecule long axes in the plane $\mathbf{z} = \mathbf{const} > 0$ relative to the molecules situated in the plane $\mathbf{z} = \mathbf{0}$.

The angle φ is determined by the twist of the cholesteric spatial helical structure around **z**-axis, and the sign of φ depends on the direction of the twist. For a left-twisted cholesteric helix, φ < 0, and for a right-twisted helix, φ > 0. The value contains items that depend on the even degrees of φ, including the constant, which does not depend on φ. It follows from Equation 5.8 that, when **q** > 0, the lower value of **f**$_A$ corresponds to a negative sign of φ, that is, φ < 0. This means that the cholesteric helix has a left-handed twist if the local spiral structure of the molecules that form the cholesteric has a right-handed twist. If the wave number of the cholesteric structure is marked as **k**, the received condition can be described as **qk** < 0.

The cholesteric formed by the phase exclusion of linear double-stranded, right-hand twisted DNA molecules that belong to the B-family has an abnormal band in the CD spectrum, which, according to the results of theoretical calculations, shows that, for the particles of CLCD, there is a specific left-hand twist of the cholesteric spatial structure. Nevertheless, it is known that the change in DNA or solvent properties can result in the formation of a right-hand twisted CLCD structure from these molecules.

This means that the foregoing theoretical approach needs to be broadened, bearing in mind the peculiarities of the interaction between these molecules. It can be assumed that if the interaction between molecules and the solvent is the reason for the appearance of dipoles on the long axes of the molecule and forming a spatial structure with certain chirality, the change of properties of the solvent can cause the change of the direction of the macroscopic cholesteric helical structure [28].

Using Equation 5.8 and integrating the whole liquid-crystal volume, its free energy, **f**, can be evaluated. The value of the modulus of the cholesteric's wave number |**k**| is calculated from the minimum condition of **f** relative to |**k**|. For different values of the parameters describing the cholesteric in the point of the free energy minimum,

$$|\mathbf{k}/\mathbf{q}| \approx 10^{-2} \sim 10^{-5} \tag{5.9}$$

In this case, the relation of the pitch value, **p**, of the spiral structure of the molecules to the pitch of the cholesteric helical spatial structure, **P**, formed by these molecules is equal to

$$\mathbf{p}/\mathbf{P} \approx 10^{-2} \sim 10^{-5} \tag{5.10}$$

The effect of the change in the sign of the abnormal band in the CD spectrum at the formation of CLCD from DNA molecules bound to an anthracycline antibiotic, daunomycin (DAU), was first described in Reference 29. The increase of the degree of binding of DAU that has positively charged amino groups in its structure to a DNA molecule is equivalent to the decrease in distance between positive charges on the surface of this molecule. The attempt to explain this phenomenon theoretically was proposed in Reference 30. When more that one DAU molecule is bound per a spiral twist of the DNA molecule, a spiral arrangement of DAU positive charges is created on the DNA surface. It is assumed in Reference 30 that, at the high DAU concentration at which the change in cholesteric helical sense is analyzed, the pitch of the spiral structure formed by added dipoles remains constant. Therefore, an additional interaction between the two helical structures of dipoles evolves and contributes to parameter λ. This parameter is included in Equations 5.5–5.7 and determines the cholesteric ordering of the DNA liquid-crystalline phase. The existence of such a contribution can even cause the change in the sense (direction) of the initial cholesteric helical spatial structure.

The electrostatic interaction between two charged spiral lines can be represented with examples from a contribution series. The first contribution corresponds to the potential energy of interaction of two charged lines and does not depend on the sign of their rotation angle—that is, it is a nonchiral potential. The second term describes the interaction between constant dipoles perpendicular to the DNA axis and directed from the axis to the positive charges. This very interaction caused by the presence of positively charged groups in DAU molecules makes an additional contribution to free energy and affects the value of the pitch of the cholesteric helical structure, **P**.

The electrostatic interaction between two macromolecules with helical dipole allocation a is described by the equation

$$U(1,2) = \sum_{k,l=1}^{M} R_{kl}^{-3} \exp(-xR_{kl}) \left[(\vec{d}_k \vec{d}_l) - 3(\vec{d}_k \vec{u}_{kl})(\vec{d}_l \vec{u}_{kl}) \right] \qquad (5.11)$$

where **M** is a total number of dipoles bound to the DNA molecule (integrated by all $\vec{d}_k$ and $\vec{d}_l$ dipoles corresponding to the macromolecules 1 and 2, respectively); $\vec{R}_{kl}$ is the vector connecting dipole $\vec{d}_k$ of molecule 1 to dipole $\vec{d}_l$ of molecule 2: $\vec{u}_{kl} = \vec{R}_{kl}/R_{kl}$.

The direction of $\vec{d}_k$ and $\vec{d}_k$ is determined by the DNA helical structure:

$$\vec{d}_k = \vec{e}_1 \cos(qz + \varphi) + \vec{e}_2 \sin(qz + \varphi) \qquad (5.12)$$

where $q = 2\pi/p$; **p** is the pitch of DNA secondary structure; $\vec{e}_1$, $\vec{e}_2$ are two orthogonal unit vectors perpendicular to the long axis of the molecule and **z** is the direction along the DNA molecule's axis.

The exponential parameter in Equation 5.11 considers the screening of the positive charge with a counterion cloud. For the ionic strength, I, corresponding to the 0.3M solution of NaCl, the Debye–Huckel screening parameter (**x**) is equal to 0.16Å^{-1}.

According to the thermodynamic theory of perturbations, the density of the system's free energy looks as follows:

$$F = F_0 - (2kT)^{-1} < W_{dd}^2 > \qquad (5.13)$$

where $W_{dd} = (1/2)\Sigma_{ij} U(i,j)$ is the total electrostatic energy of interaction between macromolecules in a unit of volume; the potential **U(i, j)** is determined by Equation 5.11; and F_0 is the free energy of DNA chains without considering the charge distribution induced by binding DAU molecules.

Without considering the details of the calculation, the final expression for λ parameter determined by the electrostatic interaction of spirally allocated dipoles bound to the DNA can be given [30]:

$$\lambda_d = -(9/16)\ (d^4\rho\sigma^2 L/kT)\ \exp(-2xR)R^{-6}\sin(2qb) \qquad (5.14)$$

where $R = (R_0^2 + b^2)^{1/2}$; $d = r_0 e$; $q = 2\pi/p$; r_0 is the radius of the DNA molecule; L is the length of the DNA molecule; ρ is the number of DNA molecules per unit volume; σ is the number of closest neighbors; R_0 is the average distance between DNA molecules in the arising of an LCD particle; p is the pitch of helical structure of the B-form of the DNA molecule, and b is the average distance between the neighboring positive charges on the DNA surface; the other parameters have the standard meanings.

The P value of a cholesteric structure that exists both in the liquid-crystalline phase and in DNA CLCD particles is determined by the value of the parameter λ_d, which depends on the concentration of the substance bound to the DNA molecule and affects the parameter b and the contribution of all the other interactions between DNA molecules to which λ_0 parameter corresponds:

$$2\pi/P = (\lambda_0 + \lambda_d)/K_{22} \qquad (5.15)$$

where P is the pitch of the cholesteric helical structure formed by DNA molecules.

At certain values of b, the quantity $\lambda_0 + \lambda_d$ turns into zero. Under these conditions, the helical twist of the cholesteric phase (dispersion) of macromolecules bound to the compound that carries positively charged groups disappears, and, consequently, the abnormal optical activity of the formed phase disappears as well. Moreover, at a certain point in the relationship between the values of λ_0 and λ_d, a change in the sign of the initial abnormal optical activity is possible.

It can be assumed that a similar mechanism exists for the change in the direction of the cholesteric helical structure in the case of CLCD particles formed by DNA molecules bound in a complex with Chi. Indeed, Chi spiral molecules contain amino groups that carry a partial positive charge under the used conditions (pH ~ 7). Because of the peculiarities of the spatial structure of Chi molecules, the amino groups can interact with DNA phosphate groups in an alternate manner [31,32]. This recalls the fact that Chi molecules that interact with DNA not only are fixed along the DNA molecules' long axis but also create a spiral structure of dipoles. At a certain distance, b, between the charged amino groups of Chi molecules located along the DNA, the corresponding dipole–dipole interaction can compensate for the contribution of the disperse interaction between adjacent molecules of the initial DNA. In this case, the sense of twist of the cholesteric helical structure in CLCD particles formed by the DNA–Chi complex can change, which, in turn, will cause change in the sign of the abnormal optical activity of these particles.

Taking into account that the molecular mass of DNA in the experiments described above was about 7×10^5 Da (which corresponds to the molecule length $L = 3.5 \times 10^{-7}$ m), and the value R_0 determined by means of small-angle x-ray scattering was ~ 25×10^{-10} m, the calculations of the parameter λ_d [14] were performed for different

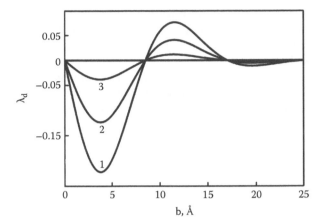

FIGURE 5.12   Theoretical curves of the variation of the λ_d parameter calculated for cholesteric dispersions of DNA–chitosan complexes, depending on the average distance **b** between positively charged amino groups in the chitosan molecules. Curves 1–3 correspond to the values of **x** parameter calculated from the solution ionic strength 0.263 (curve 1): 0.3 (curve 2); and 0.38 (curve 3).

values of parameter **x** that depend on the solution's ionic strength. The other values of the parameters from Reference 5.14 were

$$\rho = 10^{25} \text{ m}^{-3};\ \mathbf{r_0} = 10 \times 10^{-10} \text{ m};\ \sigma = 6;\ \mathbf{T} = 25°\text{C};\ \mathbf{p} = 34 \times 10^{-10} \text{ m}$$

Figure 5.12 represents theoretically calculated curves that describe the dependence of λ_d on the average distance **b** between the charged Chi amino groups in the case of the DNA–Chi complex CLCD under different conditions. The curves given in Figure 5.12 make it possible to conclude that, according to Equation 5.14, the contribution of the interaction between two dipole spirals depends on the distance **b** between the two adjacent positive charges on the DNA molecule surface, which evolved as a result of the binding of Chi to the DNA molecule (the distance **b** decreases as the degree of binding of DNA and Chi increases).

According to Equation 5.15, the pitch, **P**, of the cholesteric helix is determined by the sum $(\lambda_0 + \lambda_d)$, where the λ_0 that defines the twist of the DNA CLCD helical structure in the absence of chitosan does not depend on the distance **b** between the amino groups and has a negative sign. At high **b** values, the parameter λ_d is almost equal to zero, and the pitch, **P**, depends only on the parameter λ_0. This situation corresponds to the low concentration of Chi, when dipoles of Chi amino groups arranged significantly far from one another do not affect the **P** value of the cholesteric formed by DNA molecules bound with Chi.

As the Chi concentration increases, the distance **b** decreases, and at $\mathbf{b} < 23$ Å, the λ_d meaning significantly affects the value **P** of the cholesteric helix. Under a certain Chi concentration, the parameter λ_d becomes negative and the twisting of the initial left-handed cholesteric helical increases, that is, the decrease in the **P** of the helix takes place. At a further increase of Chi concentration, the tendency for the cholesteric **P** is the opposite. When the distance between positively charged groups **b** becomes shorter and reaches the value of half of the **p** value of the twist of the DNA secondary structure, the parameter λ_d becomes positive and increases as the parameter **b** decreases.

Therefore, there is a threshold concentration of Chi at which the direction of the CLCD particles' cholesteric helical twist can change.

According to Figure 5.9, at the concentration of NaCl equal to 0.05 M, a double turn of the sign to zero for the cholesteric helix with the decrease of the distance b between the amino groups in the content of Chi molecules takes place.

Such an effect can be explained theoretically. The conditions of the change of direction of the cholesteric helical structure of particles of CLCDs formed by the DNA–chitosan complex follow from Equation 5.15:

$$\lambda_0 > -\lambda_d \tag{5.16}$$

The helical structure of the cholesteric is right-handed twisted:

$$\lambda_0 = -\lambda_d \tag{5.17}$$

The unwinding of the helical structure takes place:

$$\lambda_0 < -\lambda_d \tag{5.18}$$

The helical structure of the cholesteric is left-hand twisted.

If the quantity $|\lambda_0|$ decreases as the ionic strength increases slower than $|\lambda_d|$, at a low ionic strength of the solution there are two **b** values at which the condition 5.17 is observed. Indeed, Figure 5.13 shows that, in solutions with low ionic strength at **b** < **b$_1$**, the condition $\lambda_0 < -\lambda_d$ is observed, and a cholesteric with left-handed helical twist is formed, that is, the circular dichroism, CD, < 0. At **b$_1$** < **b** < **b$_2$**, the parameter

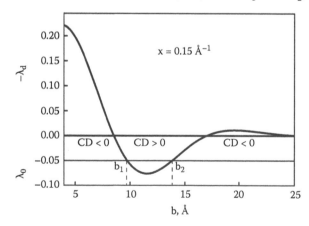

FIGURE 5.13   The graphic determination of the O-point turns for sign of cholesteric structure formed by DNA–Chi complexes in solutions of low ionic strength.

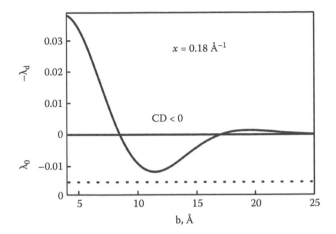

FIGURE 5.14   The graphic illustration of constant sign value of cholesteric structure formed by DNA–Chi complexes in solutions of high ionic strength.

$\lambda_0 >$ the parameter $-\lambda_d$, and a cholesteric with right-handed twist is obtained, CD $>$ 0. In the case of $\mathbf{b} > \mathbf{b}_2$, the parameter $\lambda_0 <$ the parameter $-\lambda_d$, and a cholesteric with a left-handed twist is formed, CD $<$ 0.

At high ionic strength, the foregoing condition is not observed. Figure 5.14 shows that, in this case, for all the **b** distances between positively charged amino groups fixed on the surface of DNA molecules, the inequality $\lambda_0 < -\lambda_d$ is true, which corresponds to a cholesteric with a left-handed helical twist, that is, CD $<$ 0.

A turn of the sign of the abnormal band in the CD spectrum of the DNA–Chi complex CLCDs at high ionic strength and short distances between the positive charges of Chi amino groups fixed on the DNA surface can be considered as evidence of the fact that, under these conditions, a different (unknown) mode of interaction between DNA–Chi complex molecules appears.

As at $\mathbf{b} \sim 15$ Å, where the effect of the spiral arrangement of the positively charged Chi groups is significant, the quantities $|\lambda_0|$ and $|\lambda_d|$ are comparable to each other; it is interesting to evaluate the pitch, **P** (or especially for dispersions, Pd) of the cholesteric structure formed by DNA–Chi complexes based on the value of the parameter $|\lambda_d|$. In the case of $x = 0.15$ Å^{-1}, the value of $|\lambda_d|$ is ~ 0.04 dyne/cm. Accepting that the constant $\mathbf{K}_{22}$ has a degree of 10^{-6} dyne, a simple evaluation shows that

$$\mathbf{P_d} = 2\pi\mathbf{K}_{22}/\lambda_d \sim 1.5 \times 10^{-4} \text{ cm} = 1.5 \text{ μm}$$

The calculated $\mathbf{P_d}$ value is quite close to **P**, typical of the cholesterics formed by linear double-stranded DNA molecules (Chapter 2) that was determined by the polarizing microscopy method.

It is interesting to compare the results of theoretical calculations and the experimental data that describe the properties of CLCDs formed by DNA–Chi complexes. Such comparisons make it possible to draw several important conclusions.

1. The sign of the abnormal band in the CD spectrum of the DNA–Chi complex CLCD changes at the decrease of **b**, that is, the distance between the charged amino groups in a Chi molecule. This conclusion means that the introduction of an extra number of charged groups into the structure of the DNA–Chi complex (for instance, by intercalation of some compounds between DNA nitrogen base pairs) must affect the value and even the sign of the abnormal band in the CD spectrum of these CLCDs.
2. Under certain conditions, DNA–chitosan complexes could produce LCDs having no abnormal band in the CD spectrum.

Indeed, the formation of CLCDs from DNA–Chi complexes at pH ~ 8, despite the onset of an "apparent" optical density that indicates the formation of the dispersion, is not accompanied by changes in the shape of the CD band characteristic of the initial B-form of DNA and the emergence of an abnormal CD band under these conditions. This fact is of interest from two points of view. First, it indicates once again that light-scattering itself does not cause the appearance of an abnormal band in the CD spectrum. Second, it shows that the mode of the spatial packing for the DNA–chitosan complexes in particles of dispersions is associated with a strict equilibrium of the different types of interaction forces operating between these molecules. This means that various modes of packing of DNA–Chi complex molecules in CLCD particles with different abnormal optical activity are possible.

Consequently, despite the limitations of the calculations performed, the main consequences of the theory agree well with the experimental results obtained for the DNA–chitosan CLCDs.

Therefore, the interaction of Chi with rigid, linear, double-stranded DNA molecules with low molecular mass leads to the formation of CLCDs whose properties (optical and structural) are different from the properties of classical cholesterics formed by the "pure" double-stranded DNA as a result of entropy condensation. The efficiency of the interaction between Chi and DNA depends on many factors (ionic strength, pH, etc.). The factor that regulates the spatial structure of the forming CLCDs is the distribution of the positive charges (the distance between charges, Chi molecules conformation, etc.) in the Chi polymeric chain that interacts with the DNA molecule. Such interaction not only provides the neutralization of DNA phosphate group negative charges but also makes it possible to "introduce" an extra positive charge on the surface layer of the DNA molecule, which, in turn, determines the value of the anisotropic contribution to the free energy of interaction between adjacent molecules of DNA–Chi complexes.

Chi molecules that interact with circular superhelical DNA also cause their condensation. The particles of the formed LCD do not have any specific spatial shape but can be used to deliver the genetic material into living cells [21,33–35]. The methods and techniques of the control of properties of such particles have no difference from the methods of observation of the condensed plasmid shapes formed as a result of their entropy condensation described earlier.

5.9   SUMMARY

Two important conclusions follow from the results represented in this chapter. First, in the case of high-molecular-mass DNA molecules, enthalpy condensation causes the formation of single particles that mainly have a toroidal shape with a diameter of about 1,000 Å. Second, in the case of low-molecular-mass DNA molecules, the enthalpy condensation caused by the interaction of DNA with polycations induces the formation of LC-dispersions with significantly different properties. The example of chitosan shows that the spatial structure of the formed LCD of DNA–polycation complexes is regulated by a complicated combination of such factors as polycation molecule conformation, the distance between the positive charges in its polymer chain, and the pH value and the ionic strength of the solution. Meanwhile, the physicochemical properties of CLCDs formed by the DNA–polycation complex easily change in response to the change in the properties of both polycation molecules and the solution used for the formation of these dispersions.

Consequently, the enthalpy condensation of double-stranded DNA molecules with low molecular mass depends on the combination of many factors, and only at a certain combination of solvent and polycation properties is it possible to obtain CLCDs from DNA–polycation complexes. Regardless of the uniqueness of the set of properties of a Chi molecule, this conclusion must be considered when comparing the properties of the CLCD of DNA complexes with other synthetic or natural polycations described in the works of different authors. The existence of multiple structures of CLCD formed as a result of enthalpy condensation, combined with the existence of a number of factors regulating these structures, makes the behavior of DNA–polycation complex CLCDs similar to the behavior of DNA-containing virus particles and bacteriophages.

REFERENCES

1. Oosawa, F. *Polyelectrolytes*. New York: Marcel Dekker, 1971. 160 p.
2. Allahyarov, E., Gommper, G., and Lowen, H. DNA condensation and redissolution: Interaction between overcharged DNA molecules. *J. Phys. Condens. Matter*, 2005, vol. 17, p. 1827–1840.
3. Yoshikawa, Y. and Yoshikawa, K. Diaminoalkanes with an odd number of carbon atoms induce compaction of a single double-stranded DNA chain. *FEBS Lett.*, 1995, vol. 361, p. 277–281.
4. Ichida, Y. and Yoshikawa, K. Single chain observation of collapse transition in giant DNA induced by negatively-charged polymer. *Biochem. Biophys. Res. Commun.*, 1998, vol. 242, p. 441–445.
5. Vasilevskaya, V.V., Khokhlov, A.R., Kidoaki, S., et al. Structure of collapsed persistent macromolecule: Toroids vs. spherical globule. *Biopolymers*, 1997, vol. 41, p. 51–60.
6. Yoshikawa, K., Yoshikawa, Y., and Kanbe, T. All-or-none folding transition in giant mammalian DNA. *Chem. Phys. Lett.*, 2002, vol. 354, p. 354–359.
7. Raspaud, E., Olivera de la Cruz, M., Sikorav, J.-L., et al. Precipitation of DNA by polyamines: A polyelectrolite behavior. *Biophys. J.*, 1998, vol. 74, p. 381–393.

8. Saminathan, M., Antony, T., Shirahata, A., et al. Ionic and structural specificity effect of natural and synthetic polyamines on the aggregation and resolubilization of single-, double-, and triple-stranded DNA. *Biochemistry*, 1999, vol. 38, p. 3821–3830.

9. Skuridin, S.G., Kadykov, V.A., Shashkov, V.S., et al. The formation of a compact form of DNA in solution induced by interaction with spermidine. *Mol. Biol.* (Russian Edition), 1978, vol. 12, p. 413–420.

10. Rielland, S.L. and Williams, R.E. Water-soluble lysine-containing polypeptides. III. Sequental lysine-glycine polypeptides. A circular dichroism and electron microscopy study of annealed complexes with DNA, sonicated DNA, and denaturated DNA. *Can. J. Chem.*, 1976, vol. 54, p. 3884–3894.

11. Kabanov, V.A., Sergeeev, V.G., Pyshkina, O.A., et al. Interpolyelectrolyte complexes formed by DNA and astramol poly(propyleneimine) dendrimers. *Macromolecules*, 2000, vol. 33, p. 9587–9593.

12. Sato, N., Kobayashi, H., Saga, T., et al. Tumor targeting and imaging of intraperitoneal tumors by use of antisense olido-DNA complexed with dendrimers and/or avidin in mice. *Clin. Cancer Res.*, 2001, vol. 7, p. 3606–3612.

13. Evdokimov, Yu.M., Salyanov, V.I., Krylov, A.S., et al. The formation of liquid-crystalline dispersions of DNA-Chitosan complexes under the conditions of molecular crowding. *Biophysics* (Russian Edition), 2004, vol. 49, p. 789–799.

14. Yevdokimov, Yu.M., Salyanov, V.I., Semenov, S.V., et al. Formation of liquid-crystalline dispersions of double-stranded DNA-Chitosan complexes. *Mol. Biol.* (Russian Edition), 2002, vol. 36, p. 532–541.

15. Yevdokimov, Yu.M., Salyanov, V.I., Skuridin, S.G., and Dembo, A.T. Some x-ray parameters of liquid-crystalline dispersions of Nucleic acid-Chitosan complexes. *Mol. Biol.* (Russian Edition), 2002, vol. 36, p. 706–714.

16. Yevdokimov, Yu.M., Skuridin, S.G., and Salyanov, V.I. The liquid-crystalline phases of double-stranded nucleic acids in vitro and in vivo. *Liq. Crystals*, 1988, vol. 3, p. 1443–1459.

17. Yevdokimov, Yu.M. and Salyanov, V.I.. Liquid-crystalline dispersions of complexes formed of chitosan with double-stranded nucleic acids. *Liq. Crystals*, 2003, vol. 30, p. 1057–1074.

18. Okuyama, K., Noguchi, K., Miyazawa, T., et al. Molecular and crystal structure of hydrated chitosan. *Macromolecules*, 1997, vol. 30, p. 5848–5855.

19. Cairns, P., Miles, M.J., Morris, V.J., et al. X-ray fibre diffraction studies of chitosan and chitosan gels. *Carbohydrate Res.*, 1992, vol. 235, p. 23–28.

20. Hayatsu, H., Kubo, T., Tanaka, Y., et al. Polynucleotide-chitosan complex, an insoluble but reactive form of polynucleotides. *Chem. Pharm. Bull.*, 1997, vol. 45, p. 1363–1368.

21. MacLaughlin, F.C., Murper, R.J., Wang, J., et al. Chitosan and depolymerized chitosan oligomers as condensing carriers for in vivo plasmid delivery. *J. Control. Release*, 1998, vol. 56, p. 259–272.

22. Osipov, M.A. Theory of cholesteric ordering in lyotropic liquid crystals. *Il nuovo cimento*, 1988, vol. 10, p. 1249–1262.

23. De Gennes, P.-G. *The Physics of Liquid Crystals*. London: Oxford University Press, 1974. p. 333.

24. Stephen, M.J. and Straley, J.P. Physics of liquid crystals. *Rev. Mod. Phys.*, 1974, vol. 46, p. 617–704.

25. Frèedericksz, V. and Zolina, V. Forces causing orientation of an anisotropic liquid. *Trans. Faraday Soc.*, 1933, vol. 29, p. 919–930.

26. Gruler, H., Scheffer, T., and Meier, G. Elastic constants of nematic liquid crystals. I. Theory of the normal deformation. *Z. Naturforcsh.*, A, 1972, vol. 27, p. 966–976.

27. Straley, J.P. Frank elastic constants of the hard-rod liquid crystal. *Phys. Rev.*, A, 1973, vol. 8, p. 2181–2183.

28. Kim, Y.H. A macroscopic model of polypeptide solutions. *J. Physique*, 1982, vol. 43, p. 559–565.
29. Yevdokimov, Yu.M., Salyanov, V.I., Skuridin, S.G., and Badaev, N.S. Liquid crystals and liquid-crystalline dispersions of DNA-Daunomycin complexes. *Mol. Biol.* (Russian Edition), 1995, vol. 29, p. 180–191.
30. Samori, B., Osipov, M.A., Domiani, I., et al. Transverse dipoles added to DNA chains by drug binding can induce inversion of the long-range chirality of DNA condensates. *Int. J. Biol. Macromol.*, 1993, vol. 15, p. 353–359.
31. Kubota, N. and Kikuchi, Y. Macromolecular complexes of chitosan. In *Polysaccharides*, Ed. by Dimitriu, S. New York: Marcel Dekker, 1988, p. 565–628.
32. Yevdokimov, Yu.M. Nucleic acids and chitosan. In *Chitin and Chitosan. Production, Properties and Usage*, Ed. by K.G. Skryabin. Moscow: Nauka, 2002, p. 178–200.
33. Aral, C., Ozbas-Turan, S., and Kabasakai, L. Studies of effective factors of plasmid DNA-loaded chitosan microspheres. 1. Plasmid size, chitosan concentration and plasmid addition techniques. *S.T.P. Pharma Sci.*, 2000, vol. 10, p. 83–88.
34. Ercelen, S., Zhang, X., Duportail, G., et al. Physicochemical properties of low molecular weight alkylated chitosans: A new class of potential nonviral vectors for gene delivery. *Colloids Surf. B, Biointerfaces*, 2006, vol. 51, p. 140–148.
35. Choi, C., Kim, D.-G., Jang, M., et al. DNA delivery using low molecular water-soluble chitosan nanocomplex as a biomedical device. *J. Appl. Polym. Sci.*, 2006, vol. 102, p. 3545–3551.

6 Liquid-Crystalline State of DNA Circular Molecules

6.1 PHASE EXCLUSION OF CIRCULAR MOLECULES OF NUCLEIC ACIDS

It is known that many prokaryote cells contain, in addition to the basic chromosome, small extrachromosomal DNAs that are called "plasmids." A plasmid is an independent, stable, circular (or linear) piece of DNA in a bacterial cell that is not a part of the normal cell genome and that never becomes integrated into the host chromosome, capable of autonomous replicating (independently from the basic chromosome). Though many plasmids provide significant selective replication advantages for the host cells (tolerance to antibiotics, heavy metals, etc.), most of them are cryptic, that is, do not appear in the cells' phenotype. Plasmids that may contain from several thousands to hundreds of thousands of base pairs, and produce from one to several hundreds of replicas per cell, are stably inherited in many cell generations.

In particular, the number of bacterial high-copy plasmids *Blue-Script* (2960 base pairs) in one *E. coli* cell may reach 1,000, which corresponds to DNA concentration high above the amount of DNA in a bacterial chromosome. The analysis of the curves x-ray scattering by intact *E. coli* bacterial cells carrying the *Blue-Script* plasmids has shown that there is a weak but clearly visible maximum on these curves. The position of this peak on the scattering curve coincides with the peak that is typical of the packing of linear DNA in liquid-crystalline phase *in vitro* [1] and corresponds to the distance between DNA molecules equal to 51.5 Å [2]. Meanwhile, the polarizing microscopy of thin layers of *E. coli* cells containing such plasmids proves the existence of periodic structures in the cells that reflect the spatial packing of the plasmids. The combination of the received data permits one to suppose that the *Blue-Script* plasmids in *E. coli* cells are characterized by cholesteric liquid-crystalline packing.

DNA circular molecules can accept a few spatial structural forms—the circular superhelical (supercoiled) form, an open ring form without superturns (relaxed structure), and the shape of a ring with one break ("nick") in one of the sugar-phosphate chains (the so-called nicked structure) (Figure 6.1). Such multiplicity of circular DNA forms determines the interest in the research of their condensation under different conditions, which has been examined in a number of works [3–9].

The study of packing of the various circular DNAs with an equal molecular mass is interesting, both from physicochemical and biological points of view.

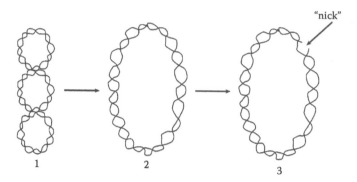

FIGURE 6.1   Schematic representation of different spatial structures of circular covalently closed DNA molecules: 1—superhelical (supercoiled) form; 2—open ring form; 3—nicked form.

6.2   FORMATION OF DISPERSIONS FROM CIRCULAR SUPERHELICAL DNA

An "apparent" optical density appears in the absorption spectrum of pBR 322 circular superhelical DNA (4363 base pairs) placed into a water–salt PEG-containing solution [10] at a certain PEG concentration, as well as in the cases of linear DNA molecules examined earlier. Development of "apparent optical density" (A_{app}) at wavelengths ($\lambda > 320$ nm) at which neither DNA nor PEG absorb testifies to the formation of a light-scattering DNA dispersion. The dependence of A_{app} on PEG concentration allows one to determine the "critical" PEG concentration (C^{cr}_{PEG}) needed to initiate the condensation of superhelical DNA.

The dependence of the "apparent" optical density (A_{app}) on the PEG concentration in solutions shows that the value of the "critical" PEG concentration necessary for the formation of a dispersion of superhelical DNA is ~110 mg/mL; this value almost coincides with the C^{cr}_{PEG} for linear double-stranded DNA molecules. Similar dependences are typical of all of the superhelical circular DNAs with different molecular mass and composition (in particular, pUC-18 DNA containing 2,700 base pairs; pGC 20 DNA—2,704 base pairs; pAB 4 DNA—7,022 base pairs; pPS-neo DNA—9,800 base pairs; pT 22 DNA—10,663 base pairs). This means that the process of formation of superhelical DNA dispersions hardly depends on the molecular mass and nucleotide composition of these molecules.

Therefore, the process formation of circular superhelical DNA dispersions takes place under such properties of the PEG-containing solution that need to initiate the formation of linear double-stranded DNA LCD particles.

Considering the presence of superturns in the structure of superhelical DNA, the question of the mode DNA molecules use to pack in their dispersion particles is interesting. An answer can be found through the same research methods that had been employed earlier to determine the properties of linear DNA LCD particles, notably circular dichroism (CD), analysis of x-ray scattering, and polarizing microscopy.

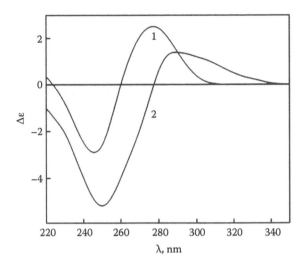

FIGURE 6.2 The CD spectra of superhelical pBR 322 DNA: (1) without PEG; (2) with 170 mg/mL PEG. C_{DNA} = 5 µg/mL; 0.3 M NaCl; 0.025 M Tris-HCl; pH 8; 0.01 M $CaCl_2$.

The formation of a dispersion by mixing water–salt solutions of pBR 322 DNA and PEG solutions ($C_{PEG} > C^{cr}_{PEG}$) is followed by the appearance of a low-intense negative band in the CD spectrum ($\Delta\varepsilon = -5$) (Figure 6.2, curve 2). Similar CD spectra are typical for dispersions obtained in PEG-containing solutions from all of superhelical DNA.

The appearance of a low-intense band in the CD spectrum at the condensation of superhelical DNA was noted in works of many authors [1,7–10].

According to the results given in Chapter 3, the amplitude of the abnormal band in the CD spectrum of CLCD formed by linear DNA molecule is related to their packing in CLCD particles, and its value depends on both the properties of the solvent and the DNA properties, in particular, on the homogeneity of their spatial structure. It can be assumed that the low-intense band in the CD spectrum of circular superhelical DNA dispersions also depends on the packing of the molecules in the forming particles.

Indeed, the analysis of thin layers of phase formed by pPS-neo circular superhelical DNA obtained as the result of low-speed centrifugation of particles of the DNA dispersion has shown that these layers are characterized by a "nonspecific" texture observed by a polarizing microscope. This "nonspecific" texture reflects the existence of anisotropy in the system, though it makes the identification of the type of superhelical DNA molecules packing in the phase difficult. (There is no systematic study of the textures of liquid crystals formed by polymers with molecular mass exceeding $1 \cdot 10^6$. Therefore, the classification of textures observed in liquid-crystalline phases formed by superhelical DNA molecules [molecular mass >$1 \cdot 10^6$] is based on comparison to the known textures of liquid crystals obtained by low-molecular-mass compounds.)

Nevertheless, the existence of a texture in thin layers of circular superhelical DNA phases indicates that there is a long-range order in the arrangement of neighboring DNA molecules (segments).

6.3 CD SPECTRA OF CIRCULAR SUPERHELICAL DNA DISPERSIONS UNDER CONDITIONS THAT MODIFY PARAMETERS OF THEIR SECONDARY STRUCTURE

It can be assumed that the ordering of the DNA circular molecules in dispersion particles depends on the morphology of DNA molecules. Two cases are possible here. First, the packing of circular DNA in dispersion particles, as well as in the case of linear double-stranded DNA molecules, is determined at the "moment of recognition" of these molecules, so the morphology of circular DNA affects both the details of packing of the molecules and the amplitude and sign of the band appearing in the CD spectrum at the phase exclusion. Second, the superhelical structure of circular DNA in dispersion particles prevents them from regular ordering. In this case, during the transition from the superhelical to the DNA open ring structure (see Figure 6.1), the appearance of a new optical activity of the particles can be expected.

To verify the first of these assumptions, the properties of the dispersion particles obtained from superhelical DNA molecules processed with compounds capable of inserting between the adjacent base pairs (intercalators) were determined [11]. These planar heterocyclic compounds of anthracene, anthraquinone, or anthracycline groups, being intercalated between the base pairs, unwind the DNA double-stranded secondary structure, which, in turn, causes the relaxation of the superturns and, even more, the change in the initial direction of superhelical turns of circular superhelical DNA. The initial circular superhelical DNA molecules were "loaded" with intercalators, which caused the progressive unwinding of superturns. Under these conditions, the DNA molecules first took the shape of an open ring and then became superhelical again, but the twist direction of superturns of the DNA molecules "loaded" with a high concentration of intercalators was opposite to the initial twist direction. This means that the intercalators can influence the morphology of circular superhelical DNA.

The process of the transition between different structural forms of circular DNA was controlled with the help of electrophoresis of DNA samples at different degree of intercalator binding. An electrophoregrams of pBR 322 circular superhelical DNA molecules complexed with an antibiotic of the anthraquinone group, mitoxanthrone, whose molecules intercalate between the DNA base pairs, are given in Figure 6.3, as an example [12]. It can be noticed that the electrophoretic mobility of pBR 322 circular superhelical DNA decreases as the antibiotic concentration increases (tracks 1–2); it reaches the mobility of the relaxed form (track 2) and increases again as the mitoxanthrone concentration increases (tracks 2–3). These data show that the binding of mitoxanthrone to circular superhelical pBR 322 DNA molecules is followed by the change in their helical turns sense.

At any condition of pBR 322 DNA molecules' being treated with intercalators, the closed structure of these molecules persists, regardless of the change in the extent of superhelicity and the sense of superhelical turns.

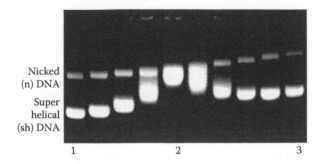

FIGURE 6.3 Electrophoregrams of pBR 322 DNA complexed with mitoxanthrone: $1 - \mathbf{r} = 0$; $2 - \mathbf{r} = 0.06$; $3 - \mathbf{r} = 0.13$. "**n**" и "**sh**"—nicked and superhelical forms of DNA, respectively. (**r**—the relation of molar concentration of bound to DNA mitoxanthrone to molar concentration of DNA nitrogen bases.)

The received pBR 322 DNA samples with different electrophoretic mobility were put into PEG-containing solution to form dispersions, and the CD spectra of these dispersions were recorded. In Figure 6.4, as an example [12] the CD spectra of pBR 322 circular superhelical DNA dispersion with (curve 1) and without (curves 2, 3) the intercalator, an antibiotic of anthracene group—bisanthrene—are given. It can be noticed that the formation of dispersion from DNA-bisanthrene complex molecules is followed by the appearance of two bands in the CD spectrum. One of the bands is located in the region of the DNA absorption; the other one is in the

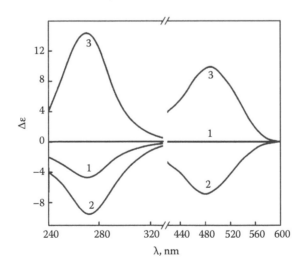

FIGURE 6.4 The CD spectra of dispersion formed by superhelical circular pBR 322 DNA and this DNA treated with bisanthrene: $1 - \mathbf{r} = 0$; 2 and $3 - \mathbf{r} = 0.042$ and 0.12, respectively. $C_{PEG} = 170$ mg/mL; 0.3 M NaCl; 0.03 M acetate buffer; pH 5.5.

region of bisanthrene absorption. Both of the bands have the same signs. Similar spectra were received for all of the dispersions of circular superhelical DNA with antibiotics and colored chemical compounds that intercalate between the DNA base pairs [11,13–15].

The amplitudes of the bands in the CD spectra of dispersion formed by superhelical DNA bound in a complex with bisanthrene are noticeably smaller than the amplitudes of bands in the CD spectrum of CLCD obtained from a bisanthrene complexed with linear double-stranded DNA molecules [12]. Figure 6.4 shows that the formation of complexes between bisanthrene and superhelical DNA [11] is followed by a change in the signs of both bands in the CD spectrum of the dispersions of these complexes. Meanwhile, the relaxation points of complexes of superhelical DNA with different intercalators—that is, the concentrations of intercalators at which the superhelical DNA is transformed to the open ring, determined by means of electrophoresis—coincide with the points where the signs of the bands in the CD spectra of the superhelical DNA dispersions bound with these complexes turn to zero. This means that there is a correlation between the direction of helical twist of the circular DNA and the sign of the bands in the CD spectra of dispersions formed by their complexes with intercalators. This conclusion is confirmed by the data in Figure 6.5.

It is noticeable that dispersion formed by the initial circular DNA molecules with left-handed superturns is characterized by the negative band in the CD spectrum. The dispersion formed by DNA molecules with right-handed superturns that evolved when the initial superhelical DNA was heated to the temperature above 50°C in the presence of formaldehyde, is characterized by the positive band in the CD spectrum. The data (insert in Figure 6.5 represents the electrophoregram of Col El DNA samples incubated with 12% formaldehyde at different temperatures; **N** and **SH**—the nicked and the superhelical forms of the DNA, respectively) indicate that the appearance of

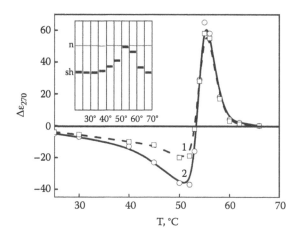

FIGURE 6.5   The temperature dependences of amplitudes of bands in the CD spectra of dispersions formed by superhelical Col El DNA (1) and pBR 322 DNA (2) incubated with 12% formaldehyde.

an open ring from a superhelical DNA molecule (see Figure 6.1) corresponds to the zero optical activity of particles of the dispersion formed by this DNA form.

The received results indicate that the sign of low-intense band in the CD spectra of circular DNA dispersions depends on the direction (sign) of the superhelical turns of these molecules. It is only possible if the DNA molecules maintain their anisotropic properties that determine the tendency of these molecules to cholesteric packing at condensation. In this case, the bands in the CD spectra of circular super-helical DNA dispersions are related to the mode of cholesteric packing of the molecules. If this is true, then the particles of the dispersions of circular superhelical DNAs and their complexes can be marked as LCD. The low intensity of the bands in the CD spectra of these LCD is determined by the "inhomogeneity" (heterogeneity) of the DNA superhelical structure that affects the efficiency of "recognition" of these molecules during the phase exclusion. Meanwhile, the disappearance of the low-intense band in the CD spectrum of LCD formed by DNA molecules that possess the open ring structure shows that the cholesteric packing of such molecules is impossible.

According to the second assumption, at the transition from superhelical DNA to the open ring structure (see Figure 6.1), an appearance of a new optical activity of LCD particles can be expected. To verify this assumption, the particles of LCD formed by superhelical DNA molecules were split with hydrolytic enzyme—micro-coccal nuclease.

Figure 6.6 (the insert represents the dependence of $\Delta\varepsilon_{270}$ value of superhelical circular DNA pPS-neo on the time of nuclease digestion) shows that, after a low extent of treatment of LCD particles, formed in a PEG-containing solution ($C_{PEG} > C^{cr}_{PEG}$) with nuclease, the pPS-neo molecules are split and form a linear molecule with the

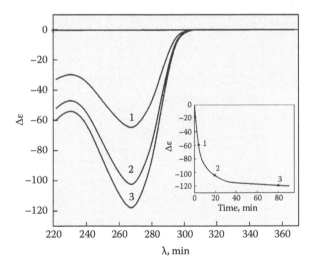

FIGURE 6.6 The CD spectra of LCD of pPS-neo DNA for different treatment times by micrococcal nuclease (MN): 1, 2, and 3–5, 20 and 30 min of MN treatment, respectively. $C_{DNA} = 15.0$ μg/mL; $C_{MN} = 0.03$ μg/mL; 170 mg/mL PEG; 0.3 M NaCl; 0.015 M Tris-HCl; pH 8; 0.001 M CaCl$_2$.

same molecular mass as the starting superhelical molecules. A sharp increase in the amplitude of a negative CD band ($\lambda = 270$ nm) (Figure 6.6, curve 2) is observed during the nuclease splitting of the superhelical DNA molecules forming the particles of the dispersion. The $\Delta\varepsilon$ value, which is only used in this case as a criterion for the dispersion of linear pPS-neo DNA molecules (molecular mass $\sim 6.5 \cdot 10^6$), reaches 100 units. This result indicates that condensation of circular DNA molecules in PEG-containing solutions does not hinder the nuclease action. The negative band in the CD spectrum indicates the appearance of a "helical" arrangement of DNA molecules forming the LCD. Such a band could result from the change in the packing of DNA molecules after the transformation of the circular form into linear form. The further splitting of pPS-neo linear DNA, which forms LCD particles in PEG-containing solution, into low-molecular-mass segments (molecular mass of $5–7 \times 10^5$ Da) results in the amplitude of the negative stripe in the CD spectrum reaching its maximum value ($\Delta\varepsilon = -120$) (Figure 6.6, curve 3).

The appearance of the abnormal band in CD spectrum in the region of DNA nitrogen bases' absorption indicates that, as a result of micrococcal nuclease digestion, the mode of packing of circular superhelical pPS-neo DNA forming the LCD particles changes sharply. The texture of a thin layer of high-molecular-mass pPS-neo DNA (molecular mass $= 6.5 \times 10^6$ Da) phase obtained from LCD particles splitted with Micrococcal nuclease for 20 min (Figure 6.7b) only slightly resembles the

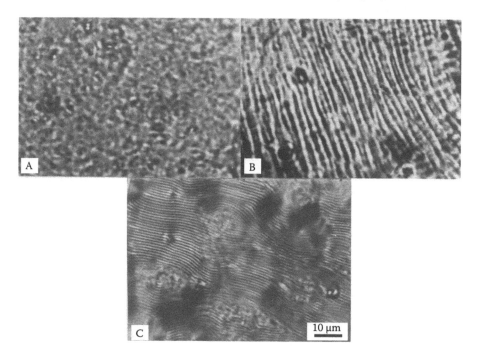

FIGURE 6.7 Textures observed in thin layers of liquid-crystalline phases formed by pPS-neo DNA. (a) Control without Micrococcal nuclease treatment; (b) nuclease treatment time –20 min; (c) 80 min (crossed polars). Solvent as in Figure 6.6.

fingerprint texture typical of classical cholesterics. Though the light and the dark lines alternate in this unusual texture with the average distance of ~3 μm between them, this texture corresponds instead to the "precholesteric" structure [16].

Nevertheless, the splitting of particles of the LCD of pBR 322 superhelical DNA with Micrococcal nuclease for 60 min, which results in the formation of linear DNA molecules with lower molecular mass (2.88×10^6 Da), causes the texture of the phase formed from such particles to become similar to the "classical" fingerprint texture [5]. Such texture, combined with the appearance of the abnormal negative band in the CD spectrum of LCD, indicates the cholesteric packing of pBR 322 DNA, both in LCD particles and in the phase fixed in the thin layer.

During the deep hydrolysis of pPS-neo linear DNA with Micrococcal nuclease, DNA fragments with molecular mass of 3×10^6 Da within LCD particles can form. The texture of phase that corresponds to such state particles (Figure 6.7c) represents a classical fingerprint texture that is typical of the cholesteric phase of linear double-stranded DNA molecules. Besides, the value of pitch, **P**, of the cholesteric helix is ~2–2.5 μm. The received value almost coincides with the pitch of the cholesteric helix typical of liquid crystals formed by linear double-stranded DNA molecules with low molecular mass [17,18].

The curves of the small-angle x-ray scattering on the phase formed as a result of low-speed centrifugation of LCD particles of the initial superspiral DNA pPS-neo (curve 1) and the phases obtained from the LCD particles of this DNA after the processing with micrococcal nuclease (curves 2–3) within different periods of time are compared in Figure 6.8.

The small meaning of maximum on x-ray scattering curves and the practical constancy of this maximum position indicate not only the retention of the high local

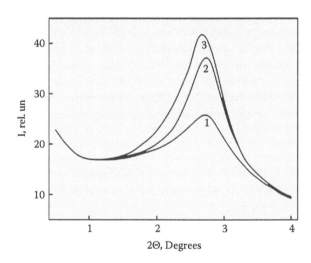

FIGURE 6.8 Small-angle x-ray scattering curves by LC phases formed by dispersions of pPS-neo DNA: (1) control without nuclease treatment; (2, 3) treatment of DNA in LCD by micrococcal nuclease for 20 and 80 min, respectively. Solvent as in Figure 6.6.

concentration of DNA in the particles but also the retention of an one-dimensional order in the packing of adjacent segments (molecules) of DNA, regardless of the change in the DNA morphology as a result of its hydrolytic enzyme digestion.

Using the results of x-ray analysis and textures of these phases, the twisting angle of the quasi-nematic layers formed by the superhelical DNA molecules can be evaluated. The value of $\mathbf{P}$ is determined by the equation $\mathbf{P} = 2\pi\mathbf{d}/\theta_0$, where $\mathbf{d}$ is the average distance between the axes of adjacent molecules of DNA in the phase, and θ_0 is the twisting angle of adjacent layers of DNA molecules in the cholesteric structure. The evaluation shows that the angle θ_0 in the case of cholesteric phases formed from LCD particles of circular superhelical pPS-neo DNA and pBR 322 DNA split with micrococcal nuclease is ~0.7°. This angle of twist of DNA layers in the obtained cholesterics is close to the twisting angle of quasi-nematic layers in CLCs and CLCDs formed by linear DNA molecules [17].

As the enzymatic splitting of the circular superhelical double-stranded DNA takes place under the condition of fixed osmotic pressure of the PEG-containing solution (i.e., when the local concentration of the forming linear fragments of this DNA does not change and remains high), the packing of a DNA molecule fragment that evolves during the hydrolysis must correspond to the cholesteric packing because of the anisotropic properties of these molecules.

Meanwhile, the classical fingerprint texture is only possible for the linear pPS-neo DNA fragments with molecular mass below 3×10^6 Da indicates the important role of the molecular mass of DNA in the process of the transition from the "nonspecific" to the cholesteric packing. This conclusion has an important theoretical meaning and directly indicates that, under the condition of phase exclusion, the cholesteric ordering of double-stranded DNA molecules is only possible in the case when the DNA molecular mass is close to 3×10^6 Da.

Therefore, the packing mode of circular superhelical DNA molecules in particles of the dispersion form because their condensation in PEG-containing water–salt solutions is very flexible and depends on the presence of turns in the DNA structure. The highly effective splitting of circular superhelical DNA molecules into linear fragments, which takes place within LCD particles, causes the change in the type of spatial packing of double-stranded DNA molecules from nonspecific to classical cholesteric type.

6.4 PACKING DENSITY AND REARRANGEMENT OF THE SPATIAL STRUCTURE OF SUPERHELICAL DNA MOLECULES IN LCD PARTICLES

Because it is known [8] that the increase in PEG concentration causes the increase in the packing density of adjacent double-stranded DNA segments and the distance between the adjacent molecules can reduce from ~50 to 25 Å, the influence of hydrolytic enzymes on the superhelical DNA molecules that form CLCD particles at different PEG concentration was investigated.

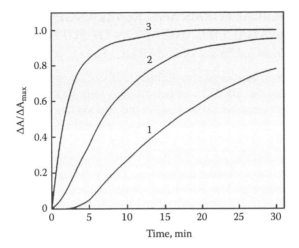

FIGURE 6.9 The dependence of the relative amplitude of the negative band ($\lambda = 270$ nm) in the CD spectra of pBR 322 DNA LCDs on the time of Micrococcal nuclease treatment. PEG concentrations (mg/mL): (1) 130; (2) 170; (3) 300; $C_{DNA} = 5$ μg/mL; $C_{MN} = 0.02$ μg/mL. Solvent as in Figure 6.6.

The curves in Figure 6.9 describe the dependence of the amplitude of the negative CD band on the duration of the enzymatic hydrolysis of pBR 322 super-helical DNA in particles of CLCDs formed at different PEG concentrations. The unexpected result is that the amplitude of the negative band increases with the PEG concentration in solution, and the maximum rate of this increase is observed in the case of the most dense packing of the DNA molecules in the LCD particles at PEG concentration of 300 mg/mL (curve 3). Nevertheless, this result corresponds to data [19–25] that show that the efficiency of the action of some enzymes (nuclease, ligase, topoisomerase, polymerase) on the condensed DNA forms increases significantly.

The increase in the efficiency of nuclease splitting of superhelical DNA molecules in LCD particles can be the result of several reasons. First, stated earlier, the phase exclusion is followed by the increase of the local DNA concentration or by the appearance in the DNA secondary structure of new nucleotide sites recognized by the Micrococcal nuclease. Under these conditions, the enzyme can contact more DNA molecules in LCD particles within its lifetime and, consequently, split them. Second, in a water–polymeric solution used for phase exclusion, the DNA–enzyme complex can be stabilized, which also causes the increase in the efficiency of the nuclease effect. The splitting of the initial superhelical DNA with a micrococcal nuclease under conditions where the superhelical DNA molecules do not form dispersions has shown that the presence of PEG does not affect DNA splitting with nuclease, that is, the presence of PEG by itself does not increase the efficiency of its effect. That is why the more probable explanation is the first assumption about the increase in the accessibility of sites recognized by enzymes as a result of superhelical DNA molecules condensation.

6.5 TOPOLOGICAL FORMS AND REARRANGEMENT OF THE SPATIAL ORGANIZATION OF SUPERHELICAL DNA MOLECULES IN LCD PARTICLES

As Figure 6.10 shows, the increase in amplitude of the abnormal band in the CD spectra, as a result of splitting the pBR 322 superhelical DNA forming LCD particles, depends on the concentration of micrococcal nuclease in the solution (curves 1–3). As the increase in the amplitude of the abnormal band in CD spectrum is caused by the change in the mode of the spatial packing of double-stranded DNA molecules, the rate of this process is determined by number of sites in DNA molecules available for splitting.

In this respect, the following question occurs: What structural form of double-stranded DNA molecules—nicked or linear—is sufficient to initiate the process of rearrangement of spatial ordering of DNA molecules in LCD particles?

Figure 6.11 shows that the dependence of the relative amplitude ($\Delta A/\Delta A_{max}$) of the negative band in the CD spectra of pBR 322 DNA LCD particles is correlated with the percentage of the linear form of the DNA that evolves as a result of its enzymatic splitting.

It is noticeable that there is a direct proportion between the content of double-stranded DNA linear molecules in LCD particles and the abnormal optical activity of these particles. This result shows that the rate of increase in abnormal negative band amplitude in the CD spectra of LCD particles is comparable to the rate of accumulation of linear DNA fragments, that is, the process of diffusion rearrangement of DNA molecules in LCD particles is quite rapid.

In the case of the splitting of the LCD particles formed by specially prepared nicked—that is, circular but not superhelical—DNA molecules with Micrococcal nuclease, there is a different situation. Here, the rate of the growth of the abnormal band in the CD spectrum of LCD particles is much lower than the rate of linear DNA accumulation.

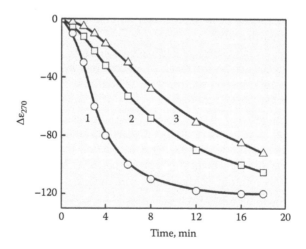

FIGURE 6.10 The dependence of the amplitude of the negative band ($\lambda = 270$ nm) in the CD spectra of superhelical pBR 322 DNA dispersion versus time of Micrococcal nuclease treatment: C_{MN} – (1) 0.125; (2) 0.062; (3) 0.031 µg/mL. Solvent as in Figure 6.6.

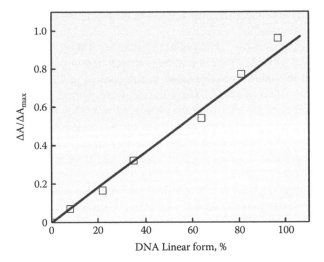

FIGURE 6.11 The dependence of the relative amplitude of the negative band ($\lambda = 270$ nm) in the CD spectra of pBR 322 DNA dispersions on the percentage of the linear form. $C_{DNA} = 10$ μg/mL; $C_{MN} = 0.01$ μg/mL; $C_{PEG} = 170$ mg/mL. Solvent as in Figure 6.6.

Moreover, experiments [8,26] have shown that the rate of restructuring of circular superhelical DNA LCD particles' spatial structure at DNA splitting and their transformation into linear form exceeds the analogous parameter in the case of nicked DNA LCD particles.

The curves given in Figure 6.12 describe the accumulation of linear pBR 322 DNA under the action of Micrococcal nuclease on the superhelical DNA in water–salt

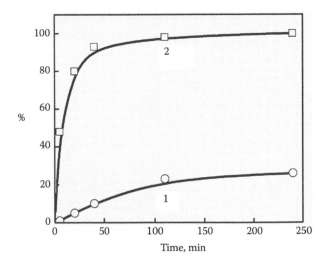

FIGURE 6.12 Accumulation of pBR 322 DNA linear form under Micrococcal nuclease action in water–salt solutions in the absence (curve 1) and in the presence of PEG (curve 2). $C_{DNA} = 5$ μg/mL; $C_{MNI} = 0.01$ μg/mL; $C_{PEG} = 170$ mg/mL. Solvent as in Figure 6.6.

solution (curve 1) and under the effect of the nuclease on the same DNA condensed in PEG solution (curve 2).

Figure 6.12 compares the curves of accumulation of the DNA linear form at the splitting of initial pBR322 DNA in a water–salt solution with Micrococcal nuclease (curve 1) and at the splitting of this DNA in the content of CLCD particles formed in the presence of PEG (curve 2). These curves were plotted using densitometry of the electrophoregrams to determine the content of the DNA linear form. The comparison of these curves shows that, in the case of the condensed state of superhelical DNA, the process of accumulation of the linear form that evolves as a result of DNA splitting with the nuclease takes place at a higher rate than in the case of the initial noncondensed pBR322 DNA.

The Micrococcal nuclease is a "single-hit" enzyme (i.e., only one site in one sugar-phosphate chain splits under its action), and at the beginning of the enzyme treatment a nicked form of DNA molecule is formed. After further processing of such DNA with Micrococcal nuclease, the second sugar-phosphate chain is split close to the first break site, which represents, by itself, a "defect" in DNA structure. Such spatial location of the two breaks in different chains of superhelical DNA molecule causes the rearrangement and transition to its linear form. Consequently, close to the beginning of the nuclease treatment of superhelical DNA molecules in LCD particles, the obtaining of both nicked and linear pBR 322 DNA structures is quite probable.

To elucidate the peculiarities of the DNA molecules' rearrangement mechanism, the curve that describes the accumulation of pBR 322 DNA linear fragments in the content of CLCD particles (Figure 6.13, curve 1) was compared to the curve 2 that describes the increase in abnormal band amplitude in the CD spectrum of the CLCD [8,26]. The comparison of the curves shows the accumulation of the linear DNA in the content of CLCD because Micrococcal nuclease treatment is followed by increase in

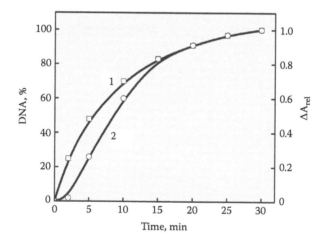

FIGURE 6.13 The dependence of relative content of linear DNA (curve 1) and that of relative amplitude of the band ($\lambda = 270$ nm) in the CD spectrum (curve 2) versus time of treatment of liquid-crystalline dispersions of superhelical pBR 322 DNA with Micrococcal nuclease. $C_{DNA} = 5$ µg/mL; $C_{MN} = 0.02$ µg/mL; other conditions as in Figure 6.12.

the amplitude of the abnormal band in the CD spectrum. Nevertheless, at the beginning of the superhelical DNA molecules' processing with nuclease, the accumulation of linear DNA in CLCD particles precedes (and thus may initiate) the process that determines an appearance of the abnormal band in the CD spectrum. Therefore, it may be assumed that the change in superhelical DNA molecules' packing, which includes the stage of linear DNA molecules' accumulation and their diffusive rearrangement in LCD particles, is similar to the process of polymer crystallization, which includes the steps of the formation of the "crystallization nucleus" and its further growth. This assumption allows one to apply the Kolmogorov–Avramy equation [27], usually used to describe the kinetics of synthetic and natural polymer crystallization, to obtain a description of the process of the increase in abnormal negative band in the CD spectrum of pBR 322 DNA CLCD. In this case, the equation looks like

$$\log \{- \ln[(\Delta A_{max} - \Delta A/\Delta A_{max})]\} = \log K + n \log t \qquad (6.1)$$

where A_{max} is the maximum amplitude of the band in the CD spectrum at $\lambda = 270$ nm at the moment t, and n and K are the constants that describe the cholesteric ordering of linear DNA fragments, which causes the appearance of abnormal optical activity.

As the increase in the abnormal band amplitude in the CD spectrum of CLCD particles is caused by the rearrangement of the DNA molecules' spatial packing, the rate of this process is determined by the number of breaks in DNA molecules, which depends on the nuclease concentration. The processing of the experimental curves shown in Figure 6.9 according to the Kolmogorov–Avramy equation leads to their linearization (Figure 6.14). The linearization of the observed kinetic curves indicates the applicability of the Kolmogorov–Avramy equation to describe the process of LCD structure rearrangement and shows that the process of the appearance

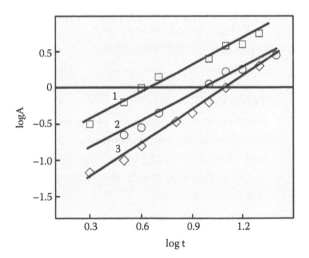

FIGURE 6.14 Kinetic curves from Figure 6.10 presented in the coordinates of the Kolmogorov–Avramy equation. 1, 2, and 3 equal 0.125, 0.062, and 0.031 μg/mL of nuclease, respectively; $C_{DNA} = 5$ μg/mL; $C_{PEG} = 170$ мг/мл; $t =$ time of nuclease treatment, min.

of abnormal band in the CD spectra of DNA LCDs is autocatalytic, which follows from at least two facts. First, it can be expected that, for an appearance of the abnormal band in the CD spectrum, the "crystallization nucleus" of the cholesteric phase with a certain diameter (~500 Å), whose optical properties cannot be detected by CD, needs to be formed (Chapter 4). Second, it can be expected that, for the appearance of the abnormal band in the CD spectrum, a certain number of linear double-stranded DNA molecules need to be orderly packed and twisted. Under any of these assumptions, the time needed for the abnormal band to appear in the CD spectrum may not coincide with the time of linear DNA molecules' accumulation in LCD particles. This very result was observed in the experiment (see Figure 6.13).

The values of **n** and **K** parameters, calculated from the Kolmogorov–Avramy equation for the case of superhelical pBR 322 DNA rearrangement caused by the action of Micrococcal nuclease, are given in the Table 6.1 [26]. There are a number of interesting facts here. First, the change in the **K** value shows that the rate of restructuring of DNA packing in LCD particles depends on the nuclease concentration. Second, the **n** and K values typical of the rearrangement of superhelical DNA packed in LCD particles differ appreciably from those ($\textbf{n} \sim 2$, $\textbf{K} \sim 8 \times 10^{-5}$) that describe the process of formation of cholesteric dispersions from both linear low-molecular-mass DNA and polynucleotides [8,28] and high-molecular-mass DNA [29]. Such a difference could be due to either of two reasons. On the one hand, there may be a continuous change in the fraction of linear DNA molecules produced as a result of Micrococcal nuclease on LCDs that can be ordered in cholesteric mode. On the other hand, the free energy accumulated in the structure of superhelical DNA molecules is probably expended to accelerate the process of restructuring of linear DNA fragments in LCD particles. Therefore, the cleavage of superhelical DNA molecules in LCDs under Micrococcal nuclease action induces the complicated process of formation and accumulation of linear DNA molecules (fragments) and their cholesteric packing; additionally, the high rate of the process of rearrangement of DNA molecules' packing in LCD may depend on the energy of superspiralization released as a result of the molecules splitting into linear fragments under the nuclease effect.

TABLE 6.1

Parameters of Kolmogorov–Avramy Equation for Rearrangement of Superhelical pBR 322 DNA Structure in CLCD Particles under Micrococcal Nuclease Treatment

Nuclease Concentration, μg/mL	n	K
0.125	1.34	$10^{-0.9}$
0.062	1.40	$10^{-1.4}$
0.031	1.59	$10^{-1.7}$

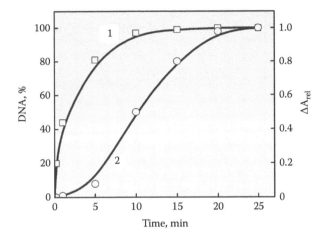

FIGURE 6.15 Relative content of the DNA linear form in LCD of nicked pBR 322 DNA (curve 1) and relative amplitude of the CD band (curve 2) versus time of Micrococcal nuclease treatment. $C_{MN} = 0.02$ μg/mL; other conditions as in Figure 6.12.

Considering the fact that, in the case of nicked (i.e., double-stranded, circular, but not superhelical DNA), the latter assumption is not correct, it is interesting to investigate the process of restructuring for LCD particles obtained from nicked DNA molecules. Hence, regarding nuclease cleavage of LCD particles formed by specially prepared nicked DNA molecules, the situation differs from the superhelical DNA case.

Figure 6.15 compares the curves for linear DNA accumulation in LCD particles formed from nicked DNA molecules (curve 1) to that of the CD band increase (curve 2) on the enzymatic hydrolysis of nicked DNA within these particles. The comparison of these curves shows that the accumulation of linear DNA proceeds significantly faster than the increase in the amplitude of the band in the CD spectrum. It can be stated that, unlike the case of superhelical DNA, the accumulation of linear fragments in LCD particles formed by nicked DNA significantly advances the rearrangement of the DNA LCD spatial structure under Micrococcal nuclease processing.

The curves of the dependence of negative band amplitude in the CD spectra of nicked DNA LCD on the time of Micrococcal nuclease cleavage are given in Figure 6.16. The curves have a distinctive S-shaped form that is typical of the formation of CLCD from low-molecular-mass, linear DNA molecules [8,28].

Comparison of Figure 6.16 and Figure 6.15 shows that, as in the case of superhelical DNA, the cholesteric packing of DNA linear fragments that evolve as nicked DNA is treated with Micrococcal nuclease—a complicated process. The processing of kinetic curves (Figure 6.16) under the Kolmogorov–Avramy equation, as well as in the case of superhelical DNA, leads to their linearization (Figure 6.17). The interpretation of the kinetic curves in the framework of the theory of polymer crystallization permits one to apply the Kolmogorov–Avramy equation in this case. The parameters of cholesteric ordering on nuclease action on nicked DNA LCD are given in Table 6.2. It should be noted that **n** and **K**, typical of the case of nicked DNA, are

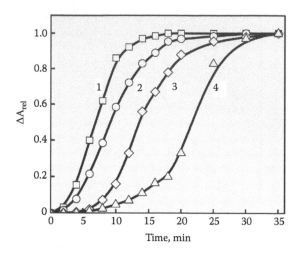

FIGURE 6.16 The dependence of the amplitude of the band ($\lambda = 270$ nm) in the CD spectra of nicked pBR 322 DNA LCD versus time of Micrococcal nuclease treatment. $C_{DNA} = 5$ µg/ mL; $C_{MN} = $ (1) 0.06, (2) 0.02, (3) 0.01, (4) 0.005 µg/mL; other conditions as in Figure 6.12.

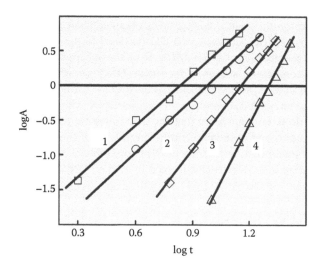

FIGURE 6.17 Kinetic curves from Figure 6.16. in the coordinates of the Kolmogorov–Avramy equation.

quite close to those for the formation of CLCD from linear DNA molecules with low molecular mass [8,28].

Comparison of the data presented in Tables 6.1 and 6.2 demonstrate a significant difference between the Kolmogorov–Avramy parameters for various topological forms of the DNA molecules. These parameters describe the mechanism and the

TABLE 6.2

Parameters of Kolmogorov–Avramy Equation for Restructuring of pBR 322 Nicked DNA LCD Spatial Structure under Micrococcal Nuclease Action

Nuclease Concentration, μg/mL	n	K
0.062	2.5	10^{-2}
0.021	2.4	$10^{-2.5}$
0.010	3.4	10^{-4}
0.005	5.0	10^{-7}

rate of formation of cholesteric structures from the linear DNA fragments accumulating in nuclease-treated dispersions. Such a difference could be explained in the framework of one of the assumptions following:

According to the first one, the free energy accumulated in superhelical DNA molecules as a result of their superspiralization is expended for the transformation of "nonspecific" into cholesteric packing of DNA molecules in the LCD particles after the splitting of these molecules. Nicked DNA has no such additional free energy; therefore, restructuring under nuclease action takes more time than for superhelical DNA. According to the second assumption, the difference in restructuring rates for superhelical and nicked DNA is due to a specific mode of the DNA "prepacking" in the LCD particles. Indeed, superhelical DNA, having more rigid molecules, may possess some kind of local order, such as precholesteric packing, in the dispersion particles that are determined at the "moment of the molecules' recognition" during phase exclusion from water–salt–PEG-containing solutions. Various topological shapes of DNA can be packed in LCD particles in different ways. Rigid, anisotropic superhelical DNA molecules, despite the existence of superturns, are packed in LCD particles with a certain degree of local helical ordering. This means that, in general, the spatial ordering appears, though there is no long-range helical twisting (which causes the existence of a low, intense band in the CD spectrum of LCD particles of superhelical DNA). That is why, as soon as neighboring DNA molecules lose their superturns and linear DNA fragments appear, cholesteric ordering of DNA molecules may arise at once in LCD particles. More flexible nicked DNA molecules are not subject to specific "prepacking" in the LCD particles. In this case, cholesteric packing requires not only the appearance of linear DNA fragments but also their mutual approaching and recognition with a subsequent helical twist, which requires time and, possibly, energy expenditures.

Both of the assumptions can explain the faster restructuring of spatial packing from "nonspecific" to cholesteric mode in the case of LCD particles formed by superhelical DNA, in comparison to similar transition in the case of particles formed by nicked DNA molecules.

Consequently, the topology of double-stranded superhelical DNA molecules fixed in the spatial structure of LCD particles influences the efficiency of their restructuring and transition to a cholesteric mode of packing of these DNA molecules.

6.6 SUMMARY

The results reviewed in this chapter are important to determine the nature and driving forces of mechanisms of high-molecular-mass DNA packing in biological objects; moreover, they make it possible to determine the reasons for the high efficiency of the enzymes under conditions of "molecular crowding," typical of living cells. The molecules of circular superhelical DNA can form dispersions because of their condensation in water–salt–PEG-containing solutions, and they pack densely in particles of these dispersions. The packing of superhelical DNA molecules differs from the classical cholesteric packing because of the existence of superturns. This packing, as well as in the case of linear DNA molecules, is flexible and depends on the combination of two parameters: the presence and sense of superhelical twists in the initial DNA molecules. Further, the morphology of circular superhelical DNA molecules acts as a "regulator" for the packing mode in LCD particles and, consequently, the sign and the amplitude of bands in the CD spectra of these particles formed by phase exclusion.

The results given in this chapter also show that the dense packing of DNA molecules in LCD particles does not inhibit the effect of such hydrolytic enzymes as micrococcal nuclease.

REFERENCES

1. Reich, Z., Wachter, E.J., and Minsky, A. Liquid-crystalline mesophases of plasmid DNA in bacteria. *Science*, 1994, vol. 264, p. 1460–1483.
2. Reich, Z., Wachter, E.J., and Minsky, A. In vitro quantitative characterization of intermolecular interactions. *J. Biol. Chem.*, 1996, vol. 270, p. 7045–7046.
3. Torbet, J. and DiCapua, E. Supercoiled DNA is interwound in liquid crystalline solution. *EMBO J.*, 1989, vol. 8, p. 4351–4356.
4. Yevdokimov, Yu.M., Salyanov, V.I., and Skuridin, S.G. Liquid-crystalline dispersions of circular superhelical DNA as a basis for biosensers. *Doklady of the USSR Acad. Sci.* (Russian Edition), 1989, vol. 307, p. 1262–1265.
5. Salyanov, V.I., Dembo, A.T., and Yevdokimov, Yu.M. Liquid-crystalline phases of circular superhelical plasmid DNA and their modification by the action of nuclease enzymes. *Liq. Crystals*, 1991, vol. 9, p. 229–238.
6. Reich, Z., Ghirlando, R., and Minsky, A. Secondary conformational polymorphism of nucleic acids as possible functional link between cellular parameters and DNA packing process. *Biochemistry*, 1991, vol. 30, p. 7828–7836.
7. Reich, Z., Levin-Zaidman, S., Gutman, S.B., et al. Supercoiling-regulated liquid-crystalline packing of topologically-constrained nucleosome-free DNA molecules. *Biochemistry*, 1994, vol. 33, p. 14177–14184.
8. Yevdokimov, Yu.M., Salyanov, V.I., and Lavrentev, P.I. Liquid crystals and liquid-crystalline dispersions of circular superhelical DNA. *Bull. USSR Acad. Sci., Phys.* (Russian Edition), 1995, vol. 59, p. 117–130.

9. Levin-Zaidman, S., Reich, Z., Wachter, E.J. et al., Flow of structural information between for DNA conformational levels. *Biochemistry*, 1996, vol. 35, p. 2985–2991.

10. Salyanov, V.I., Palumbo, M., and Yevdokimov, Yu.M. Enzyme splitting of superhelical DNA under its liquid-crystalline packing. *Mol. Biol.* (Russian Edition), 1992, vol. 26, p. 1036–1046.

11. Salyanov, V.I., Prasolov, V.S., Palumbo, M., and Yevdokimov, Yu.M. Anomalous optical activity induced at the condensation of different spatial forms of double-chain circular DNA. *Biophysics* (Russian Edition), 1989, vol. 34, p. 20–27.

12. Salyanov, V.I., Yevdokimov, Yu.M., and Palumbo, M. Investigation of properties of condensed forms of complexes of linear and circular superstranded DNA with mitoxantron and bisantrene. *Antibiot. Chemother.* (Russian Edition), 1990, vol. 35, p. 19–22.

13. Salyanov, V.I., Palumbo, M., and Yevdokimov, Yu.M. Formation of complexes of different types after Anthraquinone-DNA interaction. *Mol. Biol.* (Russian Edition), 1993, vol. 27, p. 869–879.

14. Yevdokimov, Yu.M., Salyanov, V.I., Skuridin, S.G., and Badaev, N.S. Liquid crystals and liquid-crystalline dispersions of DNA-Daunomycin complexes. *Mol. Biol.* (Russian Edition), 1995, vol. 29, p. 180–191.

15. Yevdokimov, Yu.M., Salyanov, V.I., and Berg, H. Two modes of long-range orientation of DNA bases realized upon compaction. *Nucl. Acid Res.*, 1981, vol. 9, p. 743–752.

16. Livolant, F. Ordered phases of DNA in vivo and in vitro. *Physica A*, 1991, vol. 176, p. 117–137.

17. Leforestier, A. and Livolant, F. Supramolecular ordering of DNA in the cholesteric liquid crystalline phase: An ultrastructure study. *Biophys. J.*, 1993, vol. 65, p. 56–72.

18. Yevdokimov, Yu.M., Skuridin, S.G., and Salyanov, V.I. The liquid-crystalline phases of double-stranded nucleic acids in vitro and in vivo. *Liq. Crystals*, 1988, vol. 3, p. 1443–1459.

19. Louie, D. and Serwer, P. Quantification of the effect of excluded volume on double-stranded DNA. *J. Mol. Biol.*, 1994, vol. 242, p. 547–558.

20. Zimmerman, S.B. and Minton, A.P. Macromolecular crowding: Biophysical, and physiological consequences. *Annu. Rev. Biomol. Struct.*, 1993, vol. 22, p. 27–65.

21. Tessier, D.C., Brousseau, R., and Vernet, T. Ligation of single-stranded oligodeoxyribonucleotides by T4 RNA ligase. *Anal. Biochem.*, 1986, vol. 158, p. 171–178.

22. Low, R.L., Kaguni, J.M., and Kornberg, A. Potent catenation of supercoiled and gapped DNA circles by topoisomerase I in the presence of a hydrophilic polymer. *J. Biol. Chem.*, 1984, vol. 259, p. 4576–4581.

23. Forterre, P., Mirambeau, G., Jaxel, C., et al. High positive supercoiling in vitro catalyzed by an ATP and polyethylene glycol-stimulated topoisomerase from *Sulfolobus acidocaldarius*. *EMBO J.*, 1985, vol. 4, p. 2123–2128.

24. Zimmerman, S. and Trach, S.O. Macromolecular crowding extends the range of conditions under which DNA polymerase is functional. *Biochim. Biophys. Acta*, 1988, vol. 949, p. 297–304.

25. Harrison, B. and Zimmerman, S.B. T4 polynucleotide kinase: Macromolecular crowding increases the efficiency of reaction at DNA termini. *Anal. Biochem.*, 1986, vol. 158, p. 307–315.

26. Salyanov, V.I., Lavrentev, P.I., Cernukha, B.A., and Evdokimov, Yu.M. Restructuring of spatial packing of circular DNA molecules in liquid-crystalline dispersions. *Mol. Biol.* (Russian Edition), 1994, vol. 28, p. 1283–1292.

27. Wunderlich, B. *Macromolecular Physics* (Russian Edition). Moscow: Mir, 1979, vol. 2. p. 574.

28. Skuridin, S.G., Lortkipanidze, G.B., Musaev, O.R., and Yevdokimov, Yu.M. Formation
    of liquid-crystalline microphases of two-chain nucleic acids and synthetic polynucle-
    otides of low molecular mass. *Polym. Sci., Series A* (Russian Edition), 1985, vol. 27,
    p. 2266–2273.
29. Yevdokimov, Yu.M. and Skuridin, S.G. Abnormal optical activity under intramolecular
    condensation of high-molecular mass DNA. Doklady of the USSR Acad. Sci. (Russian
    Edition), 1983, vol. 320, p. 499–502.

SECTION I SUMMARY

DNA: The results represented in Section I of this book allow one to conclude that
there are two principally different ways for condensation in the double-stranded
nucleic acid molecules (NA) that result in the formation of LCD.

*The first way of ordering rigid linear double-stranded NA molecules with low
molecular mass* is through "entropy condensation," that is, the process that takes place
at the phase exclusion of NA molecules from water–salt solutions after adding certain
water-soluble polymers to these solutions, for instance, PEG. As a result of this process, under certain conditions and provided with the high concentration of the added
polymer, the NA molecules with low molecular mass are self-assembled and form LCD
particles. (In the LCD particles, the distance between DNA molecules in quasi-nematic
layers depends on the osmotic pressure of the solvent.) The size of LCD particles evaluated theoretically is close to 500 nm; a particle contains about 10^4 NA molecules. The
nonobservance of the "critical" conditions of the formation of LCD causes the disintegration of LCD particles and the transition of NA molecules to the isotropic state.

The LCD particles obtained from double-stranded NA by this method have several peculiarities:

1. The polymer is not included in the content of the formed LCD particles.
2. There is a high local concentration of double-stranded NA in the content of
   LCD particles (from 160 to 600 mg/mL).
3. The distance, d, between adjacent NA molecules in LCD particles can be regulated from 25 to 50 Å by changing the osmotic pressure of the solution.
4. Double-stranded NA molecules mostly form CLCD with an abnormal band
   in the CD spectrum in the region of NA chromophores (nitrogen bases)
   absorption, which is caused by the geometrical and optical anisotropy of
   NA molecules.
5. The packing of NA molecules in CLCD particles is not only ordered but
   also fluid by nature because NA molecules in each quasi-nematic layer can
   both rotate around their axes and be displaced laterally.
6. Because of the fluid nature of the packing of NA in quasi-nematic layers,
   various compounds can easily diffuse into CLCD particles.
7. The CLCD particles formed in water–salt solutions maintain their physicochemical properties in a broad range of conditions.

*The second way of ordering rigid, linear, double-stranded NA molecules with
low molecular mass* is the "enthalpy condensation," that is, the phase exclusion of
the molecules from water–salt solutions caused by the attraction between the NA

molecules whose phosphate group negative charges are neutralized by counterions (the distance between the [NA-polycation] complex molecules in a quasi-nematic layer). This is determined by the energy of interaction between the (NA–polycation) complexes and remains practically constant. To realize this method, polycations that neutralize 80–90% of NA phosphate group negative charges are needed; after that, the attraction between the (NA–polycation) complex molecules is strong enough to form dispersion particles. An important stage of the enthalpy condensation process is the interaction between the approaching adjacent (NA–polycation) complex molecules. The peculiarities of the interaction depend on the surface properties of NA molecules, spatial structure of polycation molecules, and the mode of positive charges distribution in them, as well as on the properties of the solvent.

The (NA–polycation) dispersion particles obtained from double-stranded NA by this method have a few peculiarities:

1. The phase exclusion of (NA–polycation) complexes and the formation of dispersions take place when both the "critical" concentration of the polycation in the solution (the "critical" concentration is comparable to that of the initial NA molecules) and the "critical" molecular mass of polycation is reached.
2. Polycation molecules are always included in content of the formed dispersion particles.
3. The energy of interaction between the (NA–polycation) complex molecules provides the fixed distance, d, between the adjacent molecules in dispersion particles. This distance usually corresponds to the Bragg distance from 26 to 29 Å that is typical of the hexagonal packing of (NA–polycation) complexes.
4. Depending on the experimental combination of several factors, namely, the ionic strength of the solution, the structure of polycation molecules, the number of positively charged groups in polycation molecules, and so forth, various modes of packing of (NA–polycation) complexes may be realized.
5. Only in some ("lucky") cases one can realize the cholesteric packing of (NA–polycation) complex molecules in forming dispersion particles, which causes an appearance of an abnormal band in the CD spectrum in the region of NA chromophores' (nitrogen bases) absorption.

Regardless of the significant differences between different ways of condensation of double-stranded NA molecules, it has to be highlighted that the formed NA dispersion particles have the properties of both a solid substance and a liquid.

Therefore, NA molecules can exist in a liquid-crystalline state under model conditions, and moreover, a flexible packing mode of the NA molecules in LCD particles is easily changed, depending on certain conditions. The high local concentration of NA combined with the ordered arrangement of adjacent molecules of NA in LCD particles, which does not limit the diffusion of biologically active and chemical compounds inside the particles, provide both quick penetration of these compounds into the particles and a high rate of their interaction with double-stranded NA molecules, regardless of the way of condensation.

In this respect, the issue of the state and reactivity of double-stranded NA molecules under the cell conditions that will be considered below becomes highly important.

Section II

DNA LIQUID-CRYSTALLINE FORMS AND THEIR BIOLOGICAL ACTIVITY

7 Liquid-Crystalline State of DNA in Biological Objects

7.1 DNA AND BIOLOGICAL OBJECTS

The conclusion that NA molecules can exist in the liquid-crystalline state under model conditions and, depending on the properties of the solvent and NA molecules, the peculiarities of this state can vary significantly, allows questioning of the probability of the existence of the liquid-crystalline state of DNA molecules in a living cell.

Indeed, the study of an assembly of T4 bacteriophage in *E. coli* cells showed [1,2] that, at a certain stage of this assembly, the p22 protein of *E. coli* is split into two low-molecular-mass proteins, II and VII, that contain 80 and 48% of glutamine and asparagine acids, respectively. These proteins carry total negative charge in a cell and could not form stereochemical complexes with bacteriophage DNA. The concentration of proteins in the cell at the assembly site of a phage particle reaches ~500 mg/mL—that is, *E .coli* cells provide conditions similar to the conditions of phase exclusion in PEG water–salt solutions. (This is called a "PEG-like situation" in the literature). Under these conditions, water-soluble polymers cause the phase exclusion of a single bacteriophage DNA molecule and its transition to the condensed (compact) state, which was proved by direct experiments.

The toroidal shape of DNA molecules was obtained in works in which the peculiarities of the destruction of some bacteriophage heads have been studied. Under the action of chemical compounds, the bacteriophage head opened like a tulip flower, and the DNA molecule was observed in a state close to natural. In this case, the DNA molecule has a toroidal structure. (Figure 7.1 represents both the initial bacteriophage particles [g] and the toroidal DNA particles [a, b, c, d].)

This work [3] needs to be noted as the authors have, probably, been the first to obtain an electron-microscopy image of *Bacillus mycoides* N1 bacteriophage head destroyed by means of a special method proposed by Russian scientists. The destroyed bacteriophage head and toroidal DNA are clearly visible in Figure 7.2 in the right lower corner. Nevertheless, the authors of the work conducted in 1962 did not pay enough attention to this fact.

There is evidence that adjacent fragments of DNA molecule are ordered in the toroids, though the question of which state this ordering corresponds to is still open [4].

The structures called "molecular patterns" can be observed on thin sections of such biological objects as chromosomes prepared for electron microscopy. A well-known example of such patterns was discovered in the case of *Dinoflagellates* (testacean

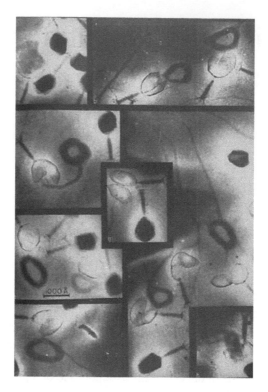

FIGURE 7.1 Electron-microscopic images of T2 DNAs' compact forms obtained after destruction of bacteriophage heads.

flagellants) sections. (Figure 7.3 distinctively demonstrates the arched structures that indicate the spatially twisted mode of DNA packing in chromosomes.)

The existence of a series of arcs on chromosome sections is confirmed in the example of materials prepared with the help of different methods (classical fixation, immobilization and thin sections, and freeze-fracture replicas, when water remains in amorphous state) and through observation of the received samples with the help of the electron microscope.

According to the interpretation of Y. Bouligand and others [5], the molecular patterns (Figure 7.3) that appear as a number of parallel lines and series of arcs correspond to different orientations of the chromosome axis relative to the section plane and connect with the existence of spatially twisted packing of DNA in chromosomes that are viscous but movable systems.

Despite the various additional speculation on the origin of these arcs, geometric and goniometric analysis lead to the conclusion that they, indeed, reflect the cholesteric packing of DNA molecules in a chromosome with a continuous left-handed twist. The patterns of cholesteric packing allow one to calculate the half helix pitch for the cholesteric formed by a DNA molecule complexed with proteins.

Cholesteric DNA packing was observed not only in the case of Dinoflagellate chromosomes [6] but also in the case of bacterial nucleotides [7,8] when the metabolic

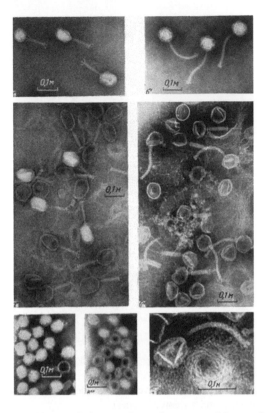

FIGURE 7.2   Electron-microscopic images for various steps of destruction of bacteriophage heads under ATP action.

activity of the bacteria is low. The arced patterns are also observed on thin sections of various species of bacteria (*E. coli, Bacillus subtilis, Rhizobium*, etc.), but they have rarely been interpreted as biological analogues of cholesteric liquid crystals. In addition, this organization does not always exist. In several species, nucleoids are decondensed during the exponential growth stage, when DNA is synthesized actively while they are condensed, and show a cholesteric organization during the inactive stationary stage. In the case of *Rhizobium*, two types of cells are observed: rapidly growing bacteria with dispersed nucleotides and slowly growing bacteria (with depressed protein synthesis and consequently transcription) with condensed nucleotides showing a series of nested arcs in their sections. Such organization is not observed in other bacteria under the conditions of normal growth, but this organization can be easily induced by adding inhibitors of protein synthesis in the culture medium.

The cholesteric organization of mitochondrial DNA has been described for several flagellate protozoa. The mitochondria of these cells contain a large amount of circular DNA molecules arranged into structures called kinetoplasts. For Bodo, this DNA is usually decondensed, but when the cells become encysted, their kinetoplast DNA condenses progressively and the series of nested arcs become visible in thin sections.

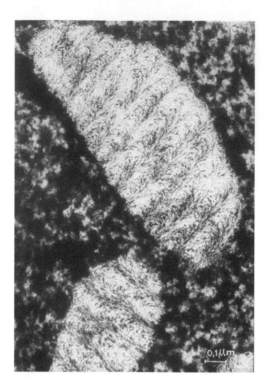

FIGURE 7.3 Typical arced pattern observed in oblique sections of two chromosomes of *Amphidium elegans* (electron-microscopic images).

A cholesteric structure is also typical of sperm nuclei from many species. During the maturation of the sperm cell, many changes in the chemical composition of chromatin, specifically the nature and the fraction of proteins bound to DNA, take place. The cholesteric structure of chromatin can be observed at the last stages of this evolution (spermatides and mature sperm nuclei), but unlike the first example, the arced patterns are rarely observed. In most mammal sperm nuclei (rabbit, rat, bull, stallion), on the freeze-fracture replicas received from thin sections, lamellar structures with a periodicity ranging from 250 to 300 Å are observed. The periodicity discovered during the analysis of intact sperm nuclei corresponds to the cholesteric structure, but the value of the half-pitch of the cholesteric in this case is relatively small to be able to observe the arch series on the 500 Å thick slices. The chromatin is apparently organized into a planar cholesteric structure with cholesteric axis perpendicular to the horizontal plane of the sperm head cell. In addition, the fact that chromatin isolated from stallion sperm nuclei has the CD spectrum typical of a cholesteric liquid-crystalline dispersion of double-stranded DNA molecules confirms the cholesteric organization of sperm cell chromatin. A similar type of chromatin organization was discovered for fishes (*Scyliorhinus caniculus*).

The probability of helical spatial organization in the case of the sperm cell nuclei of a green frog (*Rhacophorus*) and octopus (*Eledone cirrhosa*) is quite high.

TABLE 7.1
Pitch of Cholesteric Structures in Some Biological Objects

Biological Object	P, nm
Dinoflagellates	
Protocentrum micans	180–400
Protocentrum micans	72–500
Gymnodinium species	320
Protocentrum minimum	290
Peridinium cinctum	280
Polykrikos	220
Amphidinium carterae	192
Amphidinium carterae	160
Blastodinium	178
Noctiluca	150
Noctiluca	136
Gonyaulax polyedra	150
Oodinium	90
Protocentrum marialabouriae	65–68
Bacteria	
Bacillus thurigiensis	280
Rhizobium	193
Troglophytic prokaryotes	163
Bacillus subtilis	149
Escherichia coli	125
Escherichia coli	112–143
Kinetoplasts	
Bodo	133
Sperm cell nuclei	
Scorpion	>2,000
Fish	746
Pleurodele	70
Stallion	66
Bull	50–60
Rabbit	50

The data in Table 7.1 describe the parameters of cholesteric structures of some biological objects [9].

It can be noticed that the pitch, **P**, of cholesteric structures changes within the limit of 200 to 400 nm (precisely, in the case of DNA with the length of 50 nm, from 50 to 190 nm) up to more than 5,000 nm (at the average length of 2,200 to 2,500 nm).

The smallest pitch corresponds to the sperm cell nuclei (500–700 Å, depending on the sample). For the bacteria, the pitch, **P**, varies from 1,000 to 2,800 Å, for the *Dinoflagellates*—from 70 to 450 nm. In the latter groups, the pitch, **P**, varies in the samples and depends on the physiological conditions. The existence of the small pitch of cholesteric structure indicates that there are specific biological conditions that cause such very dense packing. For some sperm cell nuclei (cuttlefish, newt, salmon), the hexagonal packing of DNA molecules was described.

Therefore, in cases when DNA molecules are tightly bound to proteins (bacterial nucleotides, chromosomes of the *Protozoans*, virus heads, sperm cell heads), DNA is packed in either a crystal or liquid-crystalline mode. Taking into account the geometrical and optical anisotropy of double-stranded DNA molecules, it should be accepted that, for some biological objects, the cholesteric liquid-crystalline state of these molecules is most probably.

Considering this point of view, the question of interest is how the nucleosomes of eukaryotic cells are packed, as it is known that DNA molecules form complexes with histone proteins in the nuclei. Besides DNA is twisted upon the nucleosome particles, which causes the formation of structures of a "beads-on-a-string" type.

Interest in the liquid-crystalline structure of chromatin in living cells is based on the fact that the liquid-crystalline phases of nucleic acid complexes with polycations can change their spatial organization in response to minor changes in the properties of the solution. In this respect, the properties of an intracellular medium are determined, in particular, by the properties of histone proteins, and can determine the organization of chromatin at different stages of the cell cycle.

To determine the structure of chromosomes of higher organisms, two approaches have been used. According to the first approach, the structure of chromosomes was investigated in conditions of chromatin decondensation, when the higher orders of chromatin structure organization were automatically disintegrated. According to the second approach, chromatin was reconstructed by interaction of isolated DNA molecules with proteins of histone H1. The disadvantage of these approaches is that they were realized in diluted solutions in spite of the fact that, in cell nuclei, chromatin exists under conditions corresponding to phase exclusion. Except for several works performed with the use of the concentrated solutions, the behavior of structural elements of eukaryotic cell chromatin in conditions of phase exclusion was studied in only a limited number of works in the literature.

The behavior of chromatin core particles in concentrated solutions obtained by adding PEG (molecular mass 8,000 Da; $C_{PEG} \sim 400$ mg/mL; 0.1 M NaCl) was investigated in Reference 10. It has been shown that, under the conditions that model intracellular conditions, the nucleosome particles spontaneously order and form a phase that is liquid, though quite viscous. The calculation has shown that the concentration of particles in this phase reaches 310–485 mg/mL. This concentration is close to the concentration typical of chromosomes. For example, in the case of *Euglena* sp. chromosomes, it is equal to 433±74 mg/mL. The textures observed by the polarizing microscope indicate the formation of a liquid-crystalline hexagonal phase. Moreover, other phases could be formed by change in the concentration of nucleosome particles or the ionic strength of the solution.

Therefore, the nucleosome particles form liquid-crystalline structures even if the linker DNA fragment is split. The formation of the discotic columnar (not helically twisted) liquid-crystalline phase of the particles that includes DNA anisotropic molecules is the unexpected result of this work. Moreover, the "face-to-face" packing of cylinder-like particles of nucleosomes is contrary to the solenoid packing model proposed earlier, where the particles pack together with the 30-nm chromatin fibrils and H1 histone.

The contrast between the peculiarities of the liquid-crystalline state of nucleosome particles and the structure of native chromatin makes it possible to assume that there are unknown factors that exert different influence on the type of double-stranded DNA packing occurring under experimental conditions and on the packing of nucleosome particles under natural conditions.

The analysis of *E. coli* and stress-induced protein cocrystallization under stress conditions, where this process protects the DNA molecule from damage, indicates the fact that intracellular proteins can be one of such factors [11]. It has been shown that the process proceeds very quickly in cells, and apparently the presence of the DNA liquid-crystalline phase accelerates it. The transition to the liquid-crystalline state can cause the volume occupied by liquid-crystalline structures to become comparable to the volume of nucleotides evaluated by microscopy.

It can be assumed that the proteins within chromatin composition perform the function of a dielectric medium that affects the DNA molecules' packing type. Moreover, in the case of cells of higher organisms, there can be a hierarchy of structures determined not only by the fraction and properties of proteins but also by the spatial peculiarities of nucleosome structures. The realization of biological functions of an organism is related to the transition between these spatial structures. Nevertheless, the phase behavior of DNA in this case is also similar to the behavior of the liquid-crystalline phase because it is determined by the tendency of DNA molecules to form ordered cholesteric packing.

Another factor that may affect the formation of DNA cholesteric packing is the presence of superturns in this molecule [12]. The possibility of the existence of liquid-crystalline forms of superhelical circular double-stranded DNA molecules (plasmids) *in vitro* and *in vivo* has been studied in the literature [13–15] and illustrated by the results described in Chapter 6. Commenting on these results, it can be noted that the presence of superturns in DNA structure, which has not been considered even theoretically though it affects the type of double-stranded DNA molecules' packing at their phase exclusion, becomes a new regulatory factor.

It is also obvious that, *in vivo*, the length of DNA molecules that can be considered infinite leads to topological restrictions that do not exist in the case of liquid crystals. The liquid-crystalline organization of linear double-stranded DNA molecules of a chromosome length would take years. That is why it is necessary to consider such cell components as proteins and/or RNA when explaining the state of DNA in living cells. For example, they can change DNA molecule bending and the extent of its superhelicity, which can affect the tendency to form an ordered structure. Systematic research on the effect of various parameters, including molecule length, sequence of DNA bases, and chromatin components, will make it possible to compare the structures of liquid crystals to chromatin organization more reasonably in the future.

7.2 SUMMARY

Therefore, it is possible that liquid-crystalline dispersions are a reasonable model of genetic makeup. Considering the possibility of the liquid-crystalline state of DNA molecules in various biological objects, it is interesting to analyze the realization of some biologically relevant reactions under conditions of phase exclusion of DNA molecules.

REFERENCES

1. Laemmli, U.K., Paulson, J.R., and Hitchins, V. Maturation of the head of bacteriophage T4. A possible DNA packing mechanism: In vitro cleavage of the head proteins and structure of the core of the polyhead. *J. Supramol. Struct.*, 1974, vol. 2, p. 276–301.
2. Laemmli, U.K. Characterisation of DNA condensates induced by poly(ethylene oxide) and polylysine. *Proc. Natl. Acad. Sci. USA*, 1975, vol. 72. p. 4288–4292.
3. Poglazov, B.F., Tikhonenko, A.S., and Engelhardt, V.A. ATP action on DNA releasing from bacteriophage. *Doklady of the USSR Acad. Sci.* (Russian Edition), 1962, vol. 145, p. 4288–4292.
4. Golo, V.L., Yevdokimov, Yu.M., and Kats, E.I. Toroidal structures due to anisotropy of DNA-like molecules. *J. Biomol. Struct. Dynamics*, 1998, vol. 15, p. 757–764.
5. Bouligand, Y., Soyer, M-O., and Puiseux-Dao, S. La structure fibriliaire et l'orientation des chromosomes cher les Dinoflagelles. *Chromosoma*, 1968, vol. 24, p. 251–287.
6. Livolant, F. Cholesteric organization of DNA in vivo and in vitro. *Eur. J. Cell Biol.*, 1984, vol. 33, p. 300–311.
7. Kellenberger, E. and Arnold-Schulz-Gahmen, B. Chromatins of low-protein content: Special features of their compaction and condensation. *FEMS Microbiol. Lett.*, 1992, vol. 100, p. 361–370.
8. Kellenberger, E. About the organization of condensed and decondensed non-eukaryotic DNA and the concept of vegetative DNA (a critical review). *Biophys. Chem.*, 1988, vol. 29, p. 51–62.
9. Leforestier, A. and Livolant, F. Supramolecular ordering of DNA in the cholesteric liquid crystalline phase: An ultrastructure study. *Biophys. J.,* 1993, vol. 65, p. 56–72.
10. Leforestier, A. and Livolant, F. Liquid crystalline ordering of nucleosome core particles under macromolecular crowding conditions: Evidence for discotic columnar hexagonal phase. *Biophys. J.*, 1997, vol. 73, p. 1771–1776.
11. Wolf, S.G., Frenkel, D., Arad, T., et al. DNA protection by stress-induced biocrystallization. *Nature*, 1999, vol. 400, p. 83-85.
12. Salyanov, V.I., Dembo, A.T., and Yevdokimov Yu.M. Liquid-crystalline phases of circular superhelical plasmid DNA and their modification by the action of nuclease enzymes. *Liq. Crystals*, 1991, vol. 9, p. 229–238.
13. Reich, Z., Wachter, E.J., and Minsky, A. Liquid-crystalline mesophases of plasmid DNA in bacteria. *Science*, 1994, vol. 264, p. 1460–1483.
14. Salyanov, V.I., Palumbo, M., and Yevdokimov, Yu.M. Enzyme splitting of superhelical DNA under its liquid-crystalline packing. *Mol. Biol.* (Russian Edition), 1992, vol. 26, p. 1036–1046.
15. Minton, A.P. Excluded volume as a determinant of macromolecular structure and reactivity. *Biopolymers*, 1981, vol. 20, p. 2093–2010.

8 DNA Reactions under Conditions Causing Liquid-Crystalline Dispersions

8.1 MOLECULAR CROWDING

As was shown in previous chapters, double-stranded DNA molecules undergo condensation in solutions containing neutral polymers such as PEG or even polymers carrying negatively charged groups such as polyacrylic acid (PA) or polyglutamic acid (PG). On the whole, the effect of these polymers on DNA can be considered within the framework of variation of two physical parameters of a solution: the excluded volume, which is also called the effect of "molecular crowding," and/or water activity [1–3].

The term "macromolecular crowding" *sensu lato* denotes the conditions under which a solution exhibits properties not observed in dilute solutions where there is nonspecific (in particular, steric) interaction between macromolecules [4].

Obviously, molecular crowding makes a system thermodynamically nonideal, which in turn can alter both the conformational state of macromolecules and the efficiency of reactions involving different molecules. Indeed, molecular crowding markedly affects the chemical activity of molecules present in a given medium. According to an assessment by A. R. Minton [5], even a moderate molecular crowding changes the value of the association constant of macromolecules by several degrees of magnitude. It should be noted that Minton's calculations consider solvent as an inert structureless "background." The allowance made for solvent properties in the work by O. Berg [6] led to the results somewhat different from A. R. Minton's. It was found that there is an optimum in the extent of effects caused by molecular crowding that depends on the size (shape) of macromolecules interacting in the analyzed medium. Furthermore, at a certain shape of complexes formed between molecules, molecular crowding may not affect the efficiency of their interaction.

The condensation of DNA in water–salt solutions containing PEG, based on the mutual impermeability of DNA and PEG molecules in the examined medium, is an example of "direct" macromolecular crowding (i.e., the effect of excluded volume due to nonspecific interaction between PEG and DNA), which favors an ordered compact conformation of DNA molecules [7].

The experimental data show that macromolecular crowding has a marked effect on the efficiency of biological reactions involving a condensed or, more precisely, cholesteric liquid-crystalline form of DNA [8]. Therefore, the behavior of DNA molecules under phase exclusion conditions attracts the attention of both theorists [9,10] and experimentalists [8].

Indeed, DNA LCDs are characterized by several properties necessary for the implementation of DNA biological functions: a high local concentration, which makes it possible to accelerate the contacts between interacting molecules; the "liquid" properties of structures that provide fluidity (which opens the gate for different compounds to diffuse easily and specifically within the LC structure and make contacts with reactive groups of DNA molecules and one another); the easily restorable spatial order of the LC structure (which is necessary for some reactions, namely, for homologous pairing), combined with the ability to fluctuate in response to the minor changes in physicochemical properties of the medium; and, finally, the multiplicity of spatial forms with different structural parameters, which are typical of these dispersions.

In this respect, there are two questions of interest. First, does molecular crowding affect the efficiency of DNA condensation under the effect of polycations? Second, can enzymes work efficiently and specifically under these conditions?

To answer the first question, the condensation of DNA in PEG-containing solutions under the effect of chitosan was studied. To answer the second question, the effect of enzymes on condensed DNA formed under different conditions was researched.

8.2   CONDENSATION OF DNA UNDER THE EFFECT OF CHITOSAN IN CONDITIONS CAUSING MOLECULAR CROWDING

Figure 8.1 shows curves that allow one to estimate the "critical" Chi concentration (C^{cr}_{Chi}) necessary for the formation of CLCDs.

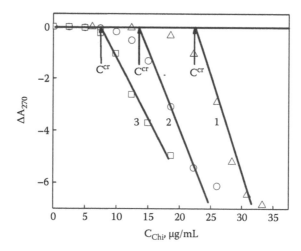

FIGURE 8.1   The dependence of amplitude of the abnormal band in CD spectrum. ($\lambda = 270$ nm) of LCDs of DNA–Chi complexes formed in PEG-containing solutions on chitosan concentration. Chitosan contains 79% amino groups; molecular mass: 8.4 kDa; $C_{DNA} = 15$ µg/mL; 0.05 M NaCl; 0.001 M Na$^+$–phosphate buffer; pH 6.86. PEG concentrations (mg/mL): 1–0; 2–50; 3–100. ΔA in optical units ($\times10^{-3}$).

The curves show that LCDs with abnormal optical activity are formed in all cases regardless of the PEG presence. Moreover, the C^{cr}_{Chi} value decreases markedly for Chi molecules containing 79% of amino groups as the PEG concentration in the solution rises. Because initial DNA molecules do not condense in the absence of Chi under the conditions used (solution ionic strength, PEG concentration in the solution) [8], the observed decrease in C^{cr}_{Chi} proves the existence of a second "indirect" effect of molecular crowding [7] associated with the increased condensation of DNA molecules in the presence of ligands interacting with these molecules.

This effect has the general character and, within the conventional framework [6,7], reflects the increased efficiency of ligand binding, that is, in our case, better neutralization of negative charges of DNA phosphate groups with positively charged groups of Chi under molecular crowding conditions. Since C^{cr} is the function of the DNA-ligand binding constant, the drop in C^{cr} indicates the growth of Chi to DNA-binding constant as the PEG concentration rises. This is also proved by the fact that the linear decrease in change in the free energy of formation of dimeric complexes with increasing PEG concentration in the solution noted in [6] favors this explanation.

Under conditions of molecular crowding, C^{cr} usually decreases as the content of amino groups in Chi molecules rises, which indicates that the higher the content of amino groups in Chi molecules, the more efficient the screening of negatively charged phosphate groups of DNA molecules. However, when the content of amino groups in Chi molecules reaches 85%, the C^{cr} decreases markedly. In this case, though, molecular crowding caused by PEG is not accompanied by a change in C^{cr}. Variations of the molecular mass of Chi could decrease the C^{cr}, although, in the range from 4,000 to 20,000, the change is not considerable.

This result demonstrates that efficient neutralization of negative charges of DNA phosphate groups with positively charged groups of Chi proceeds under the conditions of a high content of amino groups and high molecular mass. The constancy of C^{cr} with changing PEG concentration may result from the difference in the conformation of Chi molecules carrying 85% of amino groups from the conformation of other Chi samples. Such a change in conformation can be due to the different number of hydrogen bonds between the repeating n-acetyl-D-glucose residues in Chi molecules differing in the content of these residues and in molecular mass.

Within the framework of this explanation, it becomes obvious that, at a certain conformation of Chi molecules and, consequently, at a definite structure of the resulting DNA–Chi complex, the volume exclusion can result in no alteration of C^{cr} value or abnormal optical activity of the resulting CLCDs of the DNA–Chi complex. Indeed, it is shown in Reference 6 that, at a definite structure of complexes with a dumbell-like shape (depending, more precisely, on the value of their eccentricity), the volume exclusion did not lead to a change in the value of free energy of dimerization (i.e., the formation of complex between the two monomers). In this case, the value of C^{cr} may remain constant. It is possible theoretically that the equilibrium of dimerization reaction will be shifted to the formation of monomers under volume exclusion conditions.

Therefore, the value of C^{cr} is a complex function of the properties of Chi molecules and PEG concentration, and the volume exclusion caused by the rise in PEG concentration can exert various effects (depending on the properties of the used poly-cation) on the efficiency of formation of CLCD particles of DNA–Chi complexes, differing in the modes of spatial packing of the molecules of these complexes.

The dependence between the $\mathbf{C^{cr}}$ value and molecular crowding is not simple but can be explained within the framework of another approach suggested to describe the formation of CLCDs of DNA–Chi complexes [11]. According to this approach, Chi binds to DNA noncooperatively, while one Chi molecule covers a site, $\mathbf{L}$, of base pairs of DNA upon binding to make them unavailable for binding to other ligand molecules. The interaction of one Chi molecule with one binding site on DNA is characterized by the constant, $\mathbf{K}$. The maximum number of Chi molecules capable of binding to DNA molecules of $\mathbf{N}$ base pairs long is equal to $\mathbf{q_{max} = N/L}$. As Chi concentration in the solution rises, the number of Chi molecules bound in average to one DNA molecule will increase. As a whole, the solution will contain DNA molecules bound to different number of Chi molecules. After a definite extent of coverage of DNA molecules with Chi molecules and compensation of negatively charged phosphate groups of DNA, the DNA–Chi complexes form particles of LCDs. Such a model allows one to restore the realistic $\mathbf{K}$ values corresponding to Chi samples of different molecular mass.

Within the framework of the model, the dependence of $\mathbf{C^{cr}}$ on $\mathbf{K}$ value was calculated as an example (Figure 8.2).

According to the model [11], the Chi molecule (molecular mass 13,600 Da; 85% of amino groups) covers 40 base pairs upon binding: the extent of the DNA coverage resulting in transition into liquid-crystalline state is 0.8. Figure 8.2 shows that the value of $\mathbf{C^{cr}}$ decreases when $\mathbf{K}$ rises up to 10^7. As this value is reached, a "saturation effect" is observed, and the further growth of $\mathbf{K}$ does not lead to any marked change in the $\mathbf{C^{cr}}$ value. In the area of $\mathbf{K}$ values lower than 106, a sharp increase in the value of $\mathbf{C^{cr}}$ is observed.

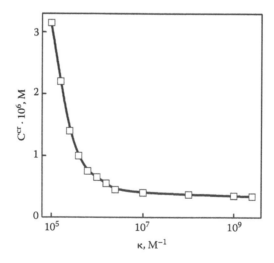

FIGURE 8.2 The theoretical dependence of C^{cr} value for chitosan on the constant of Chi to DNA binding.

At relatively high **K** values (i.e., when Chi molecules contain many amino groups and efficiently interact with DNA), the value of C^{cr} is independent of the degree of molecular crowding because C^{cr} reaches the limiting values and does not alter with the increase in **K** values. In the region of relatively low **K** values where the increase in molecular crowding is also accompanied by the growth of **K**, a decrease in the value of C^{cr} is observed because in this case there is a principal possibility for change in **K** value.

In addition, despite the attractiveness of this explanation of C^{cr} behavior for Chi samples with different properties, the model ignores the alterations of abnormal optical activity of CLCDs of DNA–Chi complexes, which is accompanied by the increase in molecular crowding.

Assuming that molecular crowding due to an increase in the PEG concentration in solution affects only the **K** value, one predicts qualitatively, using Figure 8.2, the dependences shown in Figure 8.3 obtained under conditions of increasing molecular crowding.

Figure 8.3 shows that abnormal optical activity of the resulting CLCD particles of the DNA–Chi complex increases with PEG concentration in a solution. The amplitude of the abnormal band in the CD spectrum depends on two parameters: the twist angle of neighboring quasi-nematic layers in the structure of cholesteric liquid crystal and the parameters of CLCD particles. The approximate evaluations of the size of the DNA–Chi complex dispersion particles formed in water–salt solutions show that their size is sufficiently large (4,000–5,000 Å). This allows one to suppose that a two- to threefold increase in amplitude of the band in the CD spectrum probably reflects the growth in the twist angle of quasi-nematic layers in the cholesteric

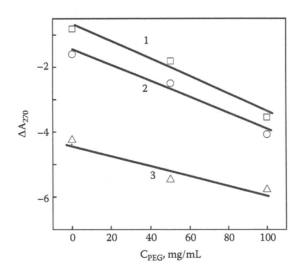

FIGURE 8.3 The dependence of amplitude of the abnormal band in the CD spectrum ($\lambda = 270$ nm) of CLCDs of DNA–Chi complexes on PEG concentration in the solution: (1) Chi: 55% amino groups; molecular mass: 5 kDa; $C_{Chi} = 30.7$ µg/mL; 0.15 M NaCl; (2) Chi: 79% amino groups; molecular mass: 8.4 kDa; $C_{Chi} = 25.9$ µg/mL; 0.15 M NaCl; (3) Chi: 55% amino groups, molecular mass; 5 kDa; $C_{Chi} = 24.7$ µg/mL; 0.05 M NaCl. $C_{DNA} = 15$ µg/mL; 0.001 M Na^{+}-phosphate buffer; pH 6.86. ΔA in optical units ($\times 10^{-3}$).

structure formed by DNA–Chi complex molecules. Within the framework of this explanation, the conformation of Chi molecules plays an important role in the interaction of Chi with DNA; the conformation should vary with the PEG concentration in such a manner that, under our conditions, it provides the spatial packing of complex molecules in the CLCD particles with the highest efficiency. This means that the amplitude of the band in the CD spectrum of the CLCDs formed by DNA–Chi complexes should increase with the PEG concentration in the solution.

Hence, the direction and the character of packing of DNA–Chi complexes in the CLCD particles will depend on the total surface charge of the molecules-forming complex, surface homogeneity, and polarity of added residues, as well as the spatial shape of molecules interacting with DNA.

One can suppose that, under certain conditions of molecular crowding (solution ionic strength, PEG concentration, Chi molecular mass, and content of amino groups), the conformation of Chi molecules can provide such a structure of the DNA–Chi complex that will cause a change in the direction of spatial packing of this complex in CLCD particles. Taking into account that the angle between quasi-nematic layers in a cholesteric liquid crystal of DNA molecules is rather small (~0.1 degree), such a change may have the character of a phase transition. Therefore, molecular crowding caused by the increase in the PEG concentration in the solution affects not only the efficiency of binding of Chi molecules to DNA but also the mode of spatial packing of resulting DNA–Chi complex molecules in particles of CLCDs.

8.3 ACTIVITY OF NUCLEOLYTIC ENZYMES UNDER CONDITIONS OF MOLECULAR CROWDING

8.3.1 EFFECT OF NUCLEASES ON SUPERHELICAL DNA-FORMING LCD PARTICLES

As shown in Chapter 6 (Section I), molecular crowding does not prohibit the action of enzymes—micrococcal nuclease on the molecules of circular superhelical DNA packed in LCD particles. The digestion with micrococcal nuclease is accompanied by the change in the mode of DNA molecules' packing in LCD particles. Meanwhile, the minimal concentration of nuclease necessary to induce the change in the type of DNA packing is rather low and equals ~ 5×10^{-5} μg/mL (i.e., ~ 5×10^{-12} M).

This is a significant result with a biological meaning. It conveys that the dense packing of double-stranded DNA molecules in LCD particles does not prohibit the nucleolytic enzymes' action. Moreover, this result is important from a practical point of view because the fact of the enzyme's action followed by the change in the DNA spatial packing within LCD particles and the appearance of an intense band in the CD spectrum allows one not only to detect the presence of nuclease in the solution but also to be able to evaluate its low concentration.

8.3.2 SPECIFICITY OF NUCLEASE ACTION UNDER CONDITIONS OF MOLECULAR CROWDING

Attention should be paid to the fact that the dense packing of superhelical DNA molecules in LCD particles can result in local change in the DNA secondary structure.

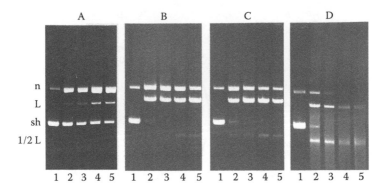

FIGURE 8.4   Electrophoregrams of pBR322 DNA samples after treatment with Micrococcal nuclease in PEG-containing water–salt solutions. PEG concentrations (mg/mL): (A) 0; (B) 130; (C) 170; (D) 300; Tracks 1–5 0, 5, 10, 20, and 30 min of nuclease treatment, respectively. $C_{DNA} = 5$ μg/mL; $C_{nuclease} = 0.02$ μg/mL; 0.3 M NaCl; 0.025 M Tris-HCl; pH 8; 0.01 M $CaCl_2$; n, L, and sh denote nicked, linear, and superhelical DNA forms, respectively.

Under these conditions, "weak" sites—that is, nucleotide sequences with distorted structure that are more efficiently recognized and split by nuclease, can appear in the DNA structure. It is also possible that, under the conditions of molecular crowding, the specificity of enzyme action can be changed.

To check the effect of superhelical circular DNA packing in LCD particles on the specificity of Micrococcal nuclease action, the detailed analysis of electrophoregrams of pBR 322 DNA samples obtained after treatment of LCD particles with enzymes under different conditions (PEG concentrations from zero to 300 mg/mL) was performed (Figure 8.4). From these results, it follows that enzymatic treatment of the DNA LCD gives rise to linear DNA fragments. The comparison of electrophoregrams given in this figure shows that the cleavage of these superhelical DNA in the content of LCD particles obtained at PEG concentrations above "critical" (110 mg/mL) yields the formation of a fragment corresponding to half the length of the pBR 322 molecule (2,180 bp; Figure 8.4 D, track 2 and 3). Control experiments showed that, in the treatment of superhelical DNA with enzymes under conditions where the LCD particles do not form, this fragment is absent.

The explanation of this effect is based on the assumption that, for the superhelical DNA molecules to be packed tightly, they must have a shape close to that of a rod. Superhelical DNA molecules, because of their "closed" structure, fold in two opposite places corresponding to the "top" and the "bottom" of the rod. It can be conjectured that loops (folds) with a small curvature radius would appear at the "defective" regions of the circular DNA molecules. The higher the packing density of superhelical DNA molecules in LCD particles, the lower the radius should be of the loops appearing in the folding places in the DNA molecules. Such folds (loops) can appear first of all in the specific sites of superhelical DNA molecules—for instance, in the sites with the inverted base sequence sufficient for the formation of a cruciform structure [12].

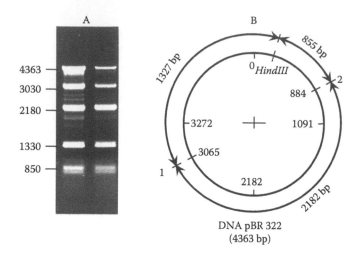

FIGURE 8.5 Electrophoregrams: A. pBR322 DNA preparations pretreated with Micrococcal nuclease in LCDs, obtained after hydrolysis with restriction enzyme Hind III. B. (1) Preferential cleavage site in the region of the cruciform; (2) new cleavage site evolving in LCDs.

Because Micrococcal nuclease preferentially attacks single-stranded stretches or DNA regions with disturbed secondary structures [13], the enzyme can efficiently cleave superhelical DNA where it folds during condensation.

To locate the sites of enzymatic hydrolysis of pBR 322 superhelical DNA in LCD particles, specific cleavage with the "single-hit" restrictase Hind III was carried out for DNA preparations with approximately equal amounts of linear form (4,363 bp) and its one-half (2,180 bp). (See Figure 8.5.) Hind III cleavage of the linear DNA molecules arising from exposure of the pBR 322 superhelical DNA in LCD particles by Micrococcal nuclease results in an appearance of three new fragments on the electrophoregram, corresponding to 3,030, 1,330, and 850 bp (Figure 8.5, where the numbers to the left are the lengths of DNA fragments in base pairs).

If the restriction site of Hind III does not correspond to the splitting site of superhelical pBR 322 DNA, cleavage of the full-sized linear pBR322 DNA molecule (4,363 bp) with the "single-hit" restrictase Hind III must cause the formation of two fragments. As seen from the electrophoregrams (Figure 8.5), these can only be the 3,030-bp and corresponding 1,330-bp fragments. Cleavage of the 2180-bp fragment by HindIII yields fragments of 1,330 and 850 bp. In addition, the electrophoregrams always contain a band corresponding to half the size of the pBR 322 DNA, indicating that Micrococcal nuclease splits the latter into two parts, equal by molecular mass but nonidentical.

The scheme constructed on the basis of these data for the cleavage of superhelical circular pBR322 DNA in LCDs (see Figure 8.5b) shows that the first (preferential) cleavage site for Micrococcal nuclease corresponds to the 3,065-bp fragment marked with point 1 on the scheme. According to the data in the literature [14], this point corresponds to the inverted sequence, that is, the place of the primary folding of superhelical circular pBR 322 DNA in the region of cruciform structure. This result confirms the hypothesis that the superhelical DNA molecules under tight packing conditions fold in a defective structural region, which is the cruciform DNA structure in the case of superhelical circular pBR 322 DNA. The place of the second fold is automatically determined by the position of the primary fold. This new splitting point is located on pBR 322 DNA molecules at 884 bp (i.e., at the directly opposite point of circular pBR322 DNA molecule [see Figure 8.5b, point 2]).

The second cleavage site appears only in superhelical DNA molecules that are tightly packed by condensation, but is not observed when pBR 322 DNA is cleaved in water–salt solutions.

Similar data were obtained in the case of LCDs of circular pT 22 DNA with subsequent nuclease treatment. The additional restrictive mapping of fragments shows obtaining fragments corresponding to half the length of superhelical circular pT 22 DNA containing 10,663 bp.

These results show that, at condensation, the superhelical circular DNA molecules adopt a shape that allows them to realize the dense packing in the forming LCD particles. Tight packing of superhelical DNA molecules in LCD particles retards the relaxation of the superhelical structure (i.e., its transition to the linear form after cleavage of the DNA at one of the defective sites), and, under these conditions, the nuclease has time to split the DNA molecule at the second folding site. The tighter the packing, the more efficient is the action of micrococcal nuclease.

Therefore, under conditions of molecular crowding, the increase in the efficiency of Micrococcal nuclease is observed. Moreover, formation of LCDs from superhelical pBR 322 DNA gives rise to an additional cleavage site for Micrococcal nuclease, that is, it alters the enzyme specificity in this system. This result may have important biological consequences because it means that the packing density of superhelical DNA in LCD particles is a factor regulating the specificity of DNA splitting with enzymes, that is, in the case of liquid-crystal packing of DNA molecules, the range of hydrolytic enzyme action can be significantly broadened.

8.4 ACTIVITY OF PROTEOLYTIC ENZYMES UNDER CONDITIONS OF MOLECULAR CROWDING

Another example confirming the maintenance of enzymatic activity under conditions of molecular crowding is the action of proteolytic enzymes on the polycation molecules bound in a complex with DNA molecules.

As shown in Chapter 5 (Section I), the interaction between polycations and double-stranded DNA molecules causes the enthalpy condensation followed by the formation, as a rule, of dispersion particles with hexagonal packing of DNA–polycation molecules. Dispersion is obtained because the DNA interaction with protamines results in a drop of the solubility of the DNA in water–salt solutions.

As an additional example, stellin B was used as a polycation [15]. This polycation was chosen because of its well-known chemical structure and physicochemical properties. Stellin B is the basic, low-molecular-mass, protein (protamine) isolated from milts of stellate sturgeon (*Acipenser stellatus*), that is, the reproductive glands

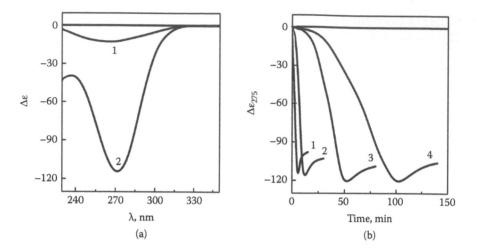

FIGURE 8.6 (a) The CD spectra of LCD formed from DNA–stellin B complex before (curve 1) and after (curve 2) trypsin treatment;. (b) the dependence of the amplitude of the band in the CD spectra ($\lambda = 275$ nm) of LCD of DNA–stellin B-complex versus time of trypsin treatment. In (b) 1—$C_{Trypsin} = 10^{-11}$ M; 2—$C_{Trypsin} = 10^{-12}$ M; 3—$C_{Trypsin} = 10^{-13}$ M; 4—$C_{Trypsin} = 10^{-14}$ M. CPEG = 170 mg/mL; 0.225 M NaCl; 0.01 M Na$^+$-phosphate buffer; pH 6.86.

of males of this fish filled with seminal fluid. It is also known that protamines are easily destructed by hydrolysis under the effect of low concentration enzymes [16]. Moreover, protamines bound to DNA *in vivo* take part in the formation of the spatial structure of gamete chromatin [17].

A sample of the LCD of the DNA–stellin B complex formed in water–salt solution was transferred to a PEG-containing solution and treated with trypsin. Figure 8.6 compares the CD spectra of this LCD before (curve 1) and after (curve 2) addition of enzyme. (Here, $r = 0.6$ is the ratio of the molar concentration of positively charged amino acid residues in the stellin B structure to the molar concentration of DNA nucleotides, i.e., the molar concentration of the negatively charged phosphate groups.) It can be seen that the LCD of the DNA–stellin B complex in PEG-containing solution with a moderate ionic strength ($\mu \leq 0.15$) does not show an intense band (curve 1). Besides, there is one maximum on the small-angle x-ray scattering curve (Figure 8.7, curve 1) on a phase obtained as a result of centrifugation of LCD prepared from the DNA–stellin B complex. The d_{Bragg} value calculated from the position of this maximum is equal to 25.07 Å, which corresponds to the average distance between DNA molecules in the hexagonal phase.

The formation of the hexagonal phase of the DNA–stellin B complex is also confirmed by the "nonspecific" texture of a thin layer of the phase (Figure 8.8A). The LC-phases in this figure were obtained from the corresponding LCD of DNA–stellin B complexes as a result of its concentration by low-speed centrifugation.

The combination of these results shows that the "enthalpy" condensation of DNA molecules actually causes the formation of LCD particles with a packing mode of DNA–stellin B complex molecules, similar to the hexagonal one.

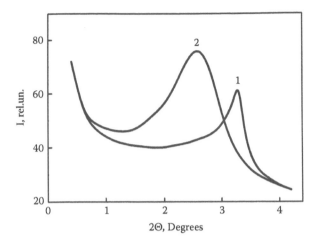

FIGURE 8.7 The x-ray diffraction curves for LC-phase formed from DNA–stellin B complex before (curve 1) and after (curve 2) trypsin treatment. C_{PEG} = 170 mg/mL; 0.225 M NaCl; 0.01 M Na$^+$-phosphate buffer; pH 6.86.

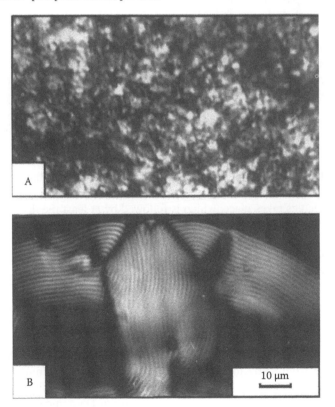

FIGURE 8.8 Textures of thin layers (~20 μm) of LC-phase formed from DNA–stellin B complex LCD before (A) and after (B) trypsin treatment (polarized light; bar corresponds to 10 μm; solvent as in Figure 8.7).

The particles of DNA–stellin B complex LCD in conditions of molecular crowding provided by PEG are affected by two structural tendencies (i.e., "structural conflict"). On the one hand, the correlation interaction between DNA–stellin B molecules tends to provide the dense hexagonal packing of DNA–stellin B complex molecules characterized not only by d_{Bragg} equal to 25.07 Å but also by the absence of an abnormal optical activity. On the other hand, the osmotic pressure of the PEG-containing solution (C_{PEG} = 170 mg/mL) provides the conditions under which the distance between adjacent DNA molecules must be high above 25 Å. This means that, in this solution, the optically active DNA molecules tend to form the CLCD possessing abnormal optical activity. The predominance of one or another structural tendency is determined by the mode of stellin B interaction with double-stranded DNA molecules (i.e., the presence of stellin B molecules, by itself, prohibits the cholesteric packing of DNA molecules in the LCDs under conditions of molecular crowding).

It is evident that, if the proteolytic enzymes maintain their activity in these conditions and can hydrolyze stellin B molecules, their disintegration can result in change in the mode of interaction between neighboring DNA molecules (or DNA molecules occupied with a very low number of stellin B fragments). Cleavage of stellin B molecules should be accompanied not only by an increase in the distance between neighboring DNA molecules but also result in their spatial mutual orientation, which unequivocally depends on the properties of the solvent. Under these conditions, the tendency of DNA molecules to the cholesteric mode of packing can be realized (i.e., the transition from hexagonal to cholesteric packing of DNA molecules in the content of the LCD can take place). This process must be followed by the change in all the physicochemical parameters of the system under study, in particular, its optical activity.

The LCD of DNA–stellin B complex in water–salt PEG solution was treated with the hydrolytic enzyme trypsin. The CD spectrum of this dispersion after enzyme processing is shown in Figure 8.6. Indeed, after trypsin treatment, the intense negative band appears in the CD spectrum of LCD of the DNA–stellin B complex (curve 2). This shows, first, that the enzymatic activity of trypsin under conditions of molecular crowding is maintained. Second, in these conditions, the enzymatic hydrolysis of stellin B molecules takes place, which is followed by at least the decrease in stellin B molecular mass. This causes a decrease in correlation interaction between the DNA–stellin B complex molecules, and the mode of spatial packing of these molecules changes. This means that enzymatic hydrolysis of stellin B permits the DNA molecules to adopt a spatial orientation that is typical of those in CLCD particles formed at C_{PEG} = 170 mg/mL. Hence, as a result of trypsin digestion of stellin B molecules, the spatial structure of LCD of DNA–stellin B complex is restructured.

Comparison of x-ray scattering curves (Figure 8.7) shows that trypsin treatment is accompanied not only by displacement of the position of the peak on the x-ray curve to smaller angles (the average distance between DNA molecules described by d_{Bragg} rises from 25.07 to 32.93 Å) but also by peak broadening (see curves 1 and 2). Meanwhile, curve 2 is only slightly different from the curve specific to the DNA CLCDs. This means that the distance between DNA molecules (and hence, their mobility) is increased. Consequently, the changes in the x-ray parameters of the

DNA–stellin B complex LC-phase confirms the restructuring of the DNA molecules in the content of particles of LCDs under the effect of trypsin.

Finally, the texture of a thin layer of the LC-phase of DNA–stellin B complex after tripsin digestion (see Figure 8.8) speaks in favor of the restructuring of the spatial organization of LCD particles. Trypsin treatment results in the appearance of, instead of the "nonspecific" texture (a), the fingerprint texture (b) typical of the cholesteric liquid crystals of DNA. The pitch, **P**, of the cholesteric formed varies from 2.5 to 3.0 μm.

The comparison of the results in the Figures 8.6–8.8 shows that that neither the density of packing of DNA molecules in the LCD particles nor the molecular crowding induced by the presence of PEG in the solution inhibits the enzymatic activity of trypsin. Moreover, these conditions in combination of the splitting of stellin B molecules in the content of DNA–stellin B complexes initiate the restructuring spatial organization of particles of the LCD of DNA–stellin B complex. This process is accompanied by the change in the shape of the CD spectrum, x-ray parameters, and type of texture. At the same time, the addition of the trypsin inhibitor di-isopropyl fluorinophosphate to the solution completely suppresses the shown changes recorded with CD spectroscopy, x-ray analysis, and polarizing microscopy.

The restructuring of spatial organization of particles of LCD formed by DNA–stellin B complex molecules takes place at a low concentration of trypsin (~10^{-14} M). This process can also be caused by the action of other hydrolytic enzymes such as *d*-chymotrypsin, pronase P, papain, and thrombin.

The observed restructuring of DNA–stellin B complexes' LCD structure under the action of proteolytic enzymes allows one to add a few remarks concerning the possible mechanism of decompactization of chromatin in sex cells. It is considered that, after fertilization of an ovule with a sperm cell, trypsin-like proteinases take part in the release of DNA from protamines [18,19]. Apparently *in vivo* as well as in the case of LCD formed by DNA–protamine complexes, the protamine molecules are hydrolyzed with the formation of a "loose" structure of nucleoprotamine. This provides favorable conditions for their further degradation and replacement with histones.

Therefore, molecular crowding probably does not exert an inhibitory influence on the biologically active compounds whose "target" are molecules of polycation within DNA–polycation complexes. Moreover, under conditions of molecular crowding, there is a possibility of the regulation of DNA molecules' spatial ordering in the content of LCDs by various biologically active compounds.

8.5   OTHER BIOCHEMICAL PROCESSES UNDER CONDITIONS OF MOLECULAR CROWDING

It can be assumed that biochemical processes such as transcription, replication, and recombination in living cells take place with the participation of DNA when its packing density is very low. However, there are contradictions to this assumption:

1. The transcriptional activity of bacterial nucleoids does not depend on their mass.
2. The chromosomes of *Dinoflagellates,* which are similar in their fine structure to bacterial chromosomes but look more condensed and have higher packing density, remain metabolically active.

To illustrate the biological activity of DNA molecules observed in various conditions providing the condensed state of these molecules, the data from different authors who studied the biologically significant reactions in the presence of PEG are collected in the Table 8.1. (The results given in Table 8.1 fit the data given earlier in describing the activity of enzymes in PEG-containing solutions.) One can attract attention to a few remarkable facts in Table 8.1.

First, it is noticeable that molecular crowding results in two effects. The first effect is the increase of efficiency of binding of both the model compounds (spermidin, protein HU) and genetically significant enzymes to DNA molecules under conditions of phase exclusion. The second effect is the increase in the rate of biological reactions. Indeed, under conditions of local DNA concentration rise because of the formation of LCDs, the diffusive limitations to the enzymatic reactions decrease, and an enzyme molecule can participate in a larger number of collisions with DNA molecules within its lifetime (i.e., the efficiency of enzymatic reactions increases). By definition, the excluded volume increases the effective concentration of macromolecules (i.e., the activity coefficient of macromolecules increases).

A. R. Minton [5] assumed that, under conditions of phase exclusion, the interaction between the molecules could be always increased regardless of their shape, while the calculated effects would be very high. The results of calculations conducted by O. Berg [6] differ from the results of A. R. Minton in two aspects. First, he has predicted the slight decrease in the binding constant caused by phase exclusion. Second, he has assumed the dependence of the phase exclusion effect on shape, which, notably, favors association. The increase in association (interaction) in the case of DNA molecules and genetically significant proteins is directly related to the issue of structure and functioning of the genome. The increase in the degree of occupation of the solution volume with inert molecules provides the formation of compact (globular) conformations of the macromolecules. It is possible that, in this state, the interaction of DNA with biologically relevant compounds can be stimulated (the assumption by [6] that intracellular molecular crowding can perform the function of an evolutionary force displacing the conformational equilibrium to compact conformations). The native proteins, ribosomes, tend to accept a compact form and realize their biological function in the compact state.

Second, based on the comparison of Figure 1.9, Figure 5.10, and Figure 5.14, evaluating the phase state of the initial double-stranded DNA molecules under laboratory conditions and the double-stranded DNA molecules formed in the conditions needed for realization of biological reactions, it can be concluded that, in most cases, the phase state of DNA molecules must correspond, definitely, to the condensed state. Precisely, under conditions of biological reactions with the participation of DNA, the

TABLE 8.1
Influence of the Crowding Effect on DNA Properties and Biologically Important Reactions

Observable Reaction	Conditions under Which Reaction Occurs	DNA Phase	Observed Effect	Remarks
DNA renaturation	PEG; 8,000 Da; 175 mg/mL	Cholesteric LC-phase	Speed up the reaction	
Cohesion of λ DNA fragments	PEG; 8,000 Da; 150 mg/mL	Cholesteric LC-phase	1,000-fold acceleration of the reaction rate	
Cohesion of λ DNA	PEG; 6,000 Da; 126 mg/mL	Cholesteric LC-phase	10-fold acceleration	The rate of the reaction corresponds to the situation *in vivo*
Ligation of single-stranded oligodeoxyribonucleotides by T4 RNA ligase	PEG; 8,000 Da; 40–300 mg/mL	Cholesteric LC-phase	The stimulation depends on conditions used	
DNA catenation by enzyme	PEG; 20,000 Da; ~70 mg/mL	Cholesteric LC-phase	The reaction without addition of PEG does not take place	
DNA supercoiling stimulated by topoisomerase I from *Sulfolobus acidocaldarius*	PEG; 6,000 Da; 100 mg/mL	Cholesteric LC-phase	Speed up the reaction	
DNA replication	PEG	—	Under low reagent concentration, the presence of PEG is absolutely required	
λdv DNA replication	PEG; 20,000 Da; 60 mg/mL	Cholesteric LC-phase	Speed up the reaction	
Nuclease and polymerase activities of DNA polymerase	PEG; 8,000 Da; 120 mg/mL	Cholesteric LC-phase	Speed up the reaction despite different conditions of inhibition	
Reaction catalyzed by T4 polynucleotide kinase	PEG; 8,000 Da; 40, 60 mg/mL	Cholesteric LC-phase	Effect depends on PEG concentration	The rate of reaction is greatly increased

continued

TABLE 8.1 (continued)
Influence of the Crowding Effect on DNA Properties and Biologically Important Reactions

Observable Reaction	Conditions under Which Reaction Occurs	DNA Phase	Observed Effect	Remarks
Exchange of DNA chains stimulated by RecA protein	PEG	—	Stimulation effect depends on reaction conditions	
Micrococcus nuclease digestion of supercoiled DNA molecules	PEG; 4,000 Da; 170 mg/mL	Cholesteric LC-phase	Efficiency of DNA digestion is increased	New site of restriction appears
DNA condensation induced by spermidin	PEG; 20,000 Da; >40 mg/mL	Cholesteric LC-phase	Efficiency of DNA condensation is increased	In absence of PEG, r ~ 7.0; in presence of PEG, r ~ 4.3
DNA condensation induced by HU-protein and other DNA-binding proteins	PEG; 8,000 Da; 80 mg/mL	Cholesteric LC-phase	Efficiency of DNA condensation is increased	Protein concentration is 1 decreased 10-fold

molecules must form cholesteric liquid-crystal dispersions (Table 8.1, "DNA Phase" column). This means that biological reactions are realized in the conditions that, at least, cause change in the DNA phase state. It is also possible that, under these conditions, the conformations of enzymes that take part in biological reactions may change [6]. The facts previously stated permit us to conclude that the cholesteric liquid-crystal form of DNA existing under conditions of molecular crowding can be considered as a biologically relevant structure [9,20,21].

The results of the work [22], where the biological activity a condensed complex obtained because the interaction of spermidin and superhelical DNA pBR 322 was investigated, speak in favor of this conclusion. It was shown that despite its dense packing this complex has a transcriptional activity in the presence of RNA-polymerase *E. coli.*

Another most interesting fact that is directly related to the DNA biological function is that, under conditions of molecular crowding, the range of properties of solvents (fraction and type of cations, etc.) in which the activity of biologically relevant enzymes is realized is significantly broadened [23–25].

Attention can also be paid to the importance of the behavior and structural state of other biopolymers in the cells. One can consider that the properties of a biological medium correspond to the conditions of molecular crowding. For instance, the total concentration of RNA and proteins in the cytoplasm of *E. coli* reaches 340 g/L, so the biological medium is thermodynamically nonideal. The thermodynamic non-idealness of the biological medium, where the intracellular concentration of dissolved substances can reach 300 g/L and take up to 20% of a cell volume, must be considered when comparing the properties of biological macromolecules observed *in vivo* and *in vitro*. It has been shown that volume exclusion can be caused not only by neutral polymers (in particular, PEG) but also by polyanions. This fact allows one to assume the possible role of such a polyanion as RNA in the cell cytoplasm. It is possible that high concentrations of RNA that depend on the functional condition of cells may affect the mode of DNA packing, which may cause the change in the rate of transcription. Hence, it can be assumed that the intracellular medium (as well as the polycations considered above) plays a dual role in the realization of intracellular processes (i.e., it can both affect the properties of genetically significant molecules and determine the type of interaction between them).

8.6 SUMMARY

Therefore, the whole set of experimental data in this chapter indicates that molecular crowding can affect the efficiency of the formation of cholesteric liquid-crystalline dispersions of double-stranded DNA and its complexes with polycations in different ways. Under conditions of molecular crowding, which one can consider as a model for intracellular conditions, the enzymatic activity of some enzymes is not only maintained but may also be different from the activity typical of water–salt solutions. Moreover, the dense packing of DNA molecules in LCD particles does not prohibit the action of enzymes and can even perform the functions of a factor regulating the specificity of DNA splitting with enzymes. This may have a significant biological meaning. The results given previously show that the liquid-crystalline cholesteric form of double-stranded DNA molecules is not inert; it maintains high biological activity. Moreover, it is assumed in Reference 26 that the liquid-crystalline, not linear, form of DNA plays the major (leading) role in the realization of biological functions of the macromolecule [27].

REFERENCES

1. Louie, D. and Serwer, P. Quantification of the effect of excluded volume on double-stranded DNA. *J. Mol. Biol.*, 1994, vol. 242, p. 547–558.
2. Zimmernan, S. Macromolecular crowding effects on macromolecular interaction: Some implications for genome structure and function. *Biochim. Biophys. Acta*, 1993, vol. 1216, p. 175–185.
3. Wiggins, P. Role water in some biological processes. *Microbiol. Rev.*, 1990, vol. 54, p. 442–449.
4. Minton, A. The influence of macromolecular crowding and macromolecular confinement on biochemical reactions in physiological media. *J. Biol. Chem.*, 2001, vol. 276, p. 10577–10580.
5. Minton, A.P. Excluded volume as a determinant of macromolecular structure and reactivity. *Biopolymers*, 1981, vol. 20, p. 2093–2010.

6. Berg, O. The influence of macromolecular crowding on thermodynamic activity: Solubility and dimerization constants for spherical and dumbbell-shaped molecules in a hard-sphere mixture. *Biopolymers*, 1990, vol. 30, p. 1027–1037.

7. Zimmerman, S.B. and Murphy, L.D. Macromolecular crowding and the mandatory condensation of DNA in bacteria. *FEBS Lett.*, 1996, vol. 390, p. 245–248.

8. Yevdokimov, Yu.M. Liquid-crystalline forms of DNA and their biological role. *Liquid Crystals and Their Application* (Russian Edition), 2003, vol. 3, p. 10–47.

9. Herzfeld, J. Entropically driven order in crowded solutions: From liquid crystals to cell biology. *Acc. Chem. Res.*, 1996, vol. 29, p. 31–37.

10. Kornyshev, A.A., Leikin, S., and Malinin, S.V. Chiral electrostatic interaction and cholesteric liquid crystals of DNA. *Eur. Phys. J.*, 2002, vol. E7, p. 83–93.

11. Nechipurenko, Yu.D., Volf, A.M., and Evdokimov, Yu.M. The distribution function describing the binding of extended ligands to DNA molecules: Possible application to the case of DNA condensation. *Biophysics* (Russian Edition), 2003, vol. 48, pp. 802–811.

12. Vologodsky, A.V. *Topology and Physical Properties of Circular DNA* (Russian Edition). Moscow: Nauka, 1988, p. 192.

13. Dingwall, C., Lomonossoff, G.P., and Laskey, R.A. High sequence specificity of micrococcal nuclease. *Nucl. Acids Res.*, 1981, vol. 9, p. 2659–3673.

14. Lilley, D.M.J. Hairpin-loop formation by inverted repeats in supercoiled DNA is a local and transmissible property. *Nucl. Acids Res.*, 1981, vol. 9, p. 1271–1289.

15. Skuridin, S.G., Hall, J., Turner, A.P.F., et al. Restructuring space ordering of (DNA-protamine) complexes in liquid crystalline dispersions under proteolytic enzyme treatment. *Liq. Crystals*, 1995, vol. 19, p. 595–602.

16. Yulkova, E.P., Rybin, V.K., and Silayev, A.B. Primary structure of stelline B. *Chem. Nat. Compounds* (Russian Edition), 1979, vol. 5, p. 700–704.

17. Suau, P. and Subirana, J.A. X-ray diffraction studies of nucleoprotamine structure. *J. Mol. Biol.*, 1977, vol. 117, p. 909–926.

18. Das, N.K. and Barker C. Mitotic chromosome condensation in sperm nucleus during post-fertilization maturation division in urechis eggs. *J. Cell Biol.*, 1976, vol. 68, p. 155–159.

19. Marushige, Y. and Marushige, K. Enzymatic unpacking of bull sperm chromatin. *Biochim. Biophys. Acta*, 1975, vol. 403, p. 180–191.

20. Fraden, S. and Kamien, R.D. Self-assembly in vivo. *Biophys. J.*, 2000, vol. 78, p. 2189–2190.

21. Holyst, R., Blazejczyk, M., Burdzy, K., et al. Reduction of dimensionality in a diffusion search process and kinetics of gene expression. *Physica A*, 2000, vol. 277, p. 71–82.

22. Baeza, I., Gariglio, P., Randel, L.M., et al. Electron microscopy and biochemical properties of polyamine-compacted DNA. *Biochemistry*, 1987, vol. 26, p. 6387–6392.

23. Hayashi, K., Nakazawa, M., Ishizaki, Y., et al. Influence of monovalent cations on the activity of T4 DNA ligase in the presence of polyethylene glycol. *Nucl. Acids Res.*, 1985, vol. 13, p. 3261–3271.

24. Hayashi, K., Nakazawa, M., Ishizaki, Y., et al. Acceleration of intermolecular ligation with E. coli DNA ligase by high concentration of monovalent cations in polyethylene glycol solutions. *Nucl. Acids Res.*, 1985, vol. 13, p. 7979–7992.

25. Zimmerman, S.B. and Harrison, B. Macromolecular crowding increases binding of DNA polymerase to DNA: An adaptive effect. *Proc. Natl. Acad. Sci. USA*, 1987, vol. 84, p. 1871–1875.

26. Sikorav, J.-L. and Church, G.M. Complementary recognition in condensed DNA: Accelerated DNA denaturation. *J. Mol. Biol.*, 1991, vol. 222, p. 1085–1108.

27. Bernal, J.D. Liquid crystals and anisotropic melts. *Trans. Faraday Soc.*, 1933, vol. 29, p. 1082–1090.

SECTION II SUMMARY

The results presented in Sections I and II of the book allow one to conclude that the liquid-crystalline state of double-stranded NA molecules has the following properties:

1. The high extent of the structural ordering of molecules that spreads over a long distance, combined with the multiplicity of ordered forms and the possibility of transition from one ordered form to other
2. A dominant cholesteric liquid–crystalline mode of packing of anisotropic, anisometrical, double-stranded NA molecules or their complexes with polycations
3. The diffusion mobility of NA molecules in the content of particles of dispersions
4. The ability of double-stranded NA molecules easily "reflecting" external factors such as temperature, changes in the properties, composition of the medium, and the presence of biologically active compounds, including enzymes, and so forth.
5. The dependence of the mode of packing of double-stranded NA molecules on the timing when (before or after the formation) the initial NA molecules were exposed to the action of the influencing factor
6. The appearance of new reactiveable sites in circular DNA molecules tightly packed in dispersion particles, which causes the change in the character of certain enzymes' action
7. The maintenance of the high biological activity of DNA molecules packed in liquid-crystal dispersion particles.

When enumerating such properties, our attention is drawn to the fact that the structural ordering and mobility of the properties are typical peculiarities of living cells.

It is reasonable to remember the statement made by J. Bernal in "A General Discussion Concerning Liquid Crystals and Anisotropic Melts" (1933) [27], which is known to everyone involved in research into biopolymeric liquid crystals: "A liquid crystal in a cell through its own structure becomes a proto organ of mechanical or electrical activity, and, when associated in specialized cells in higher animals, give rise to true organs, such as muscle or nerve. Secondly, and probably more fundamentally, the oriented molecules in liquid crystals furnish an ideal medium for catalytic action, particularly of the complex type needed to account for growth and reproduction."

This statement means that, when analyzing fine details and the peculiarities of functioning of both the whole genome and single genes, and the processes of normal and malignant growth of cells, as well as their differentiation, which are important for any kind of cells, the peculiarities of the liquid-crystalline state of double-stranded NA molecules typical of living cells need to be considered. That is why the further investigations in this field are significant and interesting.

It should be added that the fundamental principles determined at the formation of NA liquid-crystalline dispersion particles open a gateway to the practical application of these principles in the area of nanotechnology and biosensorics. Indeed, NA molecules ordered in the structure of liquid-crystalline dispersion particle may be used as building blocks for nanoconstructions, and the ability of the particles to respond to external factors allows one to use the particles as sensing units for biosensor devices.

Section III

DNA LIQUID-CRYSTALLINE DISPERSIONS IN NANOTECHNOLOGY AND BIOSENSORICS

9 Nanoconstructions Based on Nucleic Acid Molecules

9.1 THE GENERAL CONCEPT OF NANOTECHNOLOGY

This chapter covers the issues of nanodesign based on double-stranded molecules of nucleic acids fixed in the spatial structure of liquid-crystal dispersion particles and the possible practical applications of such structures.

The term *nanotechnology* was first introduced by Prof. N. Taniguchi in 1974. This new area of science that arose at the crossroad of physics, chemistry, electronics, and computing quickly developed in different countries. The terms *nanoparticles* and *nanomaterials* are known now to a broad audience of readers. Indeed, manipulations with single atoms allow one to design new "structured" materials and devices with specified unique properties. Recently, the list of terms has been broadened with such words as *nanobiotechnology* and *nanomedicine*, referring to some of the new areas of nanotechnology that use biological molecules as building blocks for creation of nanostructures.

Nanoparticles are sized from less than a nanometer to several hundreds nanometers (the word "nano" comes from the Greek word *nanos* meaning "dwarf"; this size corresponds to 10^{-9} m, nanometer, nm; i.e., the size of single atoms). High interest arose in these particles because it became obvious that single atoms (or molecules) can be used as "building blocks" to design spatial structures using an "atom-by-atom" approach. The concept of nanoparticles is quite young in the framework of modern science. It is based on two inventions of the twentieth century by Nobel Prize–winning physicists G. Binning and H. Rohrer from IBM who constructed in 1981 a scanning tunnel microscope that made it possible to "see" single atoms. The second invention was made in 1986 when the modernized microscope designed by Binning opened a way not only to observe the atoms but also to manipulate them. The paper "Positioning of Single Atoms by the Help of Scanning Tunnel Microscope" not only described the principle of operation of scanning tunnel microscope but also enumerated the prospects of this invention. In particular, the authors projected the economic benefits to companies that would begin to produce products based on "nanotechnology." In their opinion, the size and the volume of materials used for nanoproduction would not be large and, consequently, production costs would not be high.

The government of Russia, on November 18, 2004, issued a policy for the development of nanotechnology, using the following definition: "Nanotechnology is a combination of methods and approaches providing the ability to design and modify objects including components with a size less than 100 nm that have new qualities and make it possible to integrate them into rigorous functioning large-scale systems in a controllable way."

In a report by the Royal Society of London and the British Royal Academy of Engineering (2004) [1], a relatively narrow definition is given for nanotechnology: "Nanotechnology is the design, characterization, production and application of structures, devices and systems by controlling shape and size at nanometer scale."

In 2006, *Nature* provided [2] a more precise definition: "Nanotechnologies are technologies that use engineered materials or devices with the smallest functional organization on the nanometer scale *in at least one dimension*, typically ranging from 1 to ~ 100 nanometers."

All these definitions show that some properties of the created nanomaterial (or device) can be regulated by means of physical and/or chemical methods on the nanometer scale; additionally, the designed materials gain properties that were absent in the initial (bulk) materials. For example, the self-assembly of DNA molecules can cause the formation of nanotubes that can form nanowires used in nanoelectric devices. Moreover, boron-alloyed silicon wires can perform a function of highly sensitive detectors for biological and chemical substances. Nanowires modified with biotin allow one to detect the picomolar concentrations of streptavidin. These facts mean that, first of all, nanotechnologies lead to the appearance of new "functional" properties of the designed objects. Meanwhile, though chemical and/or physical methods of production of nanomaterials and devices are important within the manufacturing process, they are secondary to the functional properties of the designed nanomaterials and devices.

Taking into account the two foregoing examples, it can be noted that a DNA molecule has no "inherent tendency" to perform a function of a conductive nanowire, and neither boron nor silicon by itself can show any specific chemical substances. However, DNA nanowires and boron-alloyed silicon gain such functional properties. It happens as a result of integration of building blocks, but not because they contain DNA, boron, or silicon. Based on the same approach and using different standard blocks, various nanomaterials or devices with various functional properties can be created, and conducting wires and highly sensitive chemical detectors can be designed using other chemical or physical approaches.

Considering the functional content of nanotechnology, it becomes clear that nanotechnology is no new area of science per se, but it is an interdisciplinary science that integrates the results of fundamental sciences (such as chemistry, physics, mathematics, or biology) and applied sciences (such as material science and different areas of engineering) [3]. Nanotechnology can be considered an intersectoral area of science that includes nanoconstruction, synthesis, and research of nanomaterials or devices that have the properties considered above. Such "technical" definition of nanotechnology divides it from chemistry. It is obvious that chemistry is an integral part of nanotechnology, but these two terms are not synonyms, which often causes confusion. Chemistry includes manipulating substances on the nanometer scale; as a result of chemical processes, new products with specific chemical properties are created (for instance, with a certain melting point, pκ, or charge

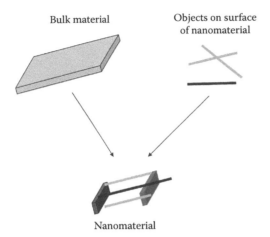

Bulk material

Objects on surface
of nanomaterial

Nanomaterial

FIGURE 9.1 (*See color insert.*) The principal scheme of "top-down" technology. Drop in size of initial material is accompanied by change in its optical properties.

distribution), which determines the way the products of these reactions can interact with other substances.

Nevertheless, as it stated earlier, nanotechnologies are determined first of all in terms of the appearance of new functional properties of nanomaterials (chemical properties can be only one of many contributions that determine these functional properties). To construct new nano-objects, several approaches, called "top-down" and "bottom-up" technologies, are employed in nanotechnology.

In the case of top-down technologies (Figure 9.1), nanomaterials are created by reducing the size of the initial material or placement of the required building blocks on the surface of some substance in a certain configuration. Notably, top-down technologies include different kinds of lithography and atomic force microscopy.

In the case of bottom-up technologies (Figure 9.2), nanomaterials are created from initial building blocks, such as atoms or molecules, which are integrated into spatial structures in a certain way.

The bottom-up technologies begin with one (or more than one) object of a certain kind (for instance, molecules) that are integrated into different ordered structures [1,2] as a result of physicochemical (biological) processes. Examples of such technology are the processes of formation of molecular assemblies (self-organization of molecules) that begin because the change occurs in chemical or physical parameters of the medium. It should be highlighted once again that, in any variant of the considered technologies, the nanostructures gain new properties that the initial objects do not have. It can be added that nanosized objects include one, two, or three-dimensional (3-D) formations, such as individual nanoparticles, nanofilms, rods and tubes, and nanoconstructed and nanocellular materials, as well as nanocomponents and nanodevices. The maximum size of the dimensions of these materials is relative, and the minimal size is determined by the size of molecules and atoms. It should be emphasized once again that, in any version of the considered approaches, the nanostructures gain new properties that the initial materials do not have.

Nanomaterials

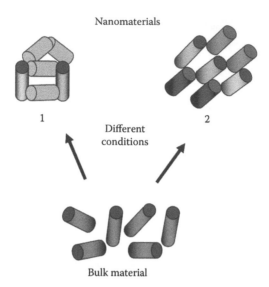

1   Different   2
conditions

Bulk material

FIGURE 9.2   *(See color insert.)* The principal scheme of "bottom-up" technology. Using various conditions, from initial material, one can create spatial structures differing by their properties.

Regardless of the variety of nanomaterials, there are a number of peculiarities typical of all of these materials. The first peculiarity is the so-called size effect [4,5]. One of the most famous examples of the size effect is the change in the gold color from yellow to red and even green as the size of the initial gold material is reduced using top-down technology (see Figure 9.1). Another example is the optical properties of spherical nanoparticles formed from any materials with semiconductor properties, such as CdS, CdSe, CdTe, ZnS, and PbS [6]. Such particles are called "quantum dots." Quantum dots have a diameter of 2 to 10 nm and contain 10–50 metal atoms; because of the so-called quantum effect, they have unique optical properties. The size and optical properties of CdSe quantum dots are compared in Figure 9.3. The reduction of the size of quantum dots is followed by a significant change in their color. It is difficult to predict the change in color of nanoparticles (though it should be noticed that, in 1982, the pioneer theoretical works in this area were performed by Soviet scientists) [7,8]. One can add that, unlike in the case of common materials, there are many factors that mask the presence of "size effect" in nanomaterials [9].

The properties of any nanoparticle depend on whether that particle contains an admixture of one or several atoms of other elements (or one type of substrate that it is sitting on). In particular, Figure 9.4 shows that, with the same size of nanoparticles (1,2,3), their properties depend on their composition [9].

These examples indicate that, in the case of nanomaterials, not only the role of the size of the particles becomes important but that they also acquire new, unique properties. The properties of an atom of an element on the surface of a nanoparticle are different from its properties within a bulk material. The consequence of a lower

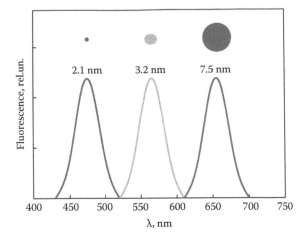

FIGURE 9.3 *(See color insert.)* The optical properties of quantum dots: three CdSe quantum dots with different diameter are shown.

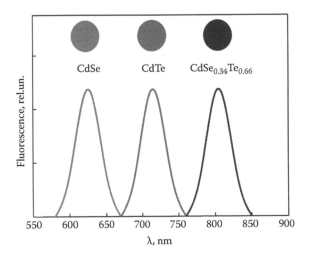

FIGURE 9.4 *(See color insert.)* The optical properties of quantum dots: three quantum dots with the same diameter (~5 nm) but different composition and the emission spectra corresponding to them are displayed.

stability of atoms on the surface is the lower melting temperature, T_m, of nanomaterials. Figure 9.5 shows the dependence of T_m of gold nanoparticles on their size [3]. It is easy to see that, as the size of the particles diminishes, the T_m value drops sharply. Melting, being a cooperative process by nature, becomes a vaguely determined phenomenon for a small amount of atoms within a nanoparticle, and, as the number of atoms in a nanoparticle reduces, the phase transition (melting) becomes blurred [4,5]. This is apparently caused by the fact that nanoparticles behave like small molecules, not like bulk material. In this case, one can speak not of phases

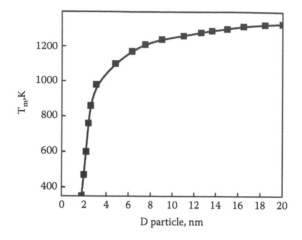

FIGURE 9.5 The dependence of T_m value of gold nanoparticles upon their size.

but of different structural isomers that exist at a certain temperature range. A nice illustration of these effects is the thermal destruction (melting) of nanostructures created on the basis of biological polymers, namely, synthetic double-stranded oligonucleotides [10].

The melting curves of synthetic double-stranded polymers that contain a different number of nucleotide pairs with the "width" of 0.34 nm are compared in Figure 9.6. (Here, T_m of a very long double-stranded molecule of poly(A) × poly(U) is 49°C. The hyperchromic effect is expressed as **100(ET – EK)/EK**, where **ET** is the optical activity at a certain temperature, and **EK** is the optical density of a denatured

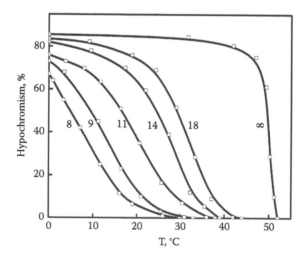

FIGURE 9.6 The change in the shape of the "melting curves" and T_m values of double-stranded oligo(A) × oligo(U) helical molecules with increase in the number of nitrogen base pairs in their content (from 8 base pairs to ∞).

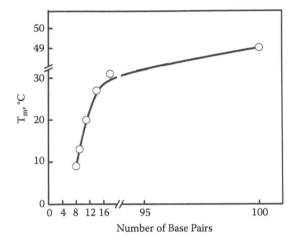

FIGURE 9.7 The dependence of T_m of double-stranded oligo(A) × oligo(U) molecules on the number of nitrogen bases in their structure.

poly(A) × poly(U) at 290 nm. Concentration of oligonucleotides is 5–11.3 mM; 50 mM Na-cocadilate buffer; pH 6.9.)

It can be noticed that, as the number of base pairs in the double-stranded structure is increased, the shape of the curve is changed from exponential to S-shaped. Moreover, the increase in the number of nucleotide pairs causes increase in T_m value (Figure 9.7).

The curve shown in this Figure 9.7 is very close by its shape to the curve of dependence of T_m for gold nanoparticles upon their size (see Figure 9.5).

While determining the connection between the properties and size of nanoparticles, not only the average size of the particles but also the mode of their size distribution should be considered important [4]. All the properties cause the appearance of nonlinear change in properties as the size of the materials drops and specific points in the dependences related to the size.

Therefore, in nanostructures that consist of a limited number of building blocks, new dominant physical phenomena appear to manifest fundamentally, such as quantum effects; statistically temporal variations of properties and their scaling, depending on the size of the structures themselves; the prevailing influence of surface behavior; absence of defects in the volume; considerable "power intensity" that determines the high chemical reactivity of structures designed; and so on. These novel phenomena are responsible for unique mechanical, electrical, magnetic, optical, chemical, and other characteristics that are opening vistas for manipulating these structures in modes unimaginable under normal conditions. Hence, manipulations at the level of individual atoms have made it possible to design new "structured" materials and devices that possess unique preset properties. That is why it can be stated that a "nanomaterial is more than size" because the most important thing here is not the size but the fact that the material gains unique properties that are functions of its size.

The combination of these data makes it possible to define nanotechnology "as the interdisciplinary science of design and the application of structured materials, devices, and systems whose functions are dependent on the nanostructure's geometry or inherent in their specific properties" [11].

As noted earlier, nanotechnology evolved from modern achievements and inventions in the area of visualization, analysis, and manipulations of nanosized structures, controllable creation of new functional materials with unique properties, and construction of nanosized devices that provided a technological breakthrough in different production areas.

The principal scheme of the integral parts of nanotechnology formed by the end of the twentieth century is illustrated by Figure 9.8. Commenting on the scheme given in this figure, it can be stated that engineering (technical) nanotechnology is directed toward the solution of such problems as (1) construction of solid substances and surfaces with a controllable molecular structure (creation of nanomaterials); (2) construction of new types of chemical compounds with controllable properties (nanoconstructing); (3) construction of nanosized self-organized or self-replicated structures; (4) fabrication of different devices (components of nanoelectronics, nano-optics, nanoenergetics, etc); and (5) integration of nanosized devices with electronic systems [12]. The creation of nanomaterials and devices with a small size, low cost, low energy consumption, and so forth, makes it possible to use them in different areas of science and technology.

The scheme given earlier shows the area of nanotechnology that is relatively new, called "nanomedicine" (it would be more precise to call it nanotechnology for medical purposes). The basic research fields in the nanomedicine area are also outlined in

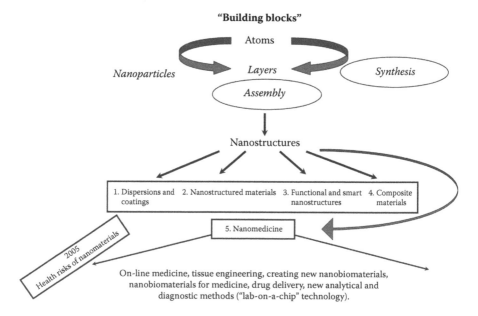

FIGURE 9.8   The scheme illustrating the basic components of nanotechnology and the main scientific research fields in the area of nanotechnology.

Figure 9.8. Proof of the importance of this area appeared in EuroNanoForum 2005, whose program was focused on the connection between nanotechnology and the health of European Union citizens. Specifically, Dr. Ottilia Saxl from the Institute of Nanotechnology in the United Kingdom, one of the organizers of the conference, said, "Medicine is the area of nanotechnology that attracts the attention of specialists from different fields. The idea that nanotechnology could help to make the treatment of many diseases more intended and targeted corresponds to the interests of both doctors and their patients."

The concept of nanotechnology making the treatment of many diseases more targeted is most attractive. Indeed, one can change the metabolism of the medicinal preparations (drugs) in the body of a certain patient in such a way that they reach their targets. This problem can be solved by application of "nanoconstructed carriers" that are also called "nanoparticles" ("nanosystems") for the targeted delivery of drugs and are created according to the biochemical peculiarities of the body of a certain patient. As the unique peculiarity of nanoparticles is the extremely developed surface, in comparison to common materials, surface, the delivery nanosystems allow one to overcome the poor solubility and poor absorption properties of the new generation drugs. Meanwhile the nanoparticles are used not only for the treatment of certain pathologies but also for diagnostics. In this respect the construction of new materials with upgraded therapeutic and diagnostic application becomes very important.

It should be noted that nanomedicine is the area of developing scientific research and applications. The most important area includes the newest field—the health risks of nanomaterials. The discussions within the framework of this field were conducted in the United States in December 2005 [13], in the Institute of Crystallography of the Russian Academy of Science in October of 2006, and in Berlin in April 2007 [14]. The participants of these discussions paid attention to the fact that the specific properties of nanoparticles (the highly developed surface, high reactivity, high catalytic activity, high mobility within an air or fluid flow, absence of metabolism, etc.) may cause unpredictable consequences as the particles enter the environment or a body of an animal or a human. This means that the elaboration of the rules for safety work with nanoparticles and the controllable application of the particles in different areas of medicine, technology, and science have started [15].

It is obvious that nanotechnologies must provide a high potential of economic growth, a high quality of life, technological and defense security, and economy of energy and resources, that is, they should fully correspond to the social demands of a society.

As the importance of nanotechnology becomes obvious to different public stratums, the awareness of the key role these technologies already play, and will play in the future, causes the creation of national nanotechnology programs in almost any developed state. These are long-term programs financed from both state and private sources. As a result, sharp competition in the area of nanotechnology evolves in different countries.

In the United States, the National Science and Technology Council (founded in 1993) formed the Interagency Working Group on Nanoscience, Engineering, and Technology (IWGN). In the 1999 session of IWGN, it was noted that, in view of the stage of development of nanotechnology in Western Europe and Japan, the

United States cannot be said to dominate this area. Considering the aspiration of the United States to be the leader in this field, one of the IWGN documents states: "The USA can not afford to be the second in this area. The country that will be the leader of development and application of nanotechnologies will have an enormous advantage in economic and military fields for decades." So, a project called the National Nanotechnology Initiative (NNI) was founded in the United States and approved in 2000 by President Bill Clinton, which caused the rapid growth of investments in the development of nanotechnologies in that country. The state financing of research conducted by different agencies within the framework of the NNI is illustrated in Table 9.1.

The data given in Table 9.1 are quite impressive: the table represents the significant volume of financing of nanotechnology research. It should be noted that, to popularize NNI in different sections of the society right after the approval of the program, some of the officials of NNI, M.C. Roco, R.S. Williams, and P. Alivisatos, in 2000

TABLE 9.1

The Financing of Different Departments in Framing the Program The National Nanotechnology Initiative

Department	2001 (Was Expended)	2006 (Expected Costs)	2007 (Planned Expenditures)	Extension during 2001–2007 (US$)	Extension during 2001–2007, %
National Science Foundation	150	344	373	223	149
US Department of Defense	125	436	345	220	179
US Department of Energy	88	207	258	170	193
US National Institutes of Health	40	175	173	133	333
US National Institute of Standards and Technology	33	76	86	53	161
US National Aeronautics and Space Administration	22	50	25	3	14
US Environmental Protection Agency	5	5	0	4	80
US Department of Agriculture	0	5	5	5	
US Department of Homeland Security	0	2	2	2	
US Justice Department	1	1	1	0	0
TOTAL	464	1,301	1,277	813	175

published a book called *Nanotechnology Research Directions: IWGN Workshop Report*. "A vision for nanotechnology research and development in the next decade," it was translated into Russian and published in 2002 [16]. In the book, the participants of the program talked about the importance of nanotechnology for various scientific and production fields in the United States. The book gives a clear concept not only about the basic scientific research fields in the area of nanotechnology but also of the main participants in the project. The number of popular publications (both positive and negative) on nanotechnology increased from 190 in 1995 to 7,000 in 2003 and, according to Lux Research, 12,000 in 2004. The journalistic activity of various offices in the area of nanotechnology and the public interest in them has attracted investments to the field.

Awaiting a huge market, the investment activity of major corporations in the world has increased sharply. Analysts from Lux Research suppose that, in the near future, the major part of the nanotechnology development funds is going to be provided by corporations. In North America, companies invest in nanotechnologies more than governmental offices (local and national)—US$1.7 billion versus US$1.6 billion, US$650 million versus US$1.3 billion in Europe. In general, there is a certain flurry of activity in nanotechnology. The prices of stocks of some companies have increased only because they added "nano" to their names, though the companies have only had indirect connection to nanotechnologies or have never conducted any research in the field.

In 2004, 1,500 companies claimed to be conducting research on nanotechnology; 1,200 of them are novice companies, more than a half of them based in the United States. Lux Research notes that this is not another bubble. The statement is based on the costs of the assets of the companies (including venture fund investments), which is still proportional to the results of the research. This means that nanotechnology innovations related to various activities can significantly change the situation in the existing production chains in the near future. The companies from the engineering field have the biggest expectations about the application of nanotechnologies. Experts are absolutely sure that the further development of nanotechnology production will make it possible to create memory microchips with a capacity of dozens of gigabytes and processors with operational frequencies of several THz.

The increasing interest of society in nanotechnology is typical not only for the United States. The investigations in the nanotechnology area are widely financed by the governments of Japan and the European Union, as well. For example, in the spring of 2004, the European Community promised to make a large investment in the development of new methods of production of microcircuits incorporating the achievements of nanotechnology. The EU countries follow the way of integration of the scientific and technology potential of the EU participants. The mechanism of integration is the 6th Frame program whose budget for the development of nanotechnology and related areas of science in 2003–2006 achieved €3.55 billion.

China is rapidly becoming a leader in terms of nanotechnology. The 5-year plan for 2001–2005 has assigned US$300 million for the development of nanotechnologies. The number of patents related to nanotechnologies is increasing sharply in this country. In 2003, China was the third by this rate in the world, with the United States and Japan being the first and second, respectively.

According to the existing forecasts for 2010–2015, science and industry world-wide need two to three million specialists in nanotechnology. As most of the leading universities are included in the process of nanotechnology development, this means that the developed countries are leading the intensive training of such specialists. Special attention is paid not only to training but also to the cooperation of academic science, private companies, and state laboratories; the importance of the interdisciplinary approach to nanotechnology issues is highlighted.

In general, the development of research in the area of nanotechnology resembles the information technology boom in 1970–1980 and the biotechnology boom in 1980–1990. These facts show that the international "nanotechnology train" is gaining acceleration.

What is the situation in Russia? First of all, the results of research conducted by Russian scientists in the fields directly related to nanotechnology correspond to the worldwide achievement and even surpass foreign efforts in some areas. Concerning the decisions taken at the governmental level, the high interest and responsibility of various agencies in the development of the national nanotechnology program should be highlighted.

The list of vital (crucial) technologies approved by the president of Russia on March 30, 2002, foresees the use of nanosized objects in some of the technologies.

Nevertheless, in 2003, at the session of the scientific, technical, and innovational security section of the Scientific Council under the Security Council of the Russian Federation, it was highlighted that nanotechnology had not yet become a priority in the scientific development field in Russia. By the first quarter of 2004, it was admitted that it had become necessary to give the problem of nanotechnology and nanomaterial development the status of a state priority and develop "the State concept of the development of works in the area of nanotechnology, nanomaterials, and nanosystems till 2019."

Starting with 2003, the training of specialists in some areas of nanotechnology became possible because the Ministry of Education and Science of Russia took the decision to create a nanotechnology educational direction outlining two special subjects: nanoelectronics and nanomaterials.

In October 2004, hearings on the development of the major directions of nanotechnology important to Russia were held in the State Duma. As a result of the hearings, all of the participants belonging to various agencies expressed interest in the development of a national nanotechnology program. The question of the specific contents of the program and its financing is still open.

As stated previously, in the principal regulations regarding the development of work in the area of nanomaterials, nanotechnology, and nanosystems until 2010, approved by the Government of Russia on October 18, 2004, an attempt was made to define such terms as *nanotechnology, nanomaterials, nanosystem hardware*, and *nanoindustry*. Though the ways in which these terms are defined in the document can be debated, this is the first attempt to develop a common language for the specialists from various agencies participating in nanotechnology research.

In February 2005, the Trade and Industry Chamber of Russia organized a discussion on nanotechnology and business. During the discussion, the development of nanotechnology as a priority scientific progress field in Russia was considered.

The participants of the session agreed to ask the president of the Trade and Industry Chamber, E.M. Primakov, to file a proposal to the government on the immediate development of the federal program on nanoindustry. On May 5, 2005, the Presidium of the Russian Medical Academy defined in its resolution that nanobiotechnology in medicine is a topical and promising field that has an important scientific and practical meaning. Establishing the foundation of a program on nanomedicine and the organization of personnel training were considered useful. In 2006, the Department of Scientific and Technical Innovations Policy under the Ministry of Science and Education of the Russian Federation funded a program of nanotechnology and material development in Russia. The program was at the discussion stage; a significant amount of money of about 100 million rubles per year is proposed to be assigned for the conduction of works in the area of nanomaterials and technologies. In 2007, Russian President V.V. Putin signed an edict establishing a nanotechnology corporation (RusNano); this program on nanotechnology includes the activities of many Russian institutes and is at the stage of formation. This makes 2007 the turning point of the foundation of nanotechnology research in Russia.

The informational support of various aspects of nanotechnology is conducted through special magazines, such as *Nano- and Microsystem Technology, Nanotechnology* (2004), *Russian Nanotechnologies* (2006), *Technologies of Living Systems* (2005), and so forth. Many publishing houses bring out monographs on various issues of nanotechnology. The nanoindustry concern, targeted at nanotechnology research, was founded; in 2004 the scientific conference "Nanotechnology for Industry" was held; a large seminar on nanomedicine has started work in Moscow.

These events show that the steps to establishing a national nanotechnology program in Russia are taken at different levels. They highlight the fact that only where there is an active state policy in the area of nanotechnologies based on consideration of fundamental knowledge of scientists in different areas and interests of potential investors and manufacturers from private companies is it possible to effectively use the intellectual, scientific, and technical potential of the country to develop a new scientific field, create the new production facilities, and improve the level of health care and security of Russia.

Nevertheless, the contents and the goals of the National nanotechnology program are not very well known in Russian society, and the issue of the level of awareness of the key role of nanotechnology in different public strata is a reason for concern for Russian scientists and manufacturers. Educating Russians on nanotechnology has countrywide implications because wider awareness will not only determine the possibility of financing specific parts of the program from various sources, including nongovernmental investments, but also will affect the efficiency of the development of Russia in the near future.

9.2 BIOLOGICAL MOLECULES AS A BACKGROUND FOR NANODESIGN

The content of a number of recent seminars and symposia held in various countries (see proceedings of the international conference "NanoTech 2005," May 8–12, Anaheim,

California; papers from the conference "NATO Workshop on Nanomaterials for Application in Medicine and Biology," October 4–5, 2006, Bonn, Germany; and the conference "Nanobio and Other Promising Biotechnologies," October 15–18, 2007, Pushchino, Russia) are indicative of rapidly expanding nanotechnological studies based on using biological macromolecules.

This interest in biological molecules is quite justified. In the process of evolution, these molecules acquired properties that are extremely attractive for nanotechnology application. The following properties can be outlined. First, the chemical diversity of such building blocks as amino acids, lipids, and nucleotides (nucleosides) cannot be compared with that of inorganics. Second, biological building blocks tend toward a spontaneous—and yet manageable, at the molecular level—formation of sophisticated spatial structures. Third, there are many approaches to assembling (polymerizing) biological building blocks that make it possible to create a variety of nanostructures. The hierarchy of self-assembling biological structures begins with monomers (i.e., nucleotides and nucleosides, amino acids, lipids, and so on) that form biopolymers (such as DNAs, RNAs, proteins, and polysaccharides); their ensembles (membranes and organelles); and, finally, cells, organs, and even organisms. Fourth, nanobiomaterials that infrequently result from self-assembling may have both improved characteristics and unique applications. This affords a good opportunity for designing a huge series of nanostructures. Hence, nanobiotechnology may be defined as a science of creating nanoconstructions (nanobiomaterials) with unique properties based on biomolecules (their complexes and ensembles). (One could say that biology is a science in which nanobiotechnology really "works.") Finally, nanobiomaterials often received by self-assembly have not only upgraded properties but also unique applications. The combination of the chemical reactivity of biopolymers and their tendency to form hierarchic nanostructures, as well as the opportunity for industrial production of biopolymers, makes biological molecules suitable for application in nanotechnology. Therefore, their use in creating artificial nanostructures based on principles furnished by nature appears to be quite logical. It would be strange not to use the possibilities provided by nature for nanotechnology. Moreover, the progress in chemical synthesis and biotechnology that allows one to combine building blocks of a different origin (i.e., to design "chimeric" molecules, containing, for instance, amino acids and synthetic organic chains) opens up fantastic vistas for designing nanomaterials and nanostructures that in principle do not exist in nature. Thereby, it can be expected that, as nanobiotechnology develops, biopolymers will be transferred from the world of biology to that of technology.

Nanodesign based on double-stranded (ds) nucleic acids (NAs) (i.e., targeted creation of complicated spatial structures with regulated properties [nanostructures, nanoconstructions, nanobiomaterials]) "building blocks" to which belong ds NA molecules or their complexes with biologically active compounds [17], is a topic of current attention among researchers from various countries.

The possibility of application of ds NA to create nanoconstructions with regulated parameters is based on a few properties characteristic of these molecules only:

1. Samples of single- and double-stranded NA molecules of preset sequence of nitrogen bases, ranging in length from just a few nucleotides to chains several tens of micrometers long, can be received on the industrial scale by means of modern biotechnological procedures. (This is not the case for any other polymer.)

2. ds NA 50–100 nm long have a high local rigidity under standard solvent conditions that allows such molecules to be used as "building blocks" without disturbing their physical properties.

3. The flexible single-stranded NA "recognizes" its complementary strand and, owing to H-bonds, forms a stable complex with the latter, thus providing an opportunity to prepare a rigid ds molecule with a preset composition.

4. The formation of branching points in the ds NA molecules combined with complementary (recognizing, "sticky") ends allows one to design planar lattices and sophisticated spatial structures.

5. The predictable and programmable type of spatial structures of rigid ds NA molecules and the mode of intermolecular interaction under various conditions as well as the intrinsic susceptibility of the NA secondary structure to external stimuli mediated by small molecules or ions open up a gate for the directed regulation of the characteristics of designed spatial constructions (i.e., to fabricate nanomechanical devices).

6. The nitrogen bases in the spatial NA constructions retain not only their capacity to interact with diverse chemical compounds and biologically active substances but also to orient these compounds in respect to the NA molecular axis, which increases an additional chemical reactivity to the whole structure.

9.3 TWO STRATEGIES OF NANODESIGN BASED ON NA MOLECULES

The strategies of creation of nanoconstructions based on NA molecules can be divided into two groups:

1. The creation of nanoconstructions by consecutive modification of initial linear NA

2. The creation of nanoconstructions using linear NA molecules (or NA complexes) fixed in a spatial structure of liquid-crystalline dispersion particles

9.3.1 NANOCONSTRUCTIONS CREATED BY MODIFICATION OF LINEAR NA MOLECULE STRUCTURE

Currently, several strategies (approaches) in the design of nanoconstructions (NaC) based on consecutive modification of an initial double-stranded NA molecule (or a

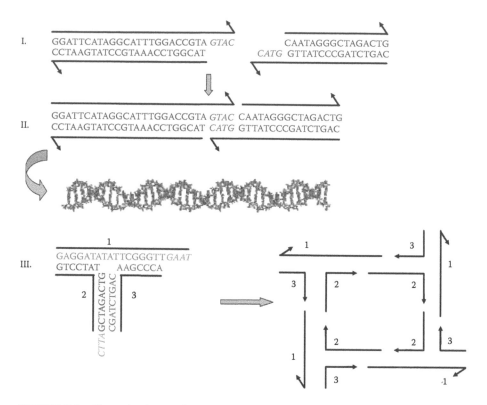

FIGURE 9.9 (*See color insert.*) Scheme of the stages of formation of a flat nanolattice from NA molecules.

synthetic nucleotide) have been described, which can arbitrarily be referred to as successive or step-by-step design. The first approach was pioneered by N. Seeman in 1982 [18,19] and is illustrated by Figure 9.9.

The initial step in Seeman's approach consists in preparing NA's fragments with single-stranded "sticky" ends (I, a and b) using biochemical procedures or direct chemical synthesis. The interaction between the sticky ends of different NA fragments leads to the formation of a single structure with two breaks in the sugar-phosphate chains. These breaks are cross-linked by enzyme–ligase, and, as a result, a rigid ds NA is formed (II). The second step is the formation of a branching point (cruciform structure, III) using NA fragments with a specific sequence of nitrogen bases. The cruciform structures may consist of three or four branches (i.e., they can differ in their spatial shapes). Sticky ends are formed in the cruciform ds NA molecule using enzymes. Cross-linking of the sticky ends of these initial molecules results in the formation of the first nanostructure—a plane (flat) 2-D nanolattice composed of ds NA molecules (on the right in Figure 9.9).

Because of mobility of the NA structure at the branching point, the resulting flate nanolattice is not very rigid. That is why, to obtain more rigid nanolattices, a few additional approaches, which may be called "supplementary" to the approach suggested by N. Seeman, have been used.

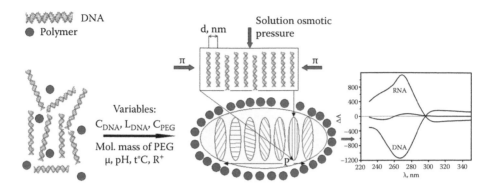

COLOR FIGURE 4.6 Scheme illustrating the formation of CLCD particles from DNA molecules at their phase exclusion from water–salt polymer-containing solution.

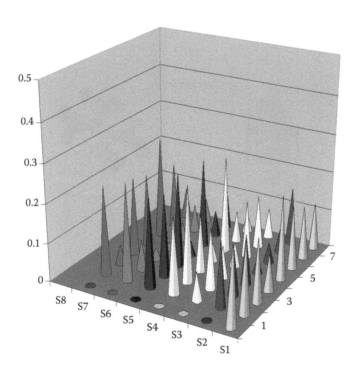

COLOR FIGURE 5.4 Chitosan binding to biochip containing single-stranded oligonucleotides. (Cone height corresponds to the concentration of chitosan bound to oligonucleotides in the given cell. Only one cell in the first row contains oligonucleotides.)

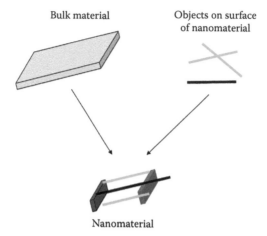

COLOR FIGURE 9.1 The principal scheme of "top-down" technology. Drop in size of initial material is accompanied by change in its optical properties.

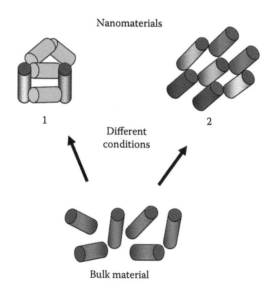

COLOR FIGURE 9.2 The principal scheme of "bottom-up" technology. Using various conditions, from initial material, one can create spatial structures differing by their properties.

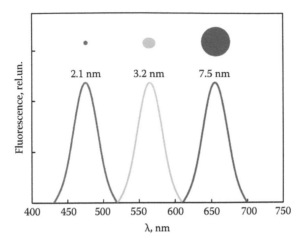

COLOR FIGURE 9.3 The optical properties of quantum dots: three CdSe quantum dots with different diameter are shown.

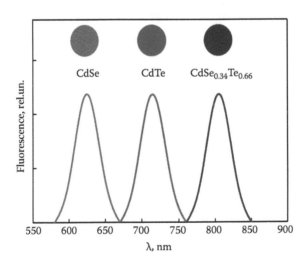

COLOR FIGURE 9.4 The optical properties of quantum dots: three quantum dots with the same diameter (~5 nm) but different composition and the emission spectra corresponding to them are displayed.

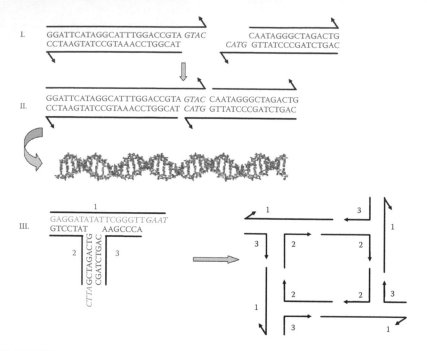

I.
GGATTCATAGGCATTTGGACCGTA *GTAC*   CAATAGGGCTAGACTG
CCTAAGTATCCGTAAACCTGGCAT   *CATG* GTTATCCCGATCTGAC

II.
GGATTCATAGGCATTTGGACCGTA *GTAC* CAATAGGGCTAGACTG
CCTAAGTATCCGTAAACCTGGCAT *CATG* GTTATCCCGATCTGAC

III.
GAGGATATATTCGGGGTT *GAAT*
GTCCTAT   AAGCCCA

COLOR FIGURE 9.9 Scheme of the stages of formation of a flat nanolattice from NA molecules.

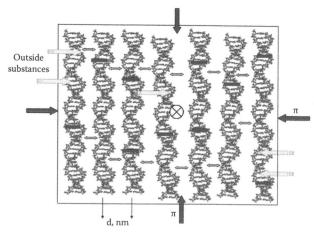

DNA molecules

DNA molecules   "Guest" molecules

COLOR FIGURE 9.12 Scheme illustrating the principal idea of "step-by-step" strategy (a 3-D nano-construction that is capable of holding "guest" molecules needs to be formed by consecutive modification of double-stranded NA molecules).

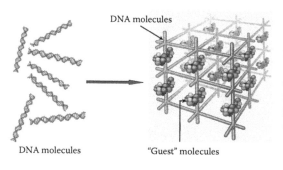

Outside substances

d, nm

π

COLOR FIGURE 9.18 Structure of DNA quasi-nematic layer in the structure of particles of liquid-crystalline dispersions obtained by phase exclusion of these molecules.

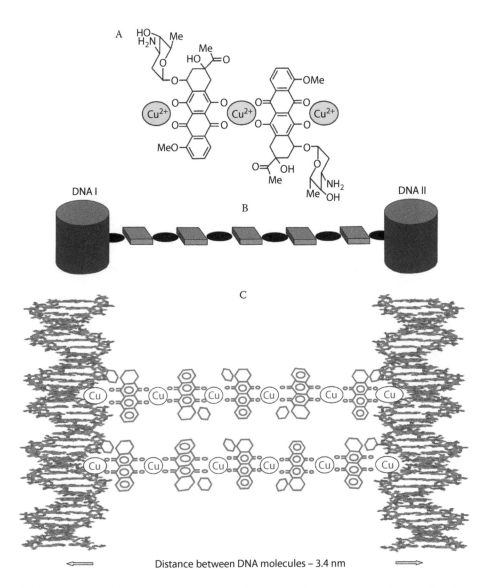

COLOR FIGURE 9.21 Structure of nanobridges formed between DNA molecules (1 and 2) fixed in the spatial structure of CLCD particles. (A) Structure of chelate complex between daunomycin molecules and cupper ions (...-[DAU-Cu²⁺-...]). (B) Schematic top view of nanobridge between DNA (I) and DNA (II) molecules. (C) Nanobridges between DNA molecules (for simplicity, nanobridges are turned on 90° degrees in respect to their standard position).

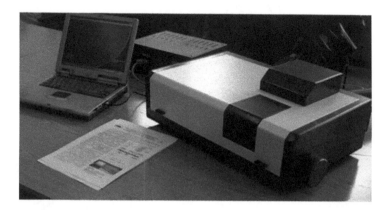

COLOR FIGURE 10.5 Portable dichrometer capable of working with biosensing units based on cholesteric DNA liquid-crystalline structures.

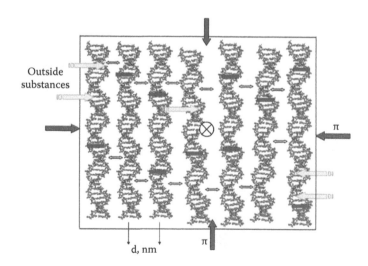

Outside substances

π

π

d, nm

COLOR FIGURE 10.6 Structure of a quasi-nematic layer formed by double-stranded DNA molecules.

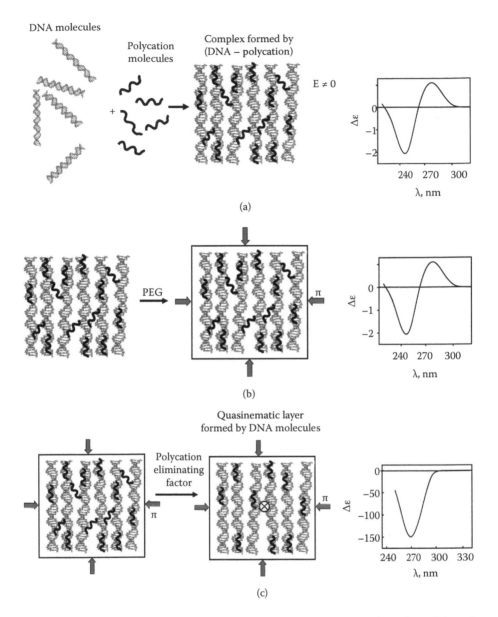

FIGURE 10.15 (*See color insert.*) The principal operation scheme of a biosensing unit based on DNA–polycation complex LCD: (a) the formation of LCD particles from (DNA–polycation) complexes is followed by a dense (hexagonal) packing of the molecules; the CD spectrum (on the right) of such a system almost coincides with the CD spectrum typical of the DNA B-form; (b) when PEG is added to the system (a), its optical and structural properties do not change (the osmotic pressure of the solution π is marked with dark arrows); (c) when factors that cause the "elimination" (destruction of the structure or extraction from the content of DNA complex) of polycation molecules are added to the system (b), the structural and optical properties of LCD particles change dramatically (under the conditions of the solvent with a fixed osmotic pressure, the DNA molecules can realize their tendency toward cholesteric packing, which is followed by the appearance of the abnormal band in the CD spectrum; on the right).

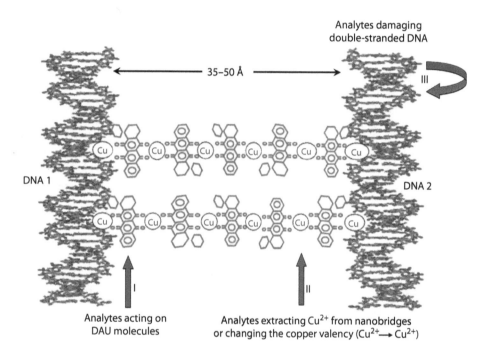

Analytes damaging
double-stranded DNA

35–50 Å

III

DNA 1

DNA 2

I

II

Analytes acting on
DAU molecules

Analytes extracting Cu^{2+} from nanobridges
or changing the copper valency ($Cu^{2+} \longrightarrow Cu^{2+}$)

COLOR FIGURE 10.18 Scheme of the nanobridges between adjacent DNA-1 and DNA-2 molecules fixed in the spatial structure of a CLCD particle (the nanobridges are turned to 90° relative to the base pairs for convenience). I–III—substances to be analyzed.

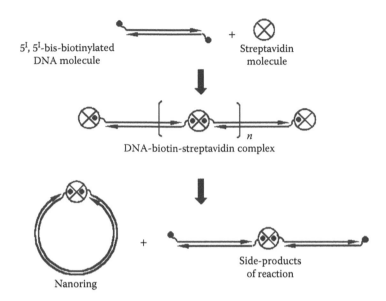

5^I, 5^I-bis-biotinylated
DNA molecule

+

Streptavidin
molecule

DNA-biotin-streptavidin complex

Nanoring

+

Side-products
of reaction

FIGURE 9.10 Steps for obtaining nanorings from double-stranded NA molecules.

In particular, the method proposed by K. Niemeyer [20,21] is based on the application of bis(biotynilated) DNA molecule and a protein that binds biotin–streptavidin (Figure 9.10, where ⊗ is streptavidin, • is biotin). The choice of this pair is based on the fact that biotin B and streptavidin S can form a strong complex BS, whose binding constant lies within the limit of 1015 M. This method allows plane nanolattice shaped as closed rings to be made using reactions of a complex formation. Nevertheless, the method has a significant disadvantage. First, a streptavidin molecule has four binding sites to biotin. The process of their interaction is probabilistic, so, besides the necessary complexes DNA–BS where each streptavidin molecule is bound to a biotin molecule, the three and tetrameric associates of biotin–streptavidin are obtained as side products of the interaction. Second, the concentration of closed nanorings of DNA–BS is not high. To increase the number of nanorings, it is necessary to control the composition of the reaction mixture after each stage and remove the side products of the reaction. Such manipulations are quite laborious, which makes the proposed method impractical.

To obtain more rigid nanolattices, a few additional approaches, which may be called "supplementary" to the approach suggested by N. Seeman, have been used. In particular, one of such techniques (by D. Bergstrom [22]) is based on the use of a synthetic molecule composed of two single-stranded, self-complementary oligonucleotides, the ends of which are bound to each other via a rigid chain of two p-(2-hydroxyethyl)-phenylethynylphenyl spacers linked to a tetrahedral carbon atom. The complementary hybridization of oligonucleotide fragments, which results in the formation of the ds structure, is accompanied by the development of a set of multiarm-star-shaped nanolattiles (Figure 9.11). The structural rigidity of these nanolattices is enabled by the sequential alternation of hydrocarbon moieties with a fixed spatial structure and ds oligonucleotides.

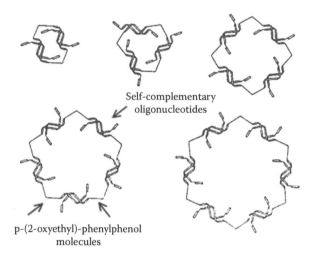

FIGURE 9.11 Multiarm-star-shaped nanoconstructions from NA molecules.

The low rigidity of flat lattice nanostructures has an advantage, too, namely, that it can be curved. Using an appropriate nucleotide sequence in starting single-stranded NAs and adequate cruciform structures, N. Seeman obtained a cube-shaped nanoconstruction (Figure 9.12) in which ds NA molecules served as the stiffening ribs [19].

An extra modification of the initial NA structure makes it possible to create nanostructures that have a shape of linked octahedrons, dodecahedrons, and so forth. Various aspects of the step-by-step strategy have been reviewed, including different plane lattices, as well as a cube-shaped nanostructure, octahedrons, dodecahedrons, and so on, in which double-stranded NA molecules served as the ribs.

The design of nanostructures using the technique just described requires large expenditures. Indeed, the following is necessary: to obtain NA fragments with specified sequences of nitrogen bases or their complexes with concrete "linkers"; to use a wide set of enzymes (restrictases and ligases) for splitting and cross-linking of NA

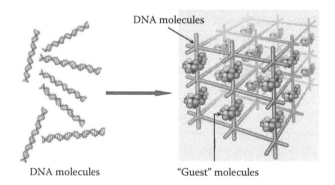

FIGURE 9.12 (*See color insert.*) Scheme illustrating the principal idea of "step-by-step" strategy (a 3-D nanoconstruction that is capable of holding "guest" molecules needs to be formed by consecutive modification of double-stranded NA molecules).

fragments between desired nitrogen bases; to select specific structures from the reaction mixture; to perform careful analysis of their properties; and to apply modern control methods (such as atomic force microscopy) in all stages of the nanodesign.

An interesting question is whether it is possible to maintain a high concentration of various compounds ("guest" molecules; see Figure 9.12) in nanoconstructions based on NA molecules. From the physicochemical point of view, the number of binding sites for "guest" molecules in a linear double-stranded NA molecule is limited. When forming cubic nanoconstructions whose stiffening ribs are the double-stranded NA molecules, the number of binding sites for "guest" molecules either stays constant or reduces. From this point of view, the creation of nanoconstructions, according to a "step-by-step" strategy capable of containing in their content a superhigh concentration of "guest" molecules, seems to be illusive.

The practical application of nanoconstructions created from single NA molecules using step-by-step technology is apparently determined by the kind of problems researchers have to solve. The most important problem in nanoconstructions is the creation of 3-D nanoconstructions with regulated properties that hold in their structure various compounds ("guest" molecules).

To create spatial nanoconstructions with built-in "guest" molecules, C. Mirkin [23], R. Letsinger [24], and P. Alivisatos [25] have almost simultaneously (1996) used a procedure based on the application of synthetic short, single-stranded NA fragments carrying SH-groups on 3′-ends as building blocks. Because of the high affinity of SH-groups to gold, NA molecules were linked with particles of colloid gold (Figure 9.13; here the distance between gold nanoparticles can be regulated).

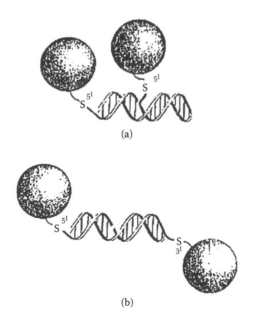

(a)

(b)

FIGURE 9.13 Hybridization of oligonucleotides bound to gold nanoparticles.

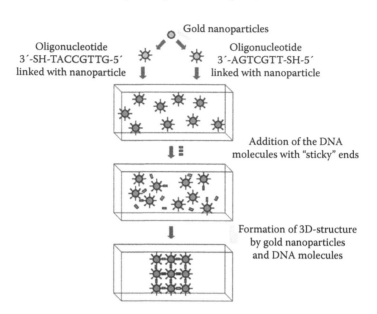

FIGURE 9.14 Scheme of formation of nanoconstructions containing gold nanoparticles linked to DNA molecules.

A fragment of double-stranded NA that has recognizing sites on 5′ or 3′ ends complementary to both types of NA molecules is added to the mixture. As a result of hybridization, a rigid double-stranded NA that holds two gold nanoparticles is formed. In principle, this means that, under certain conditions, nanoparticles can be assembled into a two-dimensional lattice (grid; Figure 9.14), that is, the local concentration of nanoparticles rises. In this case, the red color of the solution caused by surface plasmon resonance changes to blue. Hence, the process of approaching gold nanoparticles can be easily detected by the optical method. Gold nanoparticles locate at a fixed distance that depends on the length of the chain of the initial NA fragment, which is proportional to 0.34 nm (0.34 nm is the distance between the adjacent DNA base pairs) in nanoconstructions received with the use of this method. Attaching a larger number of single-stranded NA fragments to the gold nanoparticles, one can form two-dimensional nanostructures (Figure 9.15) or even three-dimensional structures, in which gold atoms alternate at regular intervals with ds NA molecules.

The advantages of this method of nanoconstructions formation are the following. First, the number of intermediate stages of work are reduced. Second, the quality of the formed construction can be controlled at the final stage. Nevertheless, the inhomogeneous distribution of gold nanoparticles modified with NA fragments in the initial solution leads to the appearance of defects in the forming grid structure.

By choosing the length and sequence of single-stranded NA fragments linked to gold nanoparticles and the number of fragments per one particle, and changing the temperature of hybridization and the subsequent cooling of the mixture, assembly of the construction can be controlled and the efficiency of formation can be increased [26].

Reference 27 describes the formation of a nanoconstruction containing in its content carbon nanotubes (CNT) and peptide nucleic acid (PNA) molecules, a

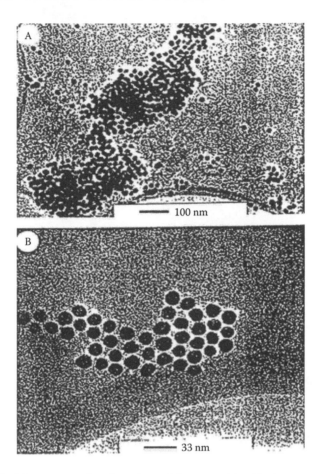

FIGURE 9.15 Nanoconstructions based on gold nanoparticles (the image was received by atomic force microscopy). A and B—different magnification.

synthetic compound (as opposed to natural NA molecules) in which the sugar-phosphate chain is replaced by a peptide backbone (Figure 9.16). In the PNA structure, the NA nitrogen bases are retained, and the distance between these bases is equal to that in natural NA. PNAs have some advantages over common NA. They are more stable in commonly used solvents and are protected from degradation with some hydrolytic enzymes. Because they contain an uncharged peptide backbone, this results in an increase in stability of the hybrid PNA–NA complexes. The attachment of a PNA molecule to a chemically modified CNT results in the formation of the CNT–PNA complex. This complex is hybridized with the complementary single-stranded fragment of the NA molecule covalently bound to the surface of a solid support to form a rigid double-stranded structure (PNA–NA) containing carbon nanotubes arranged in a particular mode. Hence, carbon nanotubes can be oriented in desired directions with the use of NA molecules arranged in a specific fashion on the surface of an electronic unit. A comprehensive review in Reference 28 covers all aspects of the formation and possible

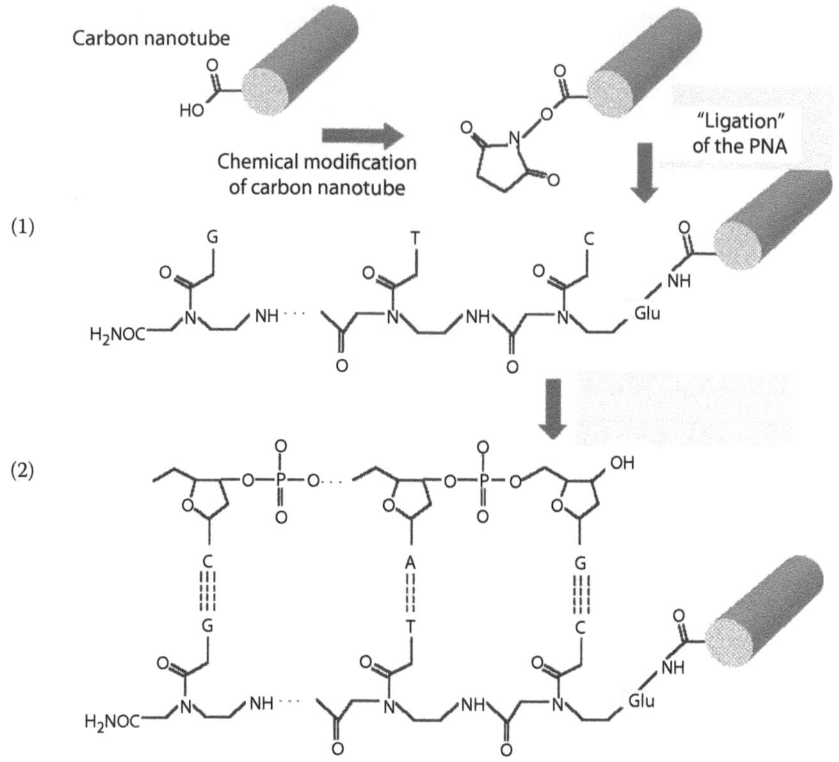

FIGURE 9.16 Carbon nanotubes bound to PNA (1) and the formation of a double-stranded structure as a result of hybridization of PNA to a fragment of single-stranded DNA (2).

practical applications of systems containing metal nanoparticles and oligonucle-otide (or NA) molecules.

"Guest" molecules in nanoconstructions can serve as targets for other biologically active compounds, that is, a biosensing unit based on nanoconstructions can be cre-ated to detect the chemical or biological substances recognized by guest molecules. In particular, the application of CNT–NA complexes as sensing units for sensors [29–30] has recently been described. The unique property of nanotubes—the ability to amplify the detected electric signal—is used in these devices.

The formation of flat DNA–protein nanoconstructions that can have a practical application has been described in Reference 32. A protein-factor of NF-kB tran-scription that has high affinity to DNA molecules was used as a connection element. Moreover, the protein was modified so that it carried an extra cysteine SH-group on the C-end. Double-stranded DNA molecules of 180 nm long were synthesized so that there were nitrogen base sequences on both of their ends that were the binding sites for NF-kB. When the protein was added to the solution containing DNA molecules, strong DNA–NF–kB complexes were formed near the ends of the DNA molecules. Then, the mixture was incubated in aerobic conditions. This led to the oxidation of SH-groups, the appearance of disulfide connections between the adjacent NF-kB molecules, and consequently to the formation of a solid two-dimensional honeycomb structure. The DNA–protein lattice constructed with nanometer accuracy was centimeter sized. It can be assumed that such two- (and three-) dimensional nanoconstructions may become the background for the further assembly of nanostructures interesting to researchers.

The use of planar lattices consisting of NA molecules as masks, followed by the deposition of current-conducting compounds as gallium arsenide or indium phosphate into these lattices, opens, in the opinion of E. Di Mauro and K. Hollenberger [33], a way for designing elements (chips) for microelectronics with an accuracy greatly exceeding the accuracy of conventional techniques.

Finally, if it is possible to realize the 3-D ordering of single spatial nanoconstructions (crystallization), it may be possible to crystallize compounds that poorly crystallize in common conditions but can be embedded into the nanoconstructions, that is, it will be possible to perform most specific refinements of the "guests." Nevertheless, such ordering by itself is a complicated task that has not yet been solved. All the variants of the considered strategy are very laborious and resource-intensive; they require persistent control of quality at each stage of the nanoconstruction process.

Comparing the techniques of nanoconstructing just described, it is necessary to consider the possible areas of application of nanoconstructions. For instance, some of the techniques are expensive because of the high cost of the enzymes used in them. However, if a higher accuracy in comparison to the existing methods is provided, the use of these techniques will be reasonable.

In conclusion, it can be stated that regardless of the first achievements of nanoconstructing based on the "step-by-step" strategy—that is, the use of single double-stranded NA molecules as building blocks—the most important task of nanoconstructing, namely, the creation of spatial constructions with controllable properties that contain in their content the molecules of the other compounds ("guest" molecules), has not been solved within the framework of this strategy. The issue of practical application of nanoconstructions based on double-stranded NA is still open.

9.3.2 THEORETICAL APPROACHES TO THE PROBLEM OF GUEST ACCUMULATION IN NANOCONSTRUCTIONS BASED ON NUCLEIC ACID MOLECULES

The 3-D nanostructure presented in Figure 9.12 provides an explanation of the basis of a different strategy for the accumulation of molecules of chemically or biologically active compounds ("guest" molecules) in nanoconstructions. According to the first approach, guest molecules can be accumulated in nanoconstruction with the use of the "free space" within the secondary structure of DNA molecules. In fact, the linear double-stranded DNA molecule can bind such compounds as intercalators (i.e., planar compounds that are inserted between nitrogen base pairs) or compounds filling one of the grooves on the surface of the macromolecule (polypeptides, proteins, etc.). The equilibrium concentration of such compounds depends on the characteristic features of the secondary structure of double-stranded DNA (the distance between base pairs, base pairs tilt, charge density of phosphate groups, etc.) and follows the known laws of adsorption. The limiting number of guest molecules can be equal, for instance, to one molecule per five base pairs, and this parameter, all other factors being the same, depends on the structure of the guest molecules. Considering the physicochemical properties of the initial molecules of double-stranded DNA, it is obvious that two ways of accumulating guest molecules in DNA-based nanoconstructions are, in principle, possible. First, the guest molecules can be attached to the initial DNA molecule, and the desired three-dimensional nanoconstruction can be formed from the DNA–guest complex using the step-by-step strategy. Evidently, the concentration of guest molecules after several biochemical and physicochemical steps of formation of nanoconstruction, which are performed under different conditions, will most likely be low. That is why the second way is more probable. According to this approach, the required 3-D nanoconstruction can be formed from the starting double-stranded DNA molecules using step-by-step technology, and then the guest molecules can be bound to the double-stranded DNA serving as stiffening ribs. Because the physicochemical characteristics of DNA molecules in nanoconstructions remain virtually unchanged, the number of guest molecules bound to double-stranded DNA molecules in nanoconstructions cannot be significantly different from the number of guests carried by the initial double-stranded DNA molecule. Therefore, the possibility of attaining superhigh concentrations of guest molecules using step-by-step technology is unlikely.

Analysis of Figure 9.12 shows that there is another possible approach to the problem of accumulation of guest molecules in double-stranded NA-based nanoconstructions. This approach is based on the concept that there is free space between NA molecules ordered in nanoconstructions and there are, a priori, no theoretical restrictions on filling this space with guest molecules. Since the distance between adjacent molecules of double-stranded NA in ordered structures of these macromolecules can be regulated within 2.5–5.0 nm, attempts can be made to attain a high concentration of guests in the free space between NA molecules. Hence, there is a possibility of a fundamentally different way of immobilizing (accumulating) of molecules of various chemical and biological compounds (guests) in ordered 3-D nanostructures. The physical chemistry of double-stranded NA and their complexes indicates the existence of various type of ordering of molecules that make it possible to fix the guest molecules in the space between NA molecules.

Chapter 3 (Section I) described [34,35] two principally different ways for spontaneous ordering of rigid, helical, double-stranded NA with low molecular mass (~1×10^6 Da), or their complexes with biologically relevant compounds in water–salt solutions and physicochemical properties of the formed liquid-crystalline dispersions. Specifically, these determine the design of NA-based nanoconstructions of various types. The scheme (Figure 9.17) describing the ordering of double-stranded DNA molecules can be reviewed to see the problems that need to be solved when creating nanoconstructions.

Approach 1—the ordering of double-stranded DNA molecules—is based on the so-called entropy condensation [36,37], that is, the process in which the change in the entropy of the system is the driving force. This condensation is achieved in the case of the phase exclusion of NA molecules from water–salt solutions on the adding of certain polymers. Under specific "critical" conditions, rigid, low-molecular-mass, double-stranded NA molecules are ordered and form LCD. The violation of the "critical" conditions—in particular, a decrease in the PEG concentration below the "critical" value—leads to disintegration of LCD particles and transition of NA molecules

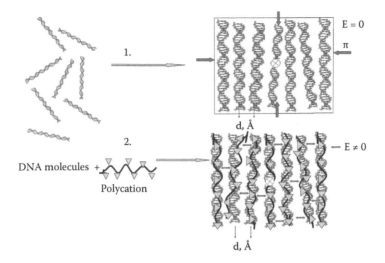

FIGURE 9.17 Scheme illustrating two possible modes of ordering of double-stranded DNA molecules. 1. Entropy condensation—phase exclusion of DNA molecules from water–salt solutions of polymers (dark arrows show that this structure is only retained at a certain osmotic pressure of the solvent [π]); 2. Enthalpy condensation—phase exclusion of DNA–polycation complexes from water–salt solutions caused by the interaction between DNA molecules and polycations.

into the isotropic state [38,39]. The packing of NA molecules in dispersion particles is fluid by nature, and NA molecules in each layer can both rotate around the axis and be displaced laterally. The theoretically estimated particle size of LCD is close to 500 nm; the particles contain approximately 10^4 NA molecules each.

NA LCD particles have several very important characteristic features for the formation of nanoconstructions:

1. The polymer (in our case, PEG) does not enter the structure of the forming dispersion particles.
2. LCD particles of double-stranded NA are characterized by a high (ranging from 160 to 600 mg/mL) local NA concentration.
3. LCD particles are characterized by the ordered arrangement of adjacent NA molecules.
4. The distance, **d**, between adjacent NA molecules in LCD particles can be regulated from 2.5 to 5.0 nm [38–40] by changing the osmotic pressure of the solution.
5. The structural elements of NA (nitrogen bases, etc.) within the LCD particles retain their high reactivity in respect to chemical and/or biologically active compounds.
6. NA molecules form cholesteric LCD (CLCD) due to their geometric and optical anisotropy. The forming cholesteric is "colored" because there are chromophores (nitrogen bases) in the content of NA molecules that absorb UV light.

The theory developed specially for colored dispersion particles [41] predicts the appearance of an abnormal band in the UV region of the CD spectrum. The latter can have different signs, depending on the parameters of the secondary structure of NA molecules. The theory does not impose restrictions on the number of molecules of colored compounds (external chromophores) that can be introduced in the LCD structure.

Hence, it can be expected that abnormal bands will appear in the CD spectra in the region of absorption of any colored compounds not only intercalating between NA base pairs but also anisotropically located in the space between adjacent NA molecules that form CLCD particles.

Approach 2 of ordering of rigid, linear low-molecular mass NA molecules and the formation of CLCD particles is based on their phase exclusion from water–salt solutions as a result of attraction (correlation interaction) between segments (molecules) of NA [42–44], in which the negative charges of the phosphate groups are neutralized by counterions (see Figure 9.17). Since the change in the enthalpy of the system is the driving force for the process, it is called "enthalpy" condensation. This process requires the presence of polycations that neutralize most (80–90%) of the negative charges of the NA phosphate groups. As the surface charge density of NA molecules decreases, the dispersion forces first equilibrate, and then become stronger than the electrostatic repulsion between adjacent molecules. This leads to the spontaneous exclusion of the resulting molecules of the NA–polycation complexes from the solution.

The most important event in this NA condensation process is the approaching of adjacent NA–polycation complexes. The peculiarities of the process depend on the properties of the NA molecules' surface, the spatial structure of polycation molecules, the mode of positive charges distribution, and the properties of the solvent [45,46]. This means that a minor modification of the NA molecules' surface induced by polycation molecules must cause the change in the properties of the forming particles of NA dispersions.

The most important features of the dispersions of NA–polycation complexes that have to be considered when creating nanoconstructions are illustrated by Figure 9.17.

1. The phase exclusion of double-stranded NA–polycation complexes is achieved when the "critical" concentration of polycation in solution comparable to the concentration of the starting NA is attained.
2. Polycation molecules always enter the composition of resulting dispersion particles.
3. As polycation molecules have a fixed spatial conformation, the energy of the interaction between the molecules of NA–polycation complexes provides a certain fixed distance, **d**, between NA–polycation complex molecules in the resulting particles; as a rule, this distance is usually close to 2.5 nm [49,50].
4. In some "lucky" cases, depending on the structure polycation molecules, it is possible to form LCD, the particles of which are characterized not only by a larger (but still fixed!) distance between NA–polycation complexes but also

by an intense band in the CD spectrum, that is, it is possible to form CLCD. Polyamines, polyamino acids, proteins (histones), dendrimers, chitosan, and so on, were used as polycations causing the formation of CLCD [47,48].

5. The formation of LCD particles of NA–polycation complexes can lead to a substantial drop in the reactivity (due to steric factors associated with arrangement of polycation on the surface of NA molecules), but other reaction sites in the polycation molecules appear. This creates the possibility of using such groups as new chemically responsive centers with high diffusive availability.

Regardless of the significant differences from the first method of ordering of NA molecules, a high local concentration of polycations in NA CLCD particles, the presence of new chemically active groups of polycations, and the ordered arrangement of the polycation molecules can provide conditions for cross-linking adjacent NA–polycation complexes in CLCD particles by chemical bridges.

The comparison of both approaches for the spontaneous ordering of NA molecules and NA complexes allows us to conclude that, at least theoretically, there is a principal possibility to use NA molecules fixed in the spatial structure of CLCD particles as building blocks with regulated properties (i.e., it is possible to create a spatial construction with tailored properties).

The use of NA CLCD particles opens a gateway for the control of the properties of both NA molecules and guest molecules in the content of nanoconstructions. Besides, the abnormal bands in the CD spectra serve as an analytical criterion to monitor the minor alterations of the structure of starting NA molecules, the CLCD particles, and the resulting nanoconstructions.

9.3.3 Nanoconstructions Based on NA Molecules Ordered by Entropy Condensation

The technology of creation of nanoconstructions developed in the Institute of Molecular Biology of the Russian Academy of Sciences [50,51], in collaboration with scientists from other institutes of the Russian Academy of Sciences and institutions of other countries, is based on the concept of self-ordering of double-stranded NA molecules (or double-stranded NA complex molecules) by phase exclusion (condensation). This means that the version of nanodesign strategy proposed by us is based on use; instead of individual NA molecules, these molecules fixed in spatial structure of their particles of CLCD spontaneously arise upon phase exclusion of NA molecules from water–salt solutions.

Meanwhile, two problems appear,

1. Can the adjacent NA molecules be linked by nanobridges containing quest molecules without disturbing the spatial structure of CLCD particles?
2. What properties will the resulting nanoconstruction demonstrate?

DNA molecules in the layers are located at the average distance, **d**, that depends on the osmotic pressure of the solution (dark arrows); the polymer does not enter the

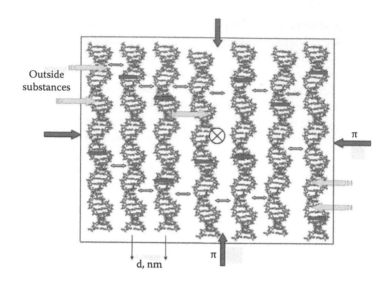

Outside
substances

π

d, nm

π

FIGURE 9.18 (*See color insert.*) Structure of DNA quasi-nematic layer in the structure of particles of liquid-crystalline dispersions obtained by phase exclusion of these molecules.

composition of the layered structure. Different compounds can easily diffuse into the layer and interact with DNA molecules. The structure of a quasi-nematic layer of CLCD particles (Figure 9.18) formed by DNA molecules explains the essence of the approach to the solution of the problem of chemically or biologically active guest molecules' accumulation based on the application of entropy condensation of DNA. This structure is retained by the osmotic pressure of the solution.

This structure shows that the distance, **d**, between adjacent NA molecules in the ordered quasi-nematic layer is equal to a few nanometers (i.e., there is a free space between adjacent NA molecules). A considerable distance between NA molecules and the "liquid" character of the packing of these molecules provide conditions for fast diffusion of "quest" molecules between both NA molecules in a single layer and NA molecules in adjacent layers in the structure of LCD particles. The retention of chemical reactivity of the structural elements of NA, high NA concentration, and an ordered arrangement of NA molecules in CLCD particles, which does not restrict the diffusion of biologically active or chemically relevant compounds, can provide a high rate of their penetration into CLCD particles and reaction with NA molecules. This means that there is, in principle, a theoretical possibility of cross-linking of adjacent NA molecules in CLCD particles by the chemical bridges containing guest molecules, and, in this case, the attaining of a high concentration of quests in the free space between NA molecules is quite possible. In this approach for accumulation of quest molecules, the concentration of quests is determined by the geometric parameters of the free space and the number of quest molecules in bridges, and is not, apparently, directly associated with the characteristic features of the secondary structure of NA molecules. Evidently, this approach to the accumulation of guest molecules will work if adjacent NA molecules are specifically fixed in the space of CLCD particles. Hence, the procedure developed for the ordering of adjacent NA molecules allows one efficient use of the space available between these molecules for positioning guest molecules.

Since the distance between NA molecules in CLCD particles is 2.5–5.0 nm, the cross-link can be called a "nanobridge," and the resulting construction containing a large number of such bridges can be called "nanoconstruction based on double-stranded NA molecules." This approach to the design of NA spatial structures can be referred to as nanodesign, the principles of this approach being fundamentally different from the above-considered "step-by-step" strategy.

The spontaneous arrangement of NA molecules or NA complexes leads to the conclusion that there is at least theoretically a principal possibility of the accumulation of molecules of various compounds in nanoconstructions based on the use of neighboring NA molecules fixed in spatial structure of CLCD particles.

The fundamental problem that needs to be solved when creating a nanoconstruction can be formulated as follows: using the initial particles of CLCD of double-stranded DNA, the adjacent DNA molecules in the CLCD structure need to be so linked with nanobridges that the mode of spatial ordering of double-stranded DNA molecules in CLCD particles is not distorted. Nanobridges will link DNA molecules located both in one layer and in adjacent quasi-nematic layers of cholesteric structure of CLCD particles, which will result in the "freezing" of DNA molecules' diffusion. Under these conditions, the formation of nanobridges can cause a dramatic change in the properties of CLCD particles. In particular, the "liquid" mode of packing of adjacent DNA molecules in CLCD particles and the diffusion mobility of DNA molecules must disappear. The structure will become "rigid" and the particle may gain the properties of a solid material, that is, the process can be accompanied by the formation of a 3-D spatial structure that can be called a "nanoconstruction" (Figure 9.19). Finally, if the process of formation is realized so that the spatial structure of CLCD particles remains constant, the abnormal optical activity will permit controlling the change in both the secondary structure of initial NA molecules and the appearance of the new molecules that form nanobridges in the nanoconstruction.

Obviously, formation of cross-links (nanobridges) between neighboring NA molecules requires the presence of terminal sites on the surface of NA molecules. If such sites are located on neighboring NA molecules, a nanobridge will cross-link the neighboring NAs. Theoretically, both metal ions specifically fixed in a groove (grooves) on the NA surface capable of forming chelates with NA nitrogen bases and the molecule of ligands additionally introduced into the system can serve as such terminal sites. Taking into account the steric location of the reactive groups (in particular, the N7 nitrogen atom of purines) in a DNA helical secondary structure, two neighboring DNA molecules can be cross-linked only when the spatial orientation of these molecules is coordinated (i.e., the positions of neighboring DNA molecules must be "phased"). This means that the formation of nanobridges is a very delicate stereochemical process that can be realized only under rather strict conditions. Besides, the theory predicts that, when constructing planar nanobridges containing chromophores specifically oriented relative to the long axes of DNA molecules in quasi-nematic layers, we should expect the appearance of an additional abnormal CD band in the absorption region of these chromophores.

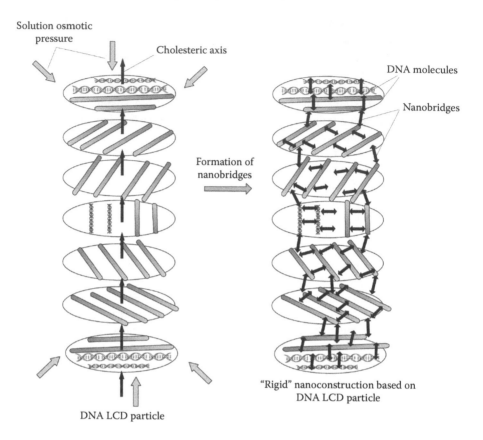

FIGURE 9.19 Scheme of "all-at-once" design: using the CLCD particles with ordered DNA molecules that exist only at a certain osmotic pressure of the solution (wide arrows), it is necessary to link DNA molecules with nanobridges without distorting the spatial structure of CLCD particles (as a result a nanoconstruction appears with the properties different from the properties of the initial CLCD particles).

The detailed analysis of the properties of compounds that can be used as components for extended nanobridges made it possible to choose the molecules of planar anthracyclines, namely, the water-soluble antibiotics of the anthracycline group [52,53]. These compounds, owing to their chemical structure, can form chelate complexes with ions of bivalent metals. The formation of chelate complexes with Cu^{2+} ions is especially efficient. Because of the peculiarities of spatial flate structure of Cu^{2+} ion, chelate complexes may contain up to 10 repeating subunits; that is, under certain conditions, anthracyclines can form extended planar structures with Cu^{2+} ions. It has been shown in Reference 52 that the presence of four reactive atoms of oxygen in the 5, 6 and 11, 12 positions of aglycone is an essential prerequisite for the formation of chelate complexes.

Considering the results of References 52 and 53, a technology of fabrication of a nanoconstruction was elaborated.

A nanoconstruction based on double-stranded DNA was obtained according to the following scheme: a CLCD was formed from double-stranded DNA molecules

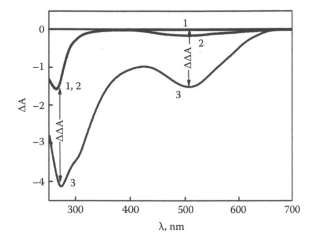

FIGURE 9.20 CD spectra of the initial DNA CLCD (1) and the CLCD DNA successively processed by DAU (2) and $CuCl_2$ solution (3). C_{DNA} = 5.5 µg/mL; Na^+-phosphate buffer, pH 6.7; ΔA in optical units ($\times 10^{-3}$); $\Delta\Delta A$ is an increase in the band amplitude as a result of nanobridge formation.

(curve 1, Figure 9.20), a solution of an anthracycline antibiotic, daunomycin (DAU), was added to it up to the formation of a external complex (curve 2), and then the formed CLCD dispersion was treated with a $CuCl_2$ solution (curve 3). Addition of $CuCl_2$ to the DNA CLCD processed with DAU and having an equilibrium value of the band amplitude at $\lambda \sim 500$ nm (curve 2) leads to the amplification of this band related to the optical properties of the chromophore of the [DAU–Cu^{2+}] complex. All other factors being equal, the amplitude of this band increases with the increase in the amplitude of the DNA abnormal negative CD band at 270 nm. Under the conditions used (the DNA molecular mass is 8×10^5 Da and the DNA concentration is 5 µg/mL), the maximum amplitude of the band at λ 500 nm is equal to 2500 (in ΔA units).

According to the theory, the amplification of the band in the chromophore absorption region shows that the chromophores of the [DAU–Cu^{2+}] complexes are spatially fixed with respect to the NA molecules in the composition of CLCD particles [41].

Although there are two different models of fixation of [DAU–Cu^{2+}] chelate complexes near the surface of DNA molecules, the experimental data [54–56] are unambiguously in favor of the model of the complex playing the role of a nanobridge between nucleic acid molecules.

The efficiency of the formation of nanobridges depends on the concentrations of both DAU molecules and copper ions [54]. Despite different values of the band amplitudes in the UV and visible regions of the CD spectrum, the obtained dependencies are S-shaped with a high degree of consistency. Amplification of the bands begins when the "critical" concentrations of both copper ions and DAU molecules are attained in the solution.

For DAU molecules, the "critical" concentration means that those DAU molecules that arise after the DAU intercalation take part in the formation of nanobridges. The "critical" concentration in the case of copper ions means that these ions induce some changes in the NA molecule, after which copper ions (or their complexes with pairs of bases and DAU molecules) become available for further chelating that is necessary for nanobridge formation. (This means that the order of the addition of components, that is, DAU and copper ions, is important for the construction of nanobridges.)

The experimental data received with the help of various physical methods including low-temperature magnetometry [55] and the results of theoretical evaluations [56] have shown directly that the [... $-Cu^{2+}-DAU-Cu^{2+}-$ ...] nanobridge (Figure 9.21) include six bivalent coppers ions. The nanobridges are flat chelate

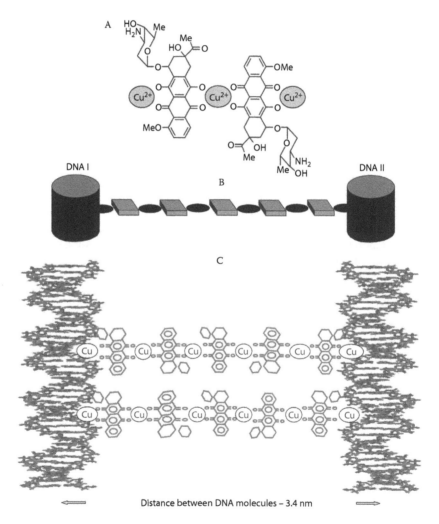

FIGURE 9.21   (*See color insert.*) Structure of nanobridges formed between DNA molecules fixed in the spatial structure of CLCD particles. (A) Structure of chelate complex between daunomycin molecules and cupper ions (...-[DAU-Cu^{2+}-...]). (B) Schematic top view of nanobridge between DNA (I) and DNA (II) molecules. (C) Nanobridges between DNA molecules (for simplicity, nanobridges are turned on 90° degrees in respect to their standard position).

complexes (Figure 9.21A) that are fixed between spatially ordered neighboring DNA molecules located both in the same layer and/or in adjacent layers (Figure 9.21B, C). This process results in the formation of a 3-D nanoconstruction.

The produced nanoconstruction has the following unique properties:

1. Unlike the initial CLCD particles, the main factor stabilizing particles of nanoconstruction is the number and "strength" of nanobridges rather than the osmotic pressure of the water–salt solution.
2. Second, a liquid packing pattern of neighboring DNA molecules in CLCD particles and DNA diffusion mobility disappeared; the structure became "rigid," and the particle acquired the properties of a solid material.
3. Third, characteristic of DNA NC is not only an abnormal optical activity appearing as an intense CD band in the DNA absorption region but also an additional abnormal optical activity in the absorption region of antibiotic chromophores [41,54].
4. The nanoconstruction is characterized by two thermal structural transitions [57]: one of them corresponds to the "CD melting" of nanobridges, and the other to the "CD melting" of DNA cholesteric. The temperature of CD melting (τ_m) of nanobridges grows with their concentration, remaining below the initial τ_m value of double-stranded DNA cholesteric.
5. Fifth, the nanoconstruction not only retained a high local concentration of DNA molecules (reaching 400 mg/mL!!!), but also acquired a high concentration of the antitumor antibiotic DAU.

The formed rigid NC structure, existing even in the absence of the osmotic pressure of polymer-containing solution, provided the possibility of immobilizing such constructs on the surface of a nuclear membrane filter and directly assessing the size of the produced particles (Figure 9.22, where two-dimensional [a,b] and 3-D [c,d] images were taken by an atomic force microscope; the dark spots are holes in the filter, $D \sim 0.15$ μm). In their shape, the particles were close to elongated cylinders; the size estimation of 400 particles demonstrated that, although the size changed in the range of 0.4–0.8 μm, the mean value was approximately 0.5 μm. This means that the size of the formed CLCD particles containing DNA molecules linked with nanobridges (i.e., the nanoconstructions existing in the absence of osmotic pressure of the solution) coincides practically with the size of the initial DNA CLCD particles theoretically calculated for the solutions with a constant osmotic pressure (Section I, Chapter 3) [38,39].

Therefore, as a result of nanodesign, it is possible not only to visualize the "rigid" DNA CLCD particles but also to receive the data describing the macroscopic parameters of these particles.

The fact of obtaining a fairly stable structure whose properties only depend on the properties of the nanobridges demonstrates that the hypothetical structure proposed for the nanoconstruction based on DNA CLCD particles (see Figure 9.19) exists in reality.

The uniqueness of the formed DNA nanoconstruction should be highlighted [58,59]. In this nanoconstruction (NaC), neighboring DNA molecules form layers with nanobridges composed of alternating antibiotic molecules and metal ions located

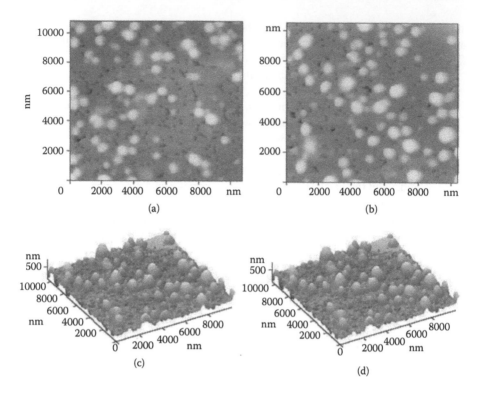

FIGURE 9.22 2-D and 3-D images of DNA particles cross-linked by nanobridges and immobilized on a nuclear membrane filter (PETP).

within and between these layers. This NaC is a 3-D formation; the diffusion mobility of neighboring NA molecules in it is fairly low and, therefore, this structure does not possess many properties characteristic of NA CLCDs. However, the structure, which is no longer "fluid," retains the optical properties of the initial cholesteric. The NaC developed is unique because it contains not only NA molecules but also antibiotic molecules and copper ions, that is, "guest" molecules. This structure accommodates a high concentration not only of NA molecules alone (up to 400 mg/mL) but antibiotic molecules (about 200 mg/mL) and copper ions. This means that the use of CLCD particles for nanodesign automatically solves the problem that remains unsolved in the "step-by-step" strategy, specifically, ordering of both neighboring NA and "guest" molecules. In addition, the cholesteric structure is formed not only by NA molecules but also "guests" (antibiotic molecules and copper ions) [60]. This fact shows that the formation of NaC sharply increases the abnormal optical activity in the absorption region of the antibiotic, a circumstance that is very important from the practical point of view since such a high optical activity makes it possible to monitor with high accuracy all changes in the properties of the NaC formed.

This statement means that the DNA nanoconstruction is not only a new type of nanobiomaterial whose properties can be regulated in the broad limits but, besides, it is easy to observe any changes of these properties.

Concerning the scientific importance of the nanoconstruction, it can be stated that a new structure formed from DNA CLCD attracts great interest on the part of both theorists [60] and experimenters [54,55,57,59].

Therefore, nanodesign taking into account the peculiarities of entropy condensation of the initial double-stranded NA leads to the formation of nanoconstructions with unique properties.

9.3.4 Nanoconstructions Based on NA Molecules Ordered as a Result of Enthalpy Condensation

As shown above (see Section I, Chapter 5), polyaminosaccharides, namely, chitosan (Chi, poly-[β (1→4)-2-amino-2-deoxi-D-glucopiranose]; Figure 9.23) attract attention as efficient agents that cause enthalpy condensation. Positively charged amino groups of chitosan efficiently neutralize the negative charges of phosphate groups of polyphosphates and nucleic acids [61,62], which is followed by the phase exclusion of complex molecules and the formation of dispersions with low-soluble particles. The interaction of Chi with low-molecular-mass double-stranded NA (DNA or poly(I) × poly(C)) leads to the formation of double-stranded NA–Chi complex CLCD that have abnormal bands in the CD spectra [63].

There is one Bragg reflection on the curve of small-angle x-ray scattering on a phase obtained as a result of low-speed sedimentation of dispersion CLCD particles (d_{Bragg} ~ 2.6 nm) [64]. The value of the Bragg reflection that describes the average distance between adjacent DNA–Chi complex molecules in the formed phase is significantly lower than the d_{Bragg} value typical of the particles of pure DNA CLCD and corresponds to the transitional area between the cholesteric and the hexagonal LC phases [38].

A number of properties of Chi molecules within DNA–Chi complexes are interesting from the nanodesign point of view. First, because of the steric structure of Chi molecules, only a certain part of the amino groups of Chi sugar residues' molecules interact with DNA in such a way that they neutralize the negative charges of DNA phosphate groups, while the other part of the amino groups is exposed to the external medium [61]. Second, the neighboring free amino and hydroxyl groups in Chi molecules can form strong chelate complexes with metal ions [65–67]. Meanwhile, the binding constant of a bivalent copper ion to chitosan may reach up to 10^{14} M [68–72], and the complex itself has a

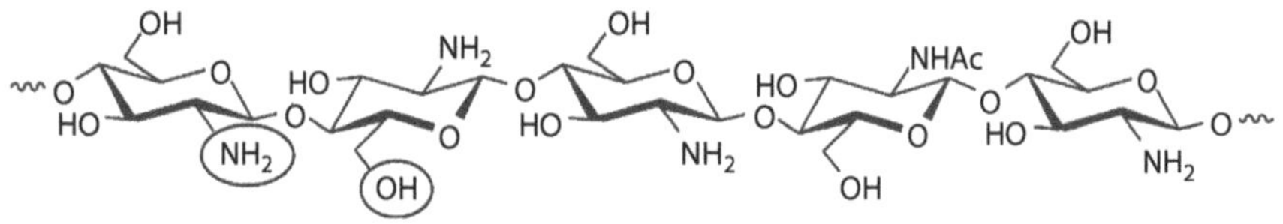

FIGURE 9.23 The structural formula of a fragment of a chitosan molecule (the chemical groups able to form chelate complexes with metal ions are marked).

planar structure [65]. Such complex can be the "start" or the "finish" site for a nanobridge that links not only DNA molecules but also Chi molecules fixed on the DNA surface, in this case. Finally, the abnormal band in the CD spectrum of CLCD formed by the DNA–Chi complex that has a negative or a positive sign, depending on a number of factors [63], allows detection of the changes in the CLCD particle's spatial structure.

Having data describing the properties of nanoconstructions based on double-stranded DNA molecules, and considering the properties of CLCD particles of DNA–Chi complexes, the first attempt to create nanobridges between adjacent Chi molecules fixed in the structure of CLCD of the DNA–Chi complex was undertaken. The received results can be summed up in the following way:

1. Regardless of the direction of the spatial twist of DNA–Chi complex molecules in CLCD particles, Chi molecules can be cross-linked by $[DAU–Cu^{2+}]_n$ complexes.

   Chi molecules interact with DNA in such a way as to allow amino groups of Chi sugar residues not only neutralize the negative charges of DNA phosphate groups but also form a specific distribution of positively charged amino groups near the DNA surface [63]. Since neighboring free amino and hydroxyl groups in Chi molecules effectively form chelate complexes with copper ions [67–72], to perform cross-linking of neighboring Chi molecules fixed in the structure of CLCD particles formed by DNA–Chi, the same approach that had been used to obtain nanobridges between pure DNA molecules was applied.

   A CLCD with an abnormal band in the CD spectrum was formed from DNA molecules via their interaction with Chi; then DAU (or $CuCl_2$) solution was added to it; and, finally, processing in a $CuCl_2$ (or DAU) solution was performed. Figure 9.24 shows that addition of $CuCl_2$ to a (DNA–Chi) CLCD (this dispersion exhibits an abnormal positive band in the CD spectrum) and its subsequent treatment with DAU leads to manifold amplification of a positive band at $\lambda \sim 500$ nm, which corresponds to the optical properties of the chromophore of the $[DAU–Cu^{2+}]$ complex.

   The increase in the amplitude of the abnormal band (amplification) is observed also at a treatment of the CLCD formed by other DNA–Chi complexes (in which Chi molecules have another content of amino groups) and characterized by an intense negative band in the CD spectrum.

   Amplification of the band in the absorption region of the $[DAU–Cu^{2+}]$ chromophore reflects, according to the theory, the spatial fixation of this chromophore with respect to Chi molecules in any CLCD formed by various DNA-Chi complexes.

   Meanwhile, the curves that reflect the dependence of the amplitudes of bands (at $\lambda \sim 310$ nm and $\lambda \sim 500$ nm) in the CD spectra of CLCD of different DNA–Chi complexes on the concentration of DAU and $CuCl_2$ are exponential, not S-shaped (as in the case of DNA CLCD).

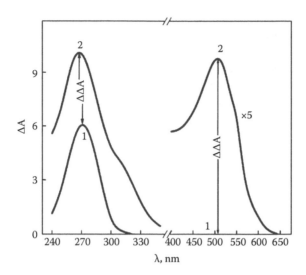

FIGURE 9.24 The CD UV- and visible spectra of a CLCD of the DNA–Chitosan complex that was treated with $CuCl_2$ (1) and then with daunomycin (2). $C_{DNA} \sim 15.5$ μg/mL; Chi sample (5 kDa) contains 65% of NH_2 groups; $C_{DAU} \times 10^{-6}$ M: 0 (1); 59.7 (2); $C_{Cu}^{2+} \times 10^{-6}$ M: 0 (1); 15 (2); 0.05 M NaCl; 0.02 M Na^+-phosphate buffer, pH 6.7; ΔA in optical units ($\times 10^{-3}$).

This indicates the availability of reactive chemical groups in Chi molecules in the structure of the (DNA–Chi) complex for the formation of complexes like $[DAU–Cu^{2+}]_n$.

The important difference of a (DNA–Chi) CLCD from a pure DNA CLCD is that the order of addition of DAU and $CuCl_2$ solutions does not affect the process of the amplification of the CD band.

2. The qualitative coincidence of the received data (Figure 9.24) to the results of the formation of nanobridges in the case of the pure DNA (see Figure 9.20) shows that the amplification of the band at $\lambda \sim 500$ nm in the CD spectra of CLCD of different DNA–Chi complexes is related to the formation of nanobridges of the $[-Cu^{2+}-DAU–Cu^{2+}-]$ type between neighboring Chi molecules that are bound into complexes with DNA molecules in CLCD particles.

3. With allowance for the fact that, as in the case of pure DNA, nanobridges are to be formed both between neighboring Chi molecules within the same layer and between neighboring Chi molecules belonging to neighboring layers, the amplification of the band at $\lambda \sim 500$ nm reflects the formation of a spatial NaC. This fact shows that the formation of $[-Cu^{2+}-DAU–Cu^{2+}-]$ nanobridges (and, therefore, nanoconstructions based on the (DNA–Chi) CLCD) has the same optical manifestation in all cases.

4. Taking into account all corrections, one can conclude with reasonable confidence that, in the case of (DNA–Chi) CLCD, the maximal band amplitude is smaller than that characteristic of NaCs formed of pure cholesteric DNA LCDs by a factor of about 3.

Such a difference may be caused by several reasons: (1) a smaller number of nanobridges; (2) a smaller physical size of nanobridges (the distance between molecules in DNA–Chi complexes is smaller than that of the distance between pure DNA molecules); and (3) the difference between the angle of inclination of nanobridges between Chi molecules in (DNA–Chi) complexes in forming CLCDs and the angle of inclination of nanobridges between pure DNA molecules. The latter reason may be related to the conformation of Chi molecules located at neighboring DNA molecules.

5. The amplitude of the abnormal band ($\lambda = 505$ nm) in the CD spectra of CLCDs of NaCs, based on different (DNA–Chi) complexes, almost does not change with an increase in temperature. This fact indicates that the spatial structure of (DNA–Chi) CLCD remains invariable, that is, the thermal stability of the NaCs based on (DNA–Chi) CLCDs significantly exceeds the stability of not only (DNA–Chi)'s (curves 1, 2) but also of the NaCs based on the pure DNA. This result may indicate not only that copper ions form strong chelate complexes with adjacent amino- and hydroxy-groups of Chi sugar residues [59,61] but also that the physical size of the $[DAU–Cu^{2+}]_n$ nanobridge in the case of nanoconstructions based on DNA–Chi complex CLCD can be small (see Figure 9.25).

6. A certain conformation of Chi molecules is needed for nanodesign. This is proved by the dependence of the amplitude of the abnormal band ($\lambda = 505$ nm) in the CD spectrum of CLCD obtained from the DNA–Chi complex processed with DAU and $CuCl_2$ on the fraction of amino groups in Chi molecules (Figure 9.26). This dependence shows that there is a minimal distance between amino groups (about 13Å, which approximately corresponds to the every-other-one distance) at which neighboring nanobridges composed of molecules of (DAU–Cu^{2+}) complexes can be located. In this case, the low abnormal optical activity indicates that, despite of the high concentration of amino- and hydroxyl groups in Chi sugar residues (potential sites of nanobridge formation), either the nanobridges located close to one another do not form because of steric reasons or the spatial orientation of adjacent nanobridges in the formed nanoconstruction is not optimal. That is why maximum optical activity is not realized under these conditions. Only with an increase in the distance between amino groups in Chi molecules can the steric structure of the Chi ensure the orientation of amino- and hydroxyl groups in sugar residues that is appropriate for the formation of optimal nanobridges in the structure of the nanoconstruction formed.

Therefore, comparison of the results given in this section of the chapter to the results on the nanodesign based on pure DNA suggest that nanobridges can be formed between neighboring Chi molecules in (DNA–Chi) complexes that are fixed in the spatial structure of CLCD particles; that is, NaCs contain molecules—not only DNA but "guests" such as Chi, DAU molecules, and copper ions. Meanwhile, the nanoconstruction has properties that are significantly different from the properties of

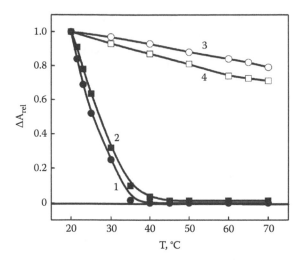

FIGURE 9.25 The temperature dependences of the relative amplitudes of the bands in the CD spectra of CLCDs formed by DNA-Chi (curve 1 and curve 2) complexes (λ = 270 nm) and nanoconstructions based on CLCDs of these complexes (curve 3 and 4, respectively; λ = 505 nm). C_{DNA} ~ 15 μg/mL; 0.05 M NaCl, 0.02 M Na$^+$-phosphate buffer, pH = 6.7. Curve 1—Chi: 65% NH$_2$; 14.6 kDa; C_{Chi} = 28.7 μg/mL; curve 2—Chi: 55% NH$_2$; 5 kDa, C_{Chi} = 28.7 μg/mL. Curve 3—Chi: 65% NH$_2$; 14.6 kDa; C_{Chi} = 28.7 μg/mL; $C_{Cu}^{2+} \times 10^{-6}$, M = 15; $C_{DAU} \times 10^{-6}$, M = 50 (3); Curve 4—Chi: 55% NH$_2$; 5 kDa; C_{Chi} 28.7 μg/mL; $C_{Cu}^{2+} \times 10^{-6}$, M = 15; $C_{DAU} \times 10^{-6}$, M = 50 (3);

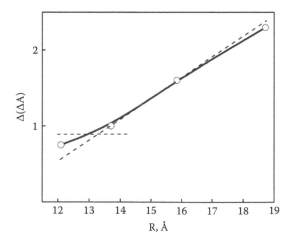

FIGURE 9.26 The dependence of the maximal amplitude of the band at λ = 505 nm in the CD spectrum of nanoconstructions based on CLCDs of the DNA–Chi complexes on the average distance, **R**, between amino groups in Chi molecules. R (Å) = 2×(5.15×100%)/NH$_2$, where NH$_2$ = the fraction of amino groups in the Chi sample. 15.5 μg/mL of DNA; $C_{DAU} \times 10^{-6}$, M = 55.

the construction based on the pure DNA CLCD (i.e., as a result of the described approach, a new type of nanoconstruction has been obtained).

It is obvious that, to form a certain spatial model of $[DAU–Cu^{2+}]_n$ between the adjacent Chi molecules in the DNA–Chi complex CLCD and, consequently, to form a spatial model of a nanoconstruction of this new type, it is necessary to perform additional research.

9.4 SUMMARY

Therefore, it can be stated that, as a result of the performed studies, new methods of immobilization (fixation) of "guest" molecules into CLCD particles of NA molecules have been proposed [73]. These methods can be modified because the nanobridges may contain other integral components and be formed between NA molecules with various polycations located on them (for instance, peptides).

The data in this chapter indicate that double-stranded NA molecules (both DNA and RNA) and complexes of these molecules ordered in the spatial structure of CLCD particles are an important polyfunctional object of nanotechnology [74]. The targeted and regulated change in the properties of these molecules provides broad possibilities to create nanobiostructures with unique properties.

REFERENCES

1. The Royal Society and The Royal Academy of Engineering: Nanoscience and Nanotechnologies: Opportunities and Uncertainties. London, 2004. p. 10.
2. Silva, G.A. Neuroscience nanotechnology: Progress, opportunities and challenges. *Nat. Rev.* 2006, vol. 7, p. 65–74.
3. Poole, Ch.P. and Owens, F.J. *Introduction to Nanotechnology* (Russian Edition). Moscow: Tekhnosfera, 2005. p. 336.
4. Andrievski, R.A. Size effects in nanomaterials: Rules and exceptions. *Proc. 1st Iran-Russia Joint Seminar and Workshop on Nanotechnology*, 2005, May 28–30, p. 1.
5. Roduner, E. Size matters: Why nanomaterials are different. *Chem. Soc. Rev.*, 2006, vol. 35, p. 583.
6. Bailey, R.E. and Nie, S. Alloyed semiconductor quantum dots: Tuning the optical properties without changing the particle size. *J. Amer. Chem. Soc.*, 2003, vol. 125, p. 7100–7106.
7. Efros, A.L. and Efros, A.L. Interband absorption of light in a semiconductor sphere. *Sov. Phys. Semicond.*, 1982, vol. 16, p. 772–775.
8. Ekimov, A.I. and Onushchenko, A.A. Quantum size effect in the optical-spectra of semiconductor microcrystals. *Sov. Phys. Semicond.*, 1982, vol. 16, p. 775–778.
9. Gusev, A.I. *Nanomaterials, Nanostructures, Nanotechnologies* (Russian Edition). Moscow: Fizmatlit, 2005. p. 416.
10. Porschke, D. Cooperative nonenzymic base recognition. II. Thermodynamics of the helix coil transition of oligoadenylic + oligouridylic acids. *Biopolymers*, 1971, vol. 10, p. 1989–2013.
11. Yevdokimov, Yu.M. and Sytchev, V.V. Nanotechnology and nucleic acids. *Technologies of Living Systems* (Russian Edition), 2007, vol. 4, p. 3–27.
12. Yevdokimov, Yu.M., Zakharov, M.A., and Skuridin, S.G. Nanotechnology based on nucleic acids. *Herald of the Russian Acad. Sci.* (Russian Edition), 2006, vol. 76, p. 112–120.

13. Report of the OECD workshop on the safety of manufactured nanomaterials. Washington DC, 2005, December 7–9, p. 149.
14. Report of the OECD workshop on the safety of manufactured nanomaterials. Current development/activities on the safety of manufactured nanomaterials. Berlin, 2007, April 25–27, p. 77.
15. Robichaud, C.O., Tanzil, D., Weilenmann, U., et al. Relative risk analysis of several manufactured nanomaterials: An insurance industry context. *Environ. Sci. Technol.*, 2005, vol. 39, p. 8985–8994.
16. *Nanotechnology Research Directions: Vision for Nanotechnology in the Next Decade.* M.C. Roco, Jr., R.S. Williams, P. Alivisatos, Eds. (Russian Edition). Moscow: Mir, 2002. p. 292.
17. Yevdokimov, Yu.M., Skuridin, S.G., Nechipurenko, Yu.D., et al. Nanoconstructions based on double-stranded nucleic acids. *Int. J. Biol. Macromol.*, 2005, vol. 36, p. 103–115.
18. Seeman, N.C. Nucleic acid junctions and lattices. *J. Theor. Biol.*, 1982, vol. 99, p. 237–247.
19. Seeman, N.C. Nanotechnology and the double helix. *Sci. Amer.*, 2004, vol. 6, p. 64–69.
20. Niemeyer, C.M., Takeshi, S., and Smith, C.L. Oligonucleotide-directed self-assembly of proteins: Semisynthetic DNA-streptavidin hybrid molecules as connectors for the generation of macroscopic arrays and the construction of supramolecular bioconjugates. *Nucl. Acids Res.*, 1994, vol. 22, p. 5530–5539.
21. Niemeyer, C.M., Adler, M., Gao, S., et al. Supramolecular nanocircles consisting of streptavidin and DNA. *Angew. Chem. Int. Ed.*, 2000, vol. 39, p. 3056–3059.
22. Shin, J. and Bergstrom, D.E. Assembly of novel DNA cycles with rigid tetrahedral linkers. *Angew. Chem. Int. Ed.*, 1997, vol. 36, p. 111–113.
23. Mirkin, C.A., Mucic, R.C., and Storhoff, L.J. A DNA based method for rationally assembling nanoparticles into macroscopic materials. *Nature*, 1996, vol. 382, p. 607–609.
24. Letsinger, R.L., Elghanian, R., Viswanadham, G., et al. Use of a steroid cyclic disulfide anchor in constructing gold nanoparticle-oligonucleotide conjugates. *Bioconjugate Chem.*, 2000, vol. 11, p. 289–291.
25. Alivisatos, A.P., Johnsson, K.P., Peng, X., et al. Organization of "nanocrystal molecules" using DNA. *Nature*, 1996, vol. 382, p. 609–611.
26. Dillenback, L.M., Goodrich, G.P., and Keating, C.D. Temperature-programmed assembly of DNA: Au nanoparticles bioconjugates. *Nano Lett.*, 2006, vol. 6, p. 16–23.
27. Williams, K.A., Veenhuizen, P.T.M., de la Torre, B.G., et al. Carbon nanotubes with DNA recognition. *Nature*, 2002, vol. 420, p. 761.
28. Katz, E. and Willner, I. Integrated nanoparticle biomolecular hybrid systems: Synthesis, properties and applications. *Angew. Chem. Int. Ed.*, 2004, vol. 43, p. 6042–6108.
29. Star, A., Tu, E., Niemann, J., et al. Label-free detection of DNA hybridization using carbon nanotube network field-effect transistors. *Proc. Natl. Acad. Sci. USA*, 2006, vol. 103, p. 921–926.
30. Hahm, J. and Lieber, C. Direct ultrasensitive electrical detection of DNA and DNA sequence variation using nanowire nanosensors. *Nano Lett.*, 2004, vol. 4, p. 51–54.
31. Tang, X., Bansaruntip, S., Nakayama, N., et al. Carbon nanotube DNA sensor and sensing mechanism. *Nano Lett.*, 2006, vol. 6, p. 1632–1636.
32. Manzanera, M., Frankel, D.J., Li, H., et al. Macroscopic 2D networks self-assembled from nanometer-sized protein/DNA complexes. *Nano Lett.*, 2006, vol. 6, p. 365–370.
33. Di Mauro, E. and Hollenberg, C.P. DNA technology in chip construction. *Adv. Mater.*, 1993, vol. 5, p. 384–386.
34. Yevdokimov, Yu.M. Spatial ordering forms of DNA and its complexes as a basis for creation of nanoconstructons for medicine and biotechnology. *Russian Nanotechnologies* (Russian Edition), 2006, vol. 1, p. 244–252.

35. Yevdokimov, Yu.M., Zakharov, M.A., and Salyanov, V.I. Liquid-crystalline dispersions of nucleic acids and their complexes as a basis for nanodesign. *Crystallography Rep.* (Russian Edition), 2006, vol. 51, p. 1221–1227.
36. Onsager, L. The effects of shape on the interaction of colloidal particles. *Ann. N.Y. Acad. Sci.*, 1949, vol. 51, p. 627–659.
37. Lerman, L.S. Intercalability, the ψ transition, and the state of DNA in Nature. *Proc. Mol. Biol. Subcell. Biol.*, 1971, vol. 2, p. 382–391.
38. Yevdokimov, Yu.M., Skuridin, S.G., and Lortkipanidze, G.B. Liquid-crystalline dispersions of nucleic acids. *Liq. Crystals*, 1992, vol. 12, p. 1–16.
39. Yevdokimov, Yu.M. Liquid-crystalline forms of nucleic acids and their biological role. *Liquid Crystals and Their Application* (Russian Edition), 2003, vol. 3, p. 10–47.
40. Livolant, F. and Leforestier, A. Condensed phases of DNA: Structures and phase transitions. *Prog. Polym. Sci.*, 1996, vol. 21, p. 1115–1164.
41. Belyakov, V.A., Orlov, V.P., Semenov, S.V., et al. Comparison of calculated and observed CD spectra of liquid crystalline dispersions formed from double-stranded DNA and from DNA complexes with coloured compounds. *Liq. Crystals*, 1996, vol. 20, p. 777–784.
42. Rouzina, I. and Bloomfield, V.A. Macroion attraction due to electrostatic correlation between screening counterions. I. Mobile surface-adsorbed ions and diffuse ion cloud. *J. Phys. Chem.*, 1996, vol. 100, p. 9977–9989.
43. Allahyarov, E., Gompper, G., and Lowen, H. DNA condensation and redissolution: Interaction between overcharged DNA molecules. *J. Phys. Condens. Matter*, 2005, vol. 17, p. 1827–1840.
44. Manning, G.S. The molecular theory of polyelectrolyte solutions with application to the electrostatic properties of polynucleotides. *Quart. Rev. Biophys.*, 1978, vol. 11, p. 179–246.
45. Kornyshev, A.A. and Leikin, S. Theory of interaction between helical molecules. *J. Chem. Phys.*, 1997, vol. 107, p. 3656–3674.
46. Kornyshev, A.A., Leikin, S., and Malinin, S.V. Chiral electrostatic interaction and cholesteric liquid crystals of DNA. *Eur. Phys. J.*, 2002, E 7, p. 83–93.
47. Saminathan, M., Antony, A., Shirahata, A., et al. Ionic and structural specificity effects of natural and synthetic polyamines on the aggregation and resolubilization of single-, double-, and triple-stranded DNA. *Biochemistry*, 1999, vol. 38, p. 3821–3830.
48. Kabanov, V.A., Sergeyev, V.G., Pyshkina, O.A., et al. Interpolyelectrolyte complexes formed by DNA and astramol poly(propylene imine) dendrimers. *Macromolecules*, 2000, vol. 33, p. 6587–9563.
49. Pelta, J., Durand, D., Doucet, J., et al. DNA mesophases induced by spermidine: Structural properties and biological implications. *Biophys. J.*, 1996, vol. 71, p. 48–63.
50. Skuridin, S.G., Dembo, A.T., Efimov, V.S., and Yevdokimov Yu.M. Liquid-crystalline phases formed by complexes of DNA with synthetic polycations. *Doklady of the USSR Acad. Sci.* (Russian Edition), 1999, vol. 365, p. 400–402.
51. Yevdokimov, Yu.M., Salyanov, V.I., and Skuridin, S.G. Molecular design for amplification of optical signal generated by DNA liquid-crystalline dispersion. *Doklady of the USSR Acad. Sci.* (Russian Edition), 1994, vol. 338, p. 827–829.
52. Yevdokimov, Yu.M., Salyanov, V.I., Buligin, L.V., et al. Liquid-crystalline structure of nucleic acids: Effect of antracycline drugs and copper. *J. Biomol. Struct. Dynamics*, 1997, vol. 15, p. 97–105.
53. Yevdokimov, Yu.M., Salyanov, V.I., Nechipurenko, Yu.D., et al. Molecular constructions (superstructures) with adjustable properties based on double-stranded nucleic acids. *Mol. Biol.* (Russian Edition), 2003, vol. 37, p. 340–355.

54. Zakharov, M.A., Sokolovskaya, L.G., Nechipurenko, Yu.D., et al. Formation of nano-constructions based on double-stranded DNA. *Biophysics* (Russian Edition), 2005, vol. 50, pp. 824–832.

55. Nikiforov, V.N., Kuznetsov, V.D., Nechipurenko, Yu.D., et al. Magnetic properties of cooper as a constituent of nanobridges formed between spatially fixed deoxyribonucleic acid molecules. *JETP Lett.*, 2005, vol. 81, p. 264–266.

56. Nechipurenko, Yu.D., Ryabokon, V.F., Semenov, S.V., and Evdokimov, Yu.M. Thermodynamic models describing the formation of "bridges" between molecules of nucleic acids in liquid crystals. *Biophysics* (Russian Edition), 2003, vol. 48, p. 635–643.

57. Zakharov, M.A., Nechipurenko, Yu.D., Lortkipanidze, G.B., and Yevdokimov, Yu.M. The thermodynamic stability of nanoconstructions based on the double-stranded DNA molecules. *Biophysics* (Russian Edition), 2005, vol. 50, p. 1036–1041.

58. Yevdokimov Yu.M., Salyanov, V.I., Gedig, E., et al. Formation of polymeric chelate bridges between double-stranded DNA molecules fixed in spatial structure of liquid-crystalline dispersions. *FEBS Lett.*, 1996, vol. 392, p. 269–273.

59. Yevdokimov, Yu.M. and Sytchev, V.V. Nanotechnology and nucleic acids. *The Open Nanosci. J.*, 2007, vol. 1, p. 19–31.

60. Golo, V.L., Kats, E.I., Volkov, Yu.S., et al. Novel cholesteric phase in dispersions of nucleic acids due to polymeric chelate bridges. *J. Biol. Phys.*, 2001, vol. 27, p. 81–93.

61. Hayatsu, H., Kobo, T., Tanaka, Y., et al. Polynucleotide-chitosan complex an insoluble but reactive form of polynucleotide. *Chem. Pharm. Bull.*, 1992, vol. 45, p. 1363–1368.

62. Danielsen, S., Varum, K., and Stokke, B.T. Influence of chitosan molecular parameters on the structure of DNA-chitosan complexes. *The 5th Int. Conf. Europ. Chitin Soc.* Trondheim, Norway, 2002, June 26–28, p. D6.

63. Yevdokimov, Yu.M. and Salyanov, V.I. Liquid crystalline dispersions of complexes formed by chitosan with double-stranded nucleic acids. *Liq. Crystals*, 2003, vol. 30, p. 1057–1074.

64. Yevdokimov, Yu.M., Salyanov, V.I., Skuridin, S.G., and Dembo, A.T. Some x-ray parameters of liquid-crystalline dispersions of nucleic acid-chitosan complexes. *Mol. Biol.* (Russian Edition), 2002, vol. 36, p. 706–714.

65. Schlick, S. Binding sites of Cu^{2+} in chitin and chitosan: An electron spin resonance study. *Macromolecules*, 1986, vol. 19, p. 192–195.

66. Braier, N.C. and Jishi, R.A. Density functional studies of Cu^{2+} and Ni^{2+} binding to chitosan. *J. Mol. Struct. (Thermochem.)*, 2000, vol. 499, p. 51–55.

67. Lee, S., Mi, F., Shen, Y., and Shyu, S. Equilibrium and kinetic studies of copper (II) ion uptake by chitosan-tripolyphosphate chelating resin. *Polymer*, 2001, vol. 42, p. 1879–1892.

68. Domard, A. pH and C.D. measurements on a fully deacetylated chitosan: Application to Cu^{II}-polymer interaction. *Int. J. Biol. Macromol.*, 1987, vol. 9. p. 98–104.

69. Monteiro, O.A.C. and Arnoldi, C. Some thermodynamic data on copper-chitin and copper-chitosan biopolymer interactions. *J. Coll. Interface Sci.*, 1999, vol. 212, p. 212–219.

70. Inoue, K., Baba, Y., and Yoshizuka, K. Adsorption of metal ions on chitosan and crosslinked copper(II)-complexed chitosan. *Bull. Chem. Soc. Jpn.*, 1993, vol. 66, p. 2915–2921.

71. Lima, I.S. and Airoldi, C.A. A thermodynamic investigation on chitosan-divalent cation interaction. *Thermochim. Acta*. 2004, vol. 421, p. 125–131.

72. Rhazi, M., Derbrieres, J., Talaimate, E., et al. Contribution to the study of the complexation of copper by chitosan and oligomers. *Polymer*, 2002, vol. 43, p. 1267–1276.

73. Yevdokimov, Yu.M., Salyanov, V.I., Kondrashina, O.V., et al. Particles of liquid-crystalline dispersions formed by (nucleic acid-rare earth element) complexes as a potential platform for neutron capture therapy. *Int. J. Biol. Macromol.*, 2005, vol. 37, p. 165–173.

74. Skuridin, S.G. and Yevdokimov, Yu.M. DNA liquid-crystalline dispersion particles as a basis for sensitive biosensor elements. *Biophysics* (Russian Edition), 2004, vol. 49, p. 468–485.

10 Biosensors Based on Nucleic Acids

10.1 GENERAL CONCEPT OF CONSTRUCTION AND OPERATION OF BIOSENSORS

The needs of modern health care, biomedicine, ecology, and the food industry determine the wide range of chemical and biologically active compounds (BAC) we need to research, from simple substances (such as salts of heavy metals and glucose) to antibiotics, antitumor drugs, enzymes, mutagens, poisons, toxins, and so forth, with the help of new, inexpensive, and specific diagnostic test systems within the framework of a new rapidly developing area of science called *analytical biotechnology* [1]. One of the achievements of analytical biotechnology is the creation of biosensors [2]. The concept of biosensors was first put forward by L. Clark and C. Lyons in 1962 [3]. Clark, an American scientist working at a clinical laboratory, decided to integrate a highly specific enzymatic system with a sensitive electrochemical transducer in a simple device called an "enzyme electrode" [3].

Figure 10.1 represents the scheme of the basic structure of biosensors known today. According to the scheme, a biosensor is an analytical device that contains a sensitive biological element connected to a transducer that transforms the received "primary" signal appearing in the system into a quantitative "secondary" signal whose value is to be proportional to the concentration of the substance to be analyzed [4].

The most important element of any biosensor is the biological sensitive element that is called a "biospecific surface," a "biodetector," or a "biosensing unit." A biodetector is generally an ensemble of biological molecules that reflects, as a result of molecular "recognition," the properties of the analyzed medium as a specific "primary" signal. The sensitivity of a biosensor depends on a number of parameters, such as the density of the biological molecules on the surface of the biodetector, the extent of amplification, and the efficiency of processing of the "primary" signal that appears in the system. This means that the sensitivity of a biosensor and the specificity of its work are determined by physicochemical processes that take place on the molecular level between both the molecules of the biodetector and the molecules of the analyzed substance.

Consequently, *the operative principle of any biosensor is based on the process of molecular "recognition,"* that is, the process that takes place on the level of single molecules that results in both the formation of a complex between the molecules with new properties and the appearance of a signal that reflects the formation of this complex.

Various objects that can be used as sensitive elements and the transducers that have been used in biosensors are given in Table 10.1.

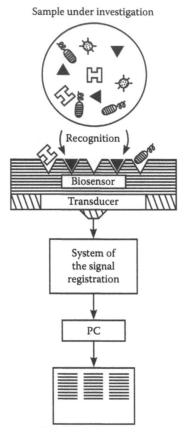

Sample under investigation

Digital information about properties
of the sample under investigation

FIGURE 10.1 Basic biosensor construction.

Biosensors allow one to detect the presence and concentration of compounds that are interesting from a practical point of view. This is especially important, for example, for ecology, health care, and veterinary medicine because biosensors open a gate to perform the ecological diagnosing of the environment and the monitoring of the health level of people and domestic animals in real time outside centralized laboratories.

The advantages of biosensors are the following:

1. The relatively high specificity of the analysis, which eliminates the preliminary sample preparation.
2. The ability to analyze a small volume sample and perform the analysis in real-time mode and within a short time of the analysis.
3. The ability to control the results of the analysis in the feedback mode, which is provided by the compatibility of biosensors with microprocessors.

TABLE 10.1

Some Potential Sensor Elements and Transducers That Can Be Used for Biosensor Creation

Biological Component	Transducer
Intact organisms	Potentiometric
Tissues	Amperometric
Organelles	Conductometric
Membranes	Impedance
Enzymes	Optical
Enzyme components	Calorimetric
Receptors	Acoustic
Antibodies	Mechanical
Nucleic acids	Chemical
Organic molecules	

4. The absence of requirements for the skills of personnel dealing with analysis, which is determined by the simplicity of the analysis.
5. The relatively low cost of the biomaterial used as the sensitive element in the biosensor.

As the operation of any biosensor is based on the process of molecular "recognition" that takes place between single molecules and is accompanied by physicochemical changes in the system, two basic functional types of biosensors can be distinguished (Figure 10.2) [1].

The first type is *affinity biosensors* (Figure 10.2a). In this case, biological molecules (receptors) of the biodetector recognize molecules of the compounds (ligands) present in the analyzed sample, discriminate these molecules, and form a tight complex with only one type of molecules. The molecular "recognition" and formation of a complex between the ligands and the receptors, resulting in complex formation between these components, are the basic processes at the molecular level occurring in this type of biosensors. "Primary" signals are provided by changes in shape, mass, color, or surface properties of the biodetector, or change in the heat content (release or absorption of heat), occurring as a result of complex formation. The most characteristic signal is emitted to an appropriate transducer that produces an amplified "secondary" signal such that the magnitude of this "secondary" signal is proportional to the concentration of the substance being analyzed.

The second type is *enzymatic biosensors* (Figure 10.2b). In this case, molecules of an enzyme in the biodetector recognize the molecules of the desired substrate present in the analyzed solution. Products (hydrogen or hydroxyl ions, etc.) appearing because of an enzymatic reaction represent the "primary" signals in the system. Again, as in the case of the first type of biosensors, the most characteristic signal is

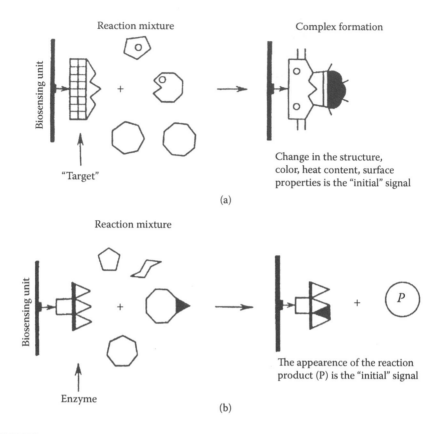

FIGURE 10.2   Scheme of operation of various types of biosensors.

detected by the corresponding transducer, amplified, and emitted as a "secondary" signal whose size is proportional to the concentration of the substrate of interest.

The enzymatic scheme of the oxidation of glucose can be represented in the following way:

$$\beta\text{-glucose} + \tfrac{1}{2}\overset{a}{O_2} + H_2O \xrightarrow{\text{enzyme } \beta\text{-glucose oxidase}} D\text{-glucono-}\delta\text{-lactone} +$$

$$+ H_2O_2 + \overset{b}{\text{release of heat}} \xrightarrow{c} D\text{-gluconic acid} + \overset{d}{H^+}$$

The scheme illustrates a number of possibilities, as indicated by letters **(a)–(d)**, which track the different parameters of the "primary" signals related to glucose concentration. For instance, signal **(a)** allows changes in oxygen concentration to be followed; **(b)** allows changes in hydrogen peroxide to be followed by colorimetry, amperometry, or the change in luminescence; heat release at **(c)** can be detected calorimetrically; finally, the appearance of hydrogen ions at **(d)** can be determined using a pH-electrode. Thus, glucose concentration can be determined using biosensors based on the analysis of a variety of "primary" signals.

Therefore, before information about the properties of the medium is obtained, and particularly the concentration of the substance being analyzed, a number of sequential events must occur in the system; the "primary" (chemical or physical) signal, which appeared as a result of molecular "recognition," is transformed by the transducer into the "secondary" signal, which is generally an electrical signal.

The specificity of a biosensor is determined by the efficiency of the reaction of the molecular "recognition" that takes place on the nanolevel, that is, on the level of single molecules forming the biodetector and the molecules of the analyzed substance. Molecular "recognition" is typical of various molecules. It is obvious that the more specific the molecular "recognition" reaction, the grater the sensitivity of a biosensor.

Biosensors using molecular "recognition" of the "receptor-ligand" or enzyme electrodes, which could in principle detect compounds with similar properties, are suitable for the rapid detection of whole groups of related substances. This is the so-called group determination, which is convenient for detecting groups of substances without particular structural specification. The sensitivity reached at group determination is within the range 10^{-7} to 10^{-15} M.

Immune biosensors based on recognition of the corresponding antigens with the molecules of monoclonal antibodies have the highest specificity. The result of high specificity is that concentrations of merely 10^{-21} M can be detected.

To increase the sensitivity of a biosensor, various means of amplification of the "primary" signal generated on the biodetector as a result of the molecular "recognition" reaction are used. In the case of enzyme electrodes, the problem has been solved by using the specific molecular peculiarities of enzyme molecules as biocatalysts, and the method of amplification of the "primary" signal is called *biocatalytic amplification* [1].

Therefore, creation of biosensors involves the decision of two problems, which are related to different fields of science. First, it is fabrication of a specific biosensing unit (biodetector) in which the molecular "recognition" of analytes is used at the highest efficiency. Such a problem can be solved within the framework of biological sciences. Second, it is the creation of an adequate recording scheme for determining the signal that appears in the system and generated by the biodetector. This problem is to be solved within the framework of technical sciences and the newest technologies.

Table 10.1 shows that, in principle, any biological structure, biological molecule (biological material), or (bio)chemical reaction can be used to fabricate biosensing units. Nevertheless, the analysis of the literature shows that most of the biosensors offered at the current moment are based on enzymes [5–8]. Regarding nucleic acids, in this case there are biosensor devices using fragments of single-stranded DNA and synthetic single-stranded polynucleotides with known nucleotide combinations as sensing units [2,9–14]. These biosensors are known as "DNA-tests" or "DNA-probes" [2] or "biological microchips (biochips)" [12,13].

The fundamental concept underlying the operation of biochips based on single-stranded DNA and RNA molecules is quite simple: the sequence of nitrogen bases of one strand of the nucleic acid "recognizes" the complementary sequence of bases on the analyzed chain and forms a hybrid complex with these bases (Figure 10.3).

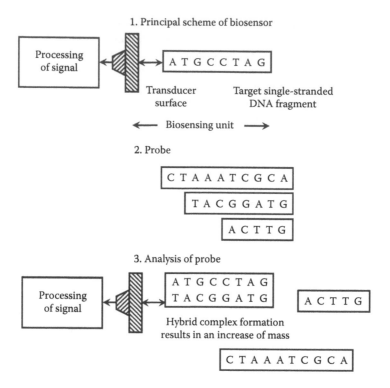

FIGURE 10.3 Generalized diagram showing the construction and operation of a biochip based on single-stranded DNA (RNA).

How to record the formation of a hybrid complex between two chains of nucleic acid molecules in a simple and efficient way has been solved [12,13,15].

DNA probes are used in medical genetics to detect DNA fragments associated with hereditary diseases such as sickle-cell anemia, phenylketonuria, and so on [2,15], as well as to analyze unknown nucleotide sequences [12]. Significant achievements were made in using DNA probes for detecting pathogenic viruses and microorganisms (bacteria, protozoa) in human and animal physiological fluids [16] and also in food [2] and the environment [17].

These foregoing facts mean that the creation of biosensors is an area of nanotechnology that employs the achievements of molecular biology and synthetic chemistry, as well as microelectronics.

10.2 DOUBLE-STRANDED DNA MOLECULE AS POLYFUNCTIONAL BIOSENSING UNIT

The structure of double-stranded DNA molecules allows one to outline a number of peculiarities that should be considered in the design of biosensors based on these molecules [1].

1. Right-handed double-stranded DNA molecules exist in different spatial forms (linear [A, B, and Z], linear, circular, superhelical, etc.) that vary noticeably in their physicochemical characteristics. A transition between these forms is induced by changes in the surrounding conditions.

2. Unlike single-stranded nucleic acids, double-stranded DNA molecules whose negatively charged phosphate groups are neutralized with counterions are rigid.

3. As a result of their own anisotropy, double-stranded DNA molecules are optically active, which appears, in particular, as a low-intensity band in the CD spectra typical of DNA molecules with different directions of helical twists or different parameters of their secondary structure.

4. Every polynucleotide chain of a double-stranded DNA molecule contains nitrogen bases (adenine, thymine, guanine, and cytosine)—chromophores that absorb in the UV region of the spectrum ($\lambda_{max} \sim 260$ nm).

5. Each chain of a nucleic acid molecule contains many chemical groups that differ in their reactivity.

6. The spatial helical structure of double-stranded DNA molecules in combination with the specific distribution of reactive groups on the surface of these molecules provides the specific "addressing" of various chemical substances and BAC that interact with nitrogen bases of nucleic acids (Figure 10.4).

The interaction of chemical substances and BAC with double-stranded DNA molecules results in different structural changes, such as, for instance, the following: increase in distance between nitrogen base pairs (Figure 10.4 [1]); formation of cross-links between adjacent nitrogen bases belonging to the same or different polynucleotide strands of the DNA molecule (Figure 10.4 [2]); cleavage of phosphodiester bonds, which disturbs the integrity of the sugar-phosphate backbone of one or both of the strands of the DNA molecule (Figure10.4 [3]); local or complete denaturation,

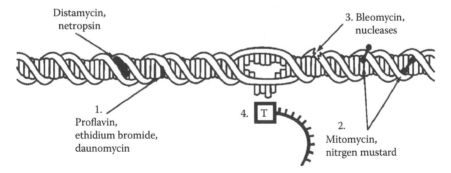

FIGURE 10.4 Modes of structural "addressing" of biologically active compounds by double-stranded nucleic acid molecules: (1) intercalation of BAC molecule between nitrogen base pairs; (2) interaction of BAC with DNA causes the formation of cross-links; (3) interaction of BAC with DNA results in the splitting of the sugar-phosphate chain; (4) interaction between BAC and DNA results in chain separation.

which is accompanied by the rupture of hydrogen bonds between nitrogen base pairs; and the unwinding of some regions of the DNA molecule or the complete separation of the polynucleotide strands (Figure 10.4 [4]).

The scheme in Figure 10.4 shows that chemical substances and BAC not only form significantly different types of complexes with DNA but also cause different changes in the physicochemical properties of the molecule. Consequently, a double-stranded DNA molecule by itself is a polyfunctional sensing unit capable of detecting chemical substances and/or BAC that interact with the DNA molecules, this or that way.

It is necessary to add that the basic concept of operation of biosensing units, which are based on double-stranded DNA or RNA molecules, differs strongly from the classical "hybridization" ideology of operation of biochips based on single-stranded fragments of nucleic acids and synthetic oligonucleotides.

Therefore, in contrast to proteins (enzymes), double-stranded DNA molecules have several attractive peculiarities. A combination of these properties opens up the possibility to use different principles of molecular "recognition" for the detection of compounds that interact with the molecules of this biopolymer.

The principal question is: "How is this potential ability of double-stranded nucleic acids to be realized?"

To design a biosensing unit that fully employs the ability of double-stranded nucleic acids to recognize and specifically "address" various chemical substances or BAC, it is convenient to make use of the particles of DNA CLCD that have physicochemical properties caused by their structural organization, which are important from an analytical biotechnology point of view. Because of very specific physicochemical properties of CLCDs, one can realize different versions of "recognition" in such biosensing units.

Within the framework of the concept of double-stranded DNA as a polyfunctional biosensing unit, various constructions of microscopic-size biosensing units based on the molecules of this biopolymer will be described in the following text. These constructions take into account both the structural and chemical capabilities of double-stranded DNA molecules and specific optical properties of DNA CLCD particles. The abnormal band in the CD spectrum of DNA CLCD, as shown in Chapter 4, Section I, was used as an analytical criterion to evaluate the extent of change in the DNA structure under the influence of BAC.

The existence of such criterion combined with the high lability of physicochemical properties of DNA as a response to the action of various factors allows one to use DNA CLCD particles as a background for microscopic-size sensing units.

A portable dichrometer was designed and fabricated at the Institute Spectroscopy of the Russian Academy of Science (RAS) to realize the potential capabilities of DNA CLCD particles as biosensing units (Figure 10.5; [18]).

10.3   CONTENT AND PRINCIPLE OF OPERATION OF AN OPTICAL BIOSENSOR BASED ON DNA LIQUID-CRYSTALLINE DISPERSIONS

An optical biosensor includes biosensing units based on DNA CLCD particles integrated to a portable dichrometer, as well as methods of detection of various classes

FIGURE 10.5 (*See color insert.*) Portable dichrometer capable of working with biosensing units based on cholesteric DNA liquid-crystalline structures.

TABLE 10.2
Some Characteristics of SCD-2 Portable Dichrometer SCD-2

Spectral range, nm	250 (200) ... 750
Spectral resolution, nm	3
Time wavelength change between limit points in spectral range, s	0.5
Probe temperature range, °C	12 ... 80
Size, mm	$500 \times 330 \times 170$
Weight, kg	14
Minimal detectable CD signal, ΔA/A	3×10^{-7}

of BAC. The most important characteristics of a portable dichrometer are shown in Table 10.2.

The control of the optical biosensor is performed with the help of a software package that makes it possible to use various modes of operation of the dichrometer and supports the library of detection methods for various classes of BAC so that the user can choose the required substance for analysis from the menu. Then, the program guides the user through the sequence of actions including the method of detection in the dialogue mode, and delivers the result in a printed form that contains information on the time and the place of the analysis, as well as type of detected substance and its fraction in the analyzed probe and in a sample.

The advantage of this biosensor is the ability to directly detect the presence and concentration of a wide range of various compounds with the help of similar biosensing units; relatively high sensitivity and selectivity in a short time and with low cost of analysis; compactness; mobility; absence of the necessity to use water and nitrogen; and the possibility of employing personnel without special qualifications.

The strategy of detecting chemical substances and BAC with the help of circular dichroism spectroscopy [19] and examples of detection of various BAC classes with the help of an optical biosensor, including constructions and operation mechanisms of the biosensors based on CLCD particles [20], are briefly described in the following text.

10.4 DNA CLCD PARTICLES AS SENSING UNITS

The high flexibility of the physicochemical properties of the DNA CLCDs with respect to different factors (Section I, Chapter 4) confirm that there are two approaches to the detection of chemical substances and BAC that cause changes in the secondary structure of double-stranded DNA molecules [21,22]. Both approaches follow from the analysis of Figure 4.6 (Section I, Chapter 4) where the formation of CLCD particles from double-stranded DNA is shown.

Figure 10.6, based on Figure 4.6, shows the structure of a quasi-nematic layer formed by double-stranded DNA molecules. At the formation of CLCDs, the chemical reactivity of structural elements (nitrogen bases, etc.) of DNA molecules remains unchanged. One can see as well that because of the "liquid" mode of DNA molecules' packing in a quasi-nematic layer, low-molecular-mass chemical or biologically active compounds (antitumor drugs, peptides, chemicals, metal cations, etc.) can readily diffuse into particles of CLCD. In addition, the conservation of the double-stranded (ds) DNA reactivity in the composition of CLCD particles, and the ordered location of neighboring DNA molecules in these particles, which does not limit the diffusion of various drugs into the particles, can ensure a high penetration rate of these compounds into the particles and, therefore, a high rate of their interaction with nucleic acid molecules.

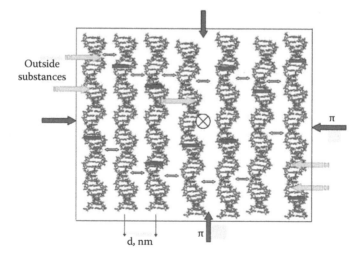

FIGURE 10.6 (*See color insert.*) Structure of a quasi-nematic layer formed by double-stranded DNA molecules.

According to the first approach, the detection of BAC is based on the fact that the mode of packing of double-stranded DNA in CLCD particles (and, consequently, the abnormal optical activity of the formed CLCD particles) is determined by the correlation interaction between adjacent double-stranded DNA molecules when they approach each other (i.e., at the "moment of formation") and depends on the peculiarities of the secondary structure of DNA molecules. A distortion of the properties of the initial DNA molecules results in a change in the mode of their spatial packing in CLCD particles and, consequently, change in the abnormal band in the CD spectrum typical of the initial DNA CLCD located in the region of absorption of nitrogen bases. Such change in the magnitude of the abnormal band can be a function of the concentration of the substance (for instance, BAC) that interacts with the starting (initial) DNA molecules. Hence, in this case, the amplitude of the abnormal band in the CD spectrum is the criterion to determine the presence of BAC.

According to the second approach, the detection of colored BAC ("external chromophores") takes into account the fact that various compounds could easily diffuse into quasi-nematic layers in the structure of CLCD particles (Figure 10.6) and form different complexes with DNA molecules. In this case, "external chromophores" are incorporated in the spatial structure of CLCD particles. It follows from the results of the theoretical calculations (Section I, Chapter 4) that the formed DNA CLCD particles have a high structural stability. The above theory imposes no limitations on the number of molecules of "external chromophores" that could be incorporated into the CLCD structure. Note that a low extent of intercorporation of "external chromophores" does not interfere with the packing mode of neighboring DNA molecules in quasi-nematic layers, leaving the overall spatial structure of DNA CLCD particles intact.

The incorporation of "external chromophores" is accompanied by an appearance of new additional, abnormal band in the CD spectrum in the region of their absorption. The amplitude of this band depends both on the mode of fixation of "external chromophores" in the DNA structure and, consequently, in the structure of quasi-nematic layers in CLCD particles and the concentration of fixed "external chromophores" (BAC). In this case, the amplitude of the new abnormal band in the CD spectrum is the criterion that determines the presence of BAC.

10.4.1   DETECTION OF BAC CAUSING DISTORTION OF DOUBLE-STRANDED DNA STRUCTURE (FIRST APPROACH)

In the first approach, the starting double-stranded DNA molecules were treated with BAC, then CLCD particles were formed from them, and their CD spectra were analyzed [21].

The decrease in the amplitude of the abnormal band in the CD spectrum typical of DNA CLCD (up to the disappearance of the band) or the change of its sign from a negative to positive one shows that the analyzed BAC distorts the secondary structure of the initial DNA molecules, which makes them unable to form classical CLCD with a negative abnormal band in the CL spectrum.

The CD spectra of the CLCD formed from DNA molecules treated by compounds and used as antitumor drugs are a matter of interest. In this respect, the coordination compounds of platinum (II) group [23,24] are of prime importance for several reasons.

1. Information on alteration in the secondary and ternary structure of DNA under the action of platinum (II) group compounds is necessary for the chemists to choose the strategy of synthesis of the new generations of anti-cancer drugs.
2. The platinum atoms reacting with DNA represents labels that can be used in the analysis of the structural characteristics of DNA.
3. The biological effects of antitumor compounds of platinum (II) group (namely, their ability to cause direct and reverse mutations or suppress the growth of the tumor cells) are currently being actively studied *in vivo* and *in vitro* in model genetic and cellular systems.

As stated earlier, the initial double-stranded DNA molecules were treated by *cis*-dichlorodiammineplatinum (II) (*cis*-Pt (II) or DDP), then CLCD particles were formed from modified DNA molecules, and their CD spectra were compared to the CD spectrum of the initial CLCD formed from native linear DNA molecules (Figure 10.7A; where r_t is the relation of molar concentration of DDP to the molar concentration of DNA nucleotides in the solution).

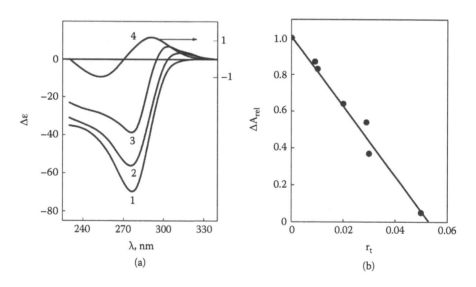

FIGURE 10.7 (a) The CD spectra of CLCD formed in PEG-containing solutions from the initial DNA molecules (1) and DNA modified by DDP (2–4); (b) the dependence of the relative amplitude of the negative band in the CD spectrum ($\lambda = 270$ nm) of DNA CLCD formed by DDP modified DNA on r_t value. In (a) curve 1—DNA B-form; curves 2–4: 2—$r_t = 0.01$; 3—$r_t = 0.03$; 4—$r_t = 0.1$. $C_{PEG} = 170$ mg/mL, molecular mass of PEG 4,000 Da; 0.5 M NaClO$_4$.

The decrease in the amplitude of the negative band in the CD spectra of CLCD formed from DNA–DDP complexes (curves 2–4) reflects the fact that the interaction of DDP with nitrogen bases is accompanied by the alteration of the parameters of the DNA secondary structure. Indeed, DDP, when interacting with DNA nitrogen bases—precisely, with N-7 and/or O-6 atoms of two purine bases (guanine bases, mainly)—may induce both changes in the distance between base pairs and their mutual orientation and may form cross-links between two guanine residues in one or in both DNA strands [25,26]. As a result of such cross-links, sites appear whose structures are dissimilar to those of native DNA. As the extent of DNA modification increases, the amplitude of the negative band in the CD spectra decreases linearly (Figure 10.7b). Zero amplitude is observed to occur at $r_t \sim 0.1$ (Figure 10.7a, curve 4). So, it can be stated that the modification of nitrogen bases causes the change in the parameters of the secondary structure of the DNA molecule, affecting the mode of packing of adjacent modified DNA molecules in CLCD particles.

The data shown in Figure 10.7 indicate that the biosensing units based on DNA CLCD particles permit one to detect very small concentrations of DDP capable of destroying the DNA secondary structure (about one atom of Pt per 100 base pairs of DNA is enough to induce change in the abnormal optical properties of DNA CLCD particles).

Therefore, Figure 10.7 not only indicates that DDP interacts with DNA nitrogen base pairs but also shows the high sensitivity of the biosensing unit based on DNA CLCD. It is important to stress that the influence of various Pt(II) compounds on the amplitude of an abnormal band in the CD spectra of the DNA CLCD correlates with differences in the biological activity of these compounds measured by biological methods [23].

In addition, similar alterations in the CD spectra of DNA CLCD can be induced by the action of various chemical substances and physical factors capable of changing the electronic structure and mutual orientation of the nitrogen bases in the initial double-stranded DNA molecules. For instance, such changes are induced by photochemical modification of nitrogen bases in the initial DNA under the combined action of UV-radiation and sodium hypophosphite [27] during the methylation of DNA *in vivo* and *in vitro* [28], as well as physical factors (such as laser irradiation) [29]. Hence, one can use the biosensing unit of the described type for the detection of BAC or physical factors causing changes in the DNA secondary structure.

10.4.2 DETECTION OF COLORED BAC ("EXTERNAL CHROMOPHORES") INTERCALATING BETWEEN DNA NITROGEN BASE PAIRS IN CLCD PARTICLES (THE SECOND APPROACH)

To realize the second approach to the detection of chemical substances and BAC, the formed DNA CLCD was processed with BAC, and then the CD spectra of the received mixtures were analyzed [22]. Obviously, for colored chemical or biologically relevant compounds forming strong complexes with DNA—that is, for compounds that are fixed in a regular mode on the surface of the DNA molecules—the same rules apply, which define the appearance of an abnormal band in the absorption

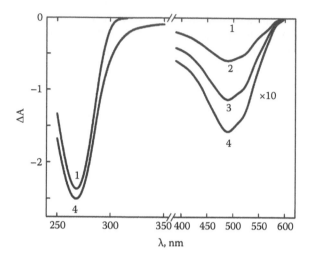

FIGURE 10.8 The experimentally measured CD spectra for the DNA CLCD particles in the absence (1) and presence (2–4) of daunomycin. Curves 1–4: $C_t = 0$; 1.2; 2.4; 3.9 × 10^{-6} M daunomycin; C_{DNA} = 12 μg/mL; 170 mg/mL PEG; 0.3 M NaCl; ΔA in optical units (2 × 10^{-3}).

region of nitrogen bases rigidly fixed in the DNA structure. For instance, an additional band in the absorption region of "external chromophores" should appear on location of these compounds between nitrogen bases of the DNA forming CLCD (Section I, Chapter 4).

Indeed, the results of theoretical calculations show that at the intercalation of such compounds as daunomycin (DAU) between DNA base pairs fixed in the structure of a CLCD particle, a negative abnormal band whose amplitude is proportional to the concentration of DAU molecules intercalating between DNA base pairs appears in the DAU absorption spectrum. These results of theoretical calculations coincide with experimental data shown in Figure 10.8.

For this case, software integrated with the portable dichrometer allows one to evaluate the concentration of DAU in the analyzed solution based on the calibration curve, considering the dependence of the amplitude of the band in the CD spectrum of CLCD on the concentration of DAU. This method was used to determine the concentration of DAU in blood samples of patients taking a DAU chemotherapy course (Figure 10.9).

An additional illustration of the second approach's practical application is shown in Figure 10.10 [19]. This figure demonstrates the CD spectra of the DNA CLCD (curve 1) and then as processed with sanguiritrin (curve 2), an antibacterial phyto-genotoxicant containing a chromophore that absorbs at λ_{max} = 317.3 nm.

According to theoretical predictions, the CD spectrum of DNA CLCD treated with sanguiritrin contains two abnormal bands. One band is located in the region of DNA absorption ($\lambda_{max} \sim 270$ nm), while the other appears in the region of sanguiritrin chromophore absorption ($\lambda_{max} \sim 337$ nm). The shape of the bands observed in the CD spectrum is similar to the shape of bands of DNA and phytogenotoxicant absorption.

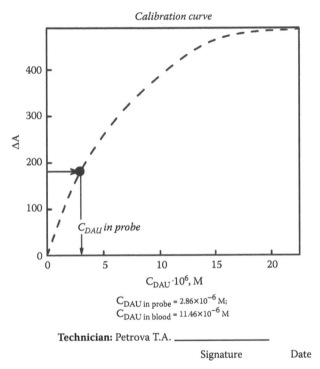

List of Analysis

Organization: Engelhardt Institute of Molecular
Biology, Russian Academy of Sciences

Drug: Daunomycin in blood

Client: Cancer Research Institute

Patient name: Starkow I.S.

Date of birth: 1951

Calibration curve

C_{DAU} *in probe*

$C_{DAU} \cdot 10^6$, M

$C_{DAU \text{ in probe}} = 2.86 \times 10^{-6}$ M;
$C_{DAU \text{ in blood}} = 11.46 \times 10^{-6}$ M

Technician: Petrova T.A. _____

Signature   Date

FIGURE 10.9   Blank of practical determination of antitumor drug daunomycin, concentration performed by means of an optical biosensor with the use of analytic software.

(The experimental displacement of the maximum of the band in the CD spectrum compared to the sanguiritrin absorption band is connected with the diffraction effect due to the "effective field" acting on molecules in the cholesteric phase, which was considered in Section I, Chapter 4.)

The appearance of the second abnormal band in the CD spectrum of DNA CLCD shows that sanguiritrin forms a complex with DNA molecules in the CLCD particles and fixed anisotropically in the quasi-nematic layers of CLCD particles.

At any concentration of sanguiritrin, both of the abnormal bands in the CD spectrum have a negative sign. According to the theory, the equal signs of the abnormal bands in the CD spectrum indicate that the orientation of sanguiritrin molecules in respect to the long axis of DNA molecules coincides with the orientation of nitrogen

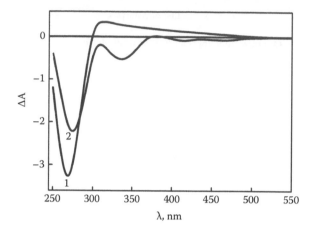

FIGURE 10.10 The CD spectra of DNA CLCD before (1) and after (2) sanguiritrin treatment. 1—$C_{Sanguiritrin} = 0$; 2—$C_{Sanguiritrin} = 7.97$ µg/mL; $C_{DNA} \sim 10$ µg/mL; $C_{PEG} = 170$ mg/mL; molecular mass of PEG = 4000 Da; 0.3 M NaCl + 10^{-2} M Na$^+$-phosphate buffer.ΔA in optical units ($\times 10^{-3}$).

bases. This is possible if sanguiritrin molecules that interact with DNA molecules forming CLCD particles intercalate between nitrogen base pairs and, hence, are fixed in the DNA structure and, consequently, in the structure of the cholesteric. Intercalating between base pairs, sanguiritrin molecules locate so that the angle of inclination of this compound about the DNA long axis is ~90°.

According to Figure 10.11 (here, each point on the curve is a result of the averaging of the optical signal generated in DNA CLCD at the analysis of three independently prepared samples with a fixed concentration of sanguiritrin) the arrows A → B → C on the insert show that the minimum concentration of sanguiritrin evaluated with the help of the biosensing unit based on DNA CLCD is ~ 0.5 µg/mL.

It can be repeated once more that the foregoing theory does not impose any limits on the number of BAC molecule that can be embedded into the structure of DNA molecules that form CLCD. Indeed, the amplitude of the abnormal band in the CD spectrum of DNA CLCD in the area of sanguiritrin absorption is directly proportional to the concentration of this compound added to the solution. Besides, from Figure 10.11, it follows that the value of $\Delta\varepsilon$ (the molar dichroism of sanguiritrin) is constant on a broad range of phytogenotoxicant concentrations.

The comparison of Figures 10.7–10.11 shows that biosensing units based on the DNA CLCD operate in the following way: nitrogen bases "recognize" molecules of drug and "address" them to the definite places in the DNA in the CLCD. A complex formation between drug and nitrogen bases is accompanied by the appearance of an optical signal. An optical signal generated in the system is "amplified" due to specific spatial properties of the cholesteric structure of the CLCD [1] and displayed as an intense band in the CD spectrum. The greater the concentration of the DNA-bound drug, the higher the amplitude of the band in the CD spectrum in the absorption region of the drug. The spectral position of intense band in the CD spectrum of a DNA CLCD, its amplitude, and sign permit one to both detect the presence of drug

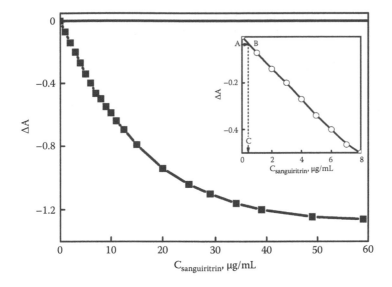

FIGURE 10.11 The dependence of the amplitude of the abnormal negative band (λ = 337 nm) in the CD spectra of DNA CLCD on sanguiritrin concentration. $C_{DNA} \sim 10$ μg/mL; $C_{PEG} = 170$ mg/mL; molecular mass of PEG = 4000 Da; 0.3 M NaCl + 10^{-2} M Na$^+$-phosphate buffer. ΔA in optical units (×10^{-3}).

(or combinations of drugs) in the solution and estimate its concentration, as well as to establish the mode of binding of the drug with the DNA molecule.

Hence, the second approach to detect biologically active compounds based on the preformed DNA CLCD can be used for analytical purposes. The minimal concentration of antibiotics capable of intercalating between the DNA nitrogen bases that can be detected by this approach is $\sim 5 \cdot 10^{-7}$ M [20].

The data in Figures 10.7–10.11 are important from a practical point of view. These data show that the abnormal optical activity of DNA CLCD can be used as an analytical criterion to detect the presence of BAC in the solution and evaluate its concentration, as well as to determine the location of BAC in the helical structure of DNA molecules. This means that DNA CLCD particles represent by themselves multifunctional polyfunctional biosensing units for detection of chemical substances and BAC whose target is the genetic material of cells. In the case of any of the described approaches to the detection of "external chromophores," BAC, and so on, the CD spectrum of CLCD makes the changes in the DNA secondary structure, caused by the effect of interaction of these compounds with the DNA nitrogen bases, "visible." Therefore, an optical biosensor that includes a portable dichrometer (see Figure 10.5) and a biosensing unit based on DNA CLCD can be used for analytical purposes.

10.4.3 Synthetic Polymer Matrixes Containing DNA CLCD as Film-Type Biosensing Units

The main difficulty in using DNA CLCDs as biosensing units is the temporal instability of their optical properties. This is explained simply by a decrease in the actual concentration of particles of DNA CLCDs in solution in long-term (more than a week) storage because of their sedimentation; in its turn, this is accompanied by an uncontrollable change in the abnormal optical properties of the dispersion.

The stabilization of DNA CLCD optical properties was achieved by preventing the sedimentation of particles by incorporating them in a synthetic polymer matrix (gel, film) [30–34]. In this case, a biosensing unit is a synthetic polymer matrix containing spatially separated particles of DNA CLCD that enable to sedimentate. The principle of functioning of such a biosensor is similar to that of "film-type indicators" [2].

Taking into account the statement concerning the mobility of the spatial structure of DNA CLCD, a key role in designing a "film-type" biosensing unit is played by the creation of such a polymer matrix that would not disturb the abnormal optical properties of DNA CLCDs.

Two methods for producing synthetic polymer matrixes (hydrogels) that contain DNA CLCD particles are described in the literature [30–34]. According to the first method, the immobilization of DNA CLCD particles in the hydrogel structure is achieved by polymerization of the hydrophilic monomer acrylamide dissolved in a water–salt solution of PEG in which a DNA CLCD was preliminarily formed [30]. A hydrogel was obtained by acrylamide polymerization fixed DNA CLCD particles almost without changing their abnormal optical properties. However, the hydrogel produced by this method has a low transparency (Figure 10.12), probably because of the incompatibility of PEG and poly(acrylamide), which limits the possibility of practical use of the gel for analytical purposes.

The second method proved to be more efficient [34]. In this method, terminal OH-groups of poly(ethylene glycol) molecules are chemically modified so that their polymerization becomes possible [33]. According to this method, hydrogels containing DNA CLCD particles were obtained using a procedure in which methacrylate macromonomers of poly(ethylene glycol) were initially used for phase exclusion of

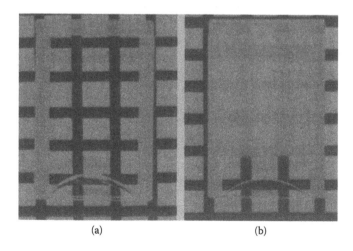

(a)   (b)

FIGURE 10.12 Images of original poly(acrylamide) gel (a) and a first synthetic polymer matrix (b) used to immobilize DNA CLCD particles (thickness of the layer—1 mm).

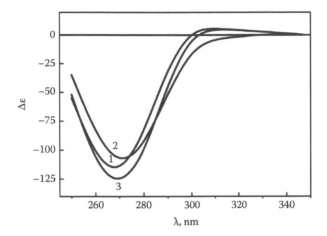

FIGURE 10.13 The CD spectra of DNA CLCD and a hydrogel containing DNA CLCD particles of this dispersion: 1—CD spectrum of DNA CLCD formed in a water–salt solution of synthetic "macromonomer" of PEG; 2—CD spectrum of DNA CLCD immobilized in a hydrogel obtained by polymerization of PEG "macromonomer"; 3—the reference CD of initial DNA CLCD formed in water–salt solution of PEG. C_{PEG} = 170 mg/mL; molecular mass of PEG = 6,000 Da; 0.3 M NaCl + 10^{-2} M Na^+-phosphate buffer.

DNA, that is, for the formation of a CLCD, and then polymerized under the action of radical initiators to yield a final film (hydrogel).

Figure 10.13 presents the CD spectra recorded before (curve 1) and after (curve 2) polymerization of a solution containing DNA CLCD and the CD spectrum of CLCD formed by DNA molecules in water–salt solution of initial PEG preparation (curve 3). Comparative analysis of the data in Figure 10.13 showed that the immobilization of DNA CLCD particles in such a hydrogel has no significant effect on the shape and amplitude of the abnormal band in the CD spectrum.

Hydrogels obtained from chemically modified PEG are optically isotropic; they have high transparency when being stored under not-too-severe conditions (humidity, temperature). These hydrogels retain the abnormal optical activity inherent in DNA CLCD for at least several months after preparation. This means that the stability of the DNA CLCD in the obtained synthetic polymer matrix is quite high.

Using examples of representatives of various groups of colored BAC (dyes, antibiotics, antitumor drugs), it was shown [34] that these compounds can diffuse into this hydrogel and interact with DNA molecules forming a CLCD, thus causing the optical effects described earlier.

Figure 10.14 shows the CD spectra of DNA CLCD in a hydrogel placed in a PEG-containing water–salt solution of DAU (the size of the hydrogel sample containing DNA CLCD: $6 \times 3 \times 20$ mm). The intense negative band with a maximum at $\lambda \sim$ 500 nm, whose shape is similar to that of the absorption band of daunomycin DAU, indicates that DAU molecules diffuse into the hydrogel in the intact native state and form a complex with DNA molecules forming the CLCD. The amplitude of the band

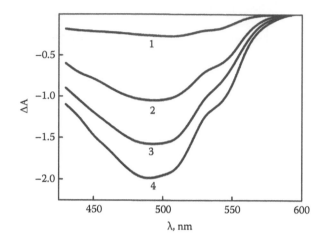

FIGURE 10.14 The CD spectra of DNA CLCD in a hydrogel placed in a PEG-containing water–salt solution of DAU and recorded within various periods of time: 1—15, 2—75, 3—120, 4—150 min. C_{DNA} = 30 µg/mL; C_{PEG} = 170 mg/mL; C_{DAU} = 1.667 x·10^{-5} M; 0.3 M NaCl + 10^{-2} M Na^+-phosphate buffer; L = 1 cm; T = 22°C; ΔA in optical units (×10^{-3}).

at $\lambda \sim 500$ nm becomes larger the more daunomycin molecules bind to DNA molecules; this finding fully corresponds with the results of the theoretical calculations previously shown. Consequently, a hydrogel based on chemically modified PEG provides the conditions for the diffusion of DAU molecules in their initial state.

Thus, the retention of the specific optical properties of DNA CLCD in combination with their attractive physicochemical properties and high stability shows that these hydrogels can be used as a good background for designing multifunctional "film-type" biosensing units.

However, practically, DNA CLCD particles, even immobilized in a gel, have a significant disadvantage. They are characterized by a dependence of the amplitude of the anomalous stripe in the CD spectrum (ΔA) on the osmotic pressure (π) of the solvent. In other words, as the gel swells in the solution of the analyzed substance, the abnormal band in the CD spectrum of DNA CLCD may drop almost to zero. This means that detection of BAC in water mediums or physiological liquids with the help of such detector is quite difficult without special precautions.

10.5 SANDWICH-TYPE BIOSENSING UNITS BASED ON (DNA–POLYCATION) LIQUID-CRYSTALLINE DISPERSIONS

It has been shown in Chapter 5 that the spatial structure of the (DNA–polycation) LCD can be regulated. Because the reorganization of the spatial structure of the (DNA–polycation) complex LCD from nonspecific to cholesteric is caused by the low concentration of the enzyme, these particles can be used as a new type of biodetectors. Such detectors are called "sandwich-type" biosensing units [35,36].

Thus, unlike biosensing units based on DNA CLCD, the "targets" for BAC action in sandwich-type biosensing units are not DNA nitrogen bases but polycation molecules that neutralize the negative charges of DNA phosphate groups. The phase exclusion of (DNA–polycation) complexes from water–salt solutions is accompanied by the formation of LCD particles with a dense packing of adjacent (DNA–polycation) molecules determined by the value of the Bragg reflection of about 25 Å that is determined by the correlation interaction between the molecules, as a rule. The abnormal band in the CD spectra of such dispersions is usually absent (Figure 10.15a). If such a dispersion is placed in a PEG-containing solution (Figure 10.15b), anisotropic DNA molecules appear to be under the effect of two "structural" tendencies. On the one hand, the correlation interaction between the (DNA–polycation) molecules tends to provide the dense, hexagonal packing of (DNA–polycation) molecules. On the other hand, in the chosen conditions (for instance, at C_{PEG} = 170 mg/mL), the osmotic pressure of the PEG-containing solution determines a different distance, longer than 25 Å, between the adjacent DNA molecules, that is, the solvent provides conditions such that the anisotropic DNA molecules tend to form a CLCD with an abnormal optical activity.

The dominance of a certain structural tendency is determined by the efficiency of the interaction between the polycation and DNA molecules, that is, the presence of polycation molecules prohibit the packing of DNA molecules under conditions of "molecular crowding" in a way typical of CLCD particles. If, under these conditions, without change in the PEG concentration, the efficiency of the correlation interaction between DNA–polycation molecules is to be reduced, for instance, because of splitting of some molecules of the polycation or extracting them from complexes with DNA, the DNA molecules can realize their ability with cholesteric packing.

Under these conditions, the transition from "hexagonal to the cholesteric" mode packing of DNA molecules will take place. This process should be accompanied by the emergence of an intense negative band in the CD spectrum in the region of DNA absorption (Figure 10.15c).

The appearance of the band can be used as an analytical criterion of the presence of BAC that causes the destruction or competitive displacement of polycation molecules from the (DNA–polycation) complexes. Using the described method, a sensing unit of a new type can be created [37].

The polycation molecules used for designing biosensing units of this type should have the following physicochemical properties:

1. The polycation should not have noticeable intrinsic absorption in the UV region of the spectra.
2. Polycation molecules should neutralize the negative charges of phosphate groups of DNA.
3. The interaction of the polycation with DNA should not cause marked changes in the secondary structure parameters of this macromolecule.
4. Polycation molecules should contain reactive groups as targets for a biologically active compound being determined.

Apparently, with allowance for these physicochemical properties of a polycation, the different versions of molecular design of sandwich-type biosensing units can be realized.

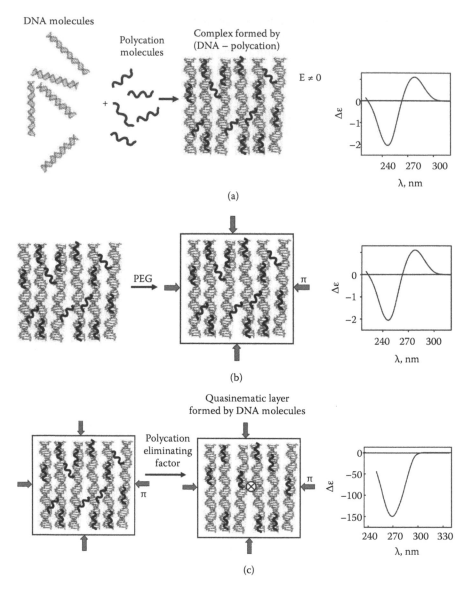

FIGURE 10.15 (*See color insert.*) The principal operation scheme of a biosensing unit based on DNA–polycation complex LCD: (a) the formation of LCD particles from (DNA–polycation) complexes is followed by a dense (hexagonal) packing of the molecules; the CD spectrum (on the right) of such a system almost coincides with the CD spectrum typical of the DNA B-form; (b) when PEG is added to the system (a), its optical and structural properties do not change (the osmotic pressure of the solution π is marked with dark arrows); (c) when factors that cause the "elimination" (destruction of the structure or extraction from the content of DNA complex) of polycation molecules are added to the system (b), the structural and optical properties of LCD particles change dramatically (under the conditions of the solvent with a fixed osmotic pressure, the DNA molecules can realize their tendency toward cholesteric packing, which is followed by the appearance of the abnormal band in the CD spectrum; on the right).

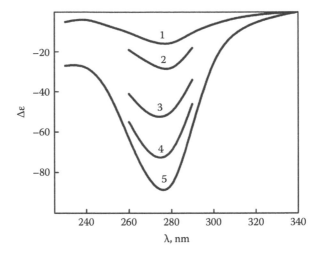

FIGURE 10.16 The CD spectra of PEG-containing water–salt solutions of LCD formed by (DNA–polyconidin) complex (r = 0.91) at various concentrations of heparin: 1—0 μg/mL; 2—0.2 μg/mL; 3—0.6 μg/mL; 4—1.0 μg/mL; 5—2.5 μg/mL. C_{PEG} = 170 mg/mL; 0.3 M NaCl + 10^{-2} M Na^+-phosphate buffer.

As an example, the following experimental data characterizing the analytical potential of a sandwich-type biosensing unit based on the (DNA–polyconidin) LCD are presented [38]. (Polyconidin is a synthetic antagonist of heparin, that is, a polyamine capable of inducing DNA condensation in water–salt salt solutions of moderate ionic strength to yield LCDs without abnormal optical activity. In our case, it is a quaternary ammonium salt of a monodisperse conidin oligomer with a polymerization degree of 25.)

In Figure 10.15, the LCD of (DNA–polyconidin) complex obtained in a water–salt solution with a moderate ionic strength has been placed in a PEG-containing water–salt solution and treated with heparin. Figure 10.16 compares the CD spectrum of the initial LCD of the (DNA–polyconidin) complex (curve 1) to the CD spectra of the same dispersion (curves 2–5) after adding heparin. (Here, **r** is the relation of the molar concentration of positively charged polyconidin groups to the molar concentration of negatively charged DNA phosphate groups in the solution.) Addition of heparin is accompanied by the emergence of a negative band in the CD spectrum of (DNA–polyconidin) LCD with a maximum at λ = 280 nm. With an increase in the heparin concentration in the solution, the amplitude of this band increases.

The dependence of the amplitude of the band at λ = 280 nm in the CD spectrum of (DNA–polyconidin) LCD on the concentration of heparin (Figure 10.17, where the arrow marks the concentration of heparin under which there is a directly proportional dependence between $\Delta\varepsilon_{280}$ value and the concentration of heparin) shows that the $\Delta\varepsilon_{280}$ value is directly proportional to the heparin concentration in the range from 0 to 1.2–1.3 μg/mL. Because the optical signal generated by the LCD of the (DNA–polyconidin) complex in a PEG-containing water–salt solution after the addition of heparin is directly proportional to the heparin concentration, the straight-line

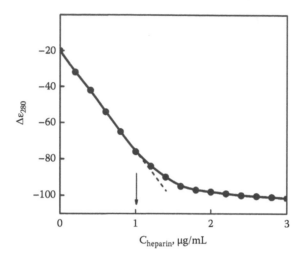

FIGURE 10.17 The dependence of the negative band amplitude in the CD spectra of PEG-containing water–salt solution of (DNA–polyconidin) LCD on heparin concentration. C_{DNA} = 15 µg/mL; $C_{Polyconidin}$ = 6.35 µg/mL; C_{PEG} = 170 mg/mL; 0.3 M NaCl + 10^{-2} M Na$^+$-phosphate buffer.

portion of the plot in Figure 10.17 can be used as a calibration line. Then, one can determine low (<1 µg/mL) heparin concentrations in a sample to be analyzed with high accuracy. Therefore, an abnormal band in the CD spectrum of DNA LCD, combined with the dependence between the amplitude of the band and the concentration of the added heparin, can be used to detect the presence of low heparin concentrations in the solution. Note that the determination of very low (physiological) heparin concentrations is of prime importance to practical health care since a change in the physiological heparin concentration in patients' blood very often indicates the preclinical phase of developing pathological states, mainly those of immune genesis. This is of extreme significance for taking preventive measures and predicting the clinical course of a pathological process under the action of treatment being performed. The minimal heparin concentration determinable with a bioanalytical system involving a portable dichrometer capable of working with sandwich-type liquid-crystalline biosensing units is ~0.5 µg/mL [39].

10.6 DNA NANOCONSTRUCTION AS A SENSING UNIT (NEW TYPE OF BIODETECTORS)

The broadening of the number of BAC and drugs detected by means of liquid-crystalline biodetectors determines the design of an absolutely new type of liquid-crystalline DNA structures whose analytical capabilities are much higher than those of the classical CLCD of linear double-stranded DNA. The strategy for obtaining such structures requires the development and use of modern physicochemical methods and biotechnological approaches. These approaches include, for instance,

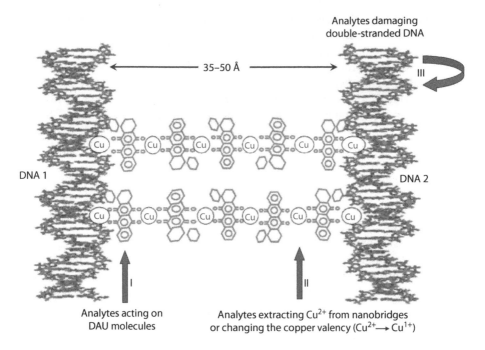

FIGURE 10.18 (*See color insert.*) Scheme of the nanobridges between adjacent DNA-1 and DNA-2 molecules fixed in the spatial structure of a CLCD particle (the nanobridges are turned to 90° relative to the base pairs for convenience). I–III—substances to be analyzed.

nanodesign, which allows one to create nanoconstructions (NaCs) based on DNA with controllable properties (Chapter 9).

As shown before (Chapter 9), an NaC is an ensemble of the DNA molecules with a low molecular mass ($<1 \cdot 10^6$ Da) fixed in the structure of a CLCD particle because of the osmotic pressure of the solvent and linked by nanobridges (Figure 10.18) [40].

The analysis of the structure in Figure 10.18 shows that the stability of DNA NaC is determined not by the osmotic pressure of the PEG-containing solution but by the number and properties of the nanobridges. This peculiarity determines the ability of DNA NaC to exist outside the "critical" conditions of DNA CLCD formation [41].

Getting back to Figure 10.18, attention is focused on the fact that NaC of DNA includes a "sensitive element," which is represented by each nanobridge. Because of the specific spatial conformation of a nanobridge formed between DAU molecules and Cu^{2+} ions, it is obvious that the nanobridge can be destroyed under the action of various chemical and physical factors. In this case, the destruction of nanobridges between DNA molecules in NaC causes the disintegration of the whole NaC. This process must be followed by the decrease (up to the total disappearance) of the abnormal optical activity of NaC, expressed as an abnormal band in the CD spectrum in the region of absorption of chromophores in the content of nanobridges. Under certain conditions, the extent of the decrease of the amplitude of the abnormal band in the CD spectrum of NaC is directly related to the concentration of the agent that destroys the nanobridge structure. This means that NaC can be used as a biodetector that reacts to the presence of substances that destroy the nanobridges, while the change in the optical properties of this biodetector can be easily recorded by the portable dichrometer. Considering these facts and the data presented, which describe the physicochemical properties of NaC based on DNA CLCD (Chapter 9), it can be said that NaC is a microscopic-size biodetector that makes it possible to detect the presence of any compound that affects the stability of the nanobridges. Such a biodetector can be called an "integral liquid-crystalline microchip" [42].

It has been shown on model compounds (in particular, ascorbic acid or BSA [43]), which change the valency of Cu^{2+} ions ($Cu^{2+} \rightarrow Cu^{1+}$) or "extract" them from the structure of nanobridges, respectively, that DNA NaC can be used as a new type of integral liquid-crystalline microchip. The limits of detection of analyzed substances ($\sim 10^{-7} - 10^{-8}$ M) comparable to the limits of detection by means of classical biochemical methods have been achieved.

As the disintegration of DNA NaC is caused by relatively low concentrations of model substances, it was interesting to use this type of liquid-crystalline biodetectors (biosensing units) in medical biotechnology to perform the preliminary screening of BAC to determine and choose pharmaceutical substances with phytogenotoxicity that have noticeable ability to form complexes with Cu^{2+} ions [44].

Figure 10.19 shows the CD spectra of DNA NaC before (curve 1) and after (curve 2) hyporamin treatment. (Hyporamin is an antibacterial and antiviral drug that is a dry purified extract from the leaves of sea-buckthom [*Hippophae rhamnoides L*] from the *Eleagnaceae* family.) Adding hyporamin is accompanied by the disappearance

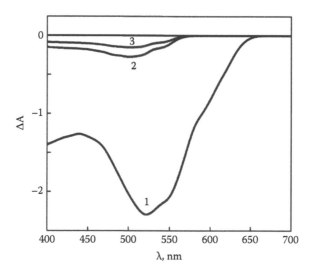

FIGURE 10.19 The CD spectra of DNA NaC before (1) and after (2) hyporamin treatment. Curve 1–$C_{Hyporamin} = 0$; curve 2–$C_{Hyporamin} = 3.953$ µg/mL. $C_{DNA} \sim 5,6$ µg/mL; $C_{PEG} = 170$ mg/mL, molecular mass of PEG = 4,000 Da; $C_{DAU} \sim 27.18 \times 10^{-6}$ M; $C_{Cu}^{2+} \sim 1 \times 10^{-5}$ M; 0.3 M NaCl + 10^{-2} M Na$^+$-phosphate buffer. ΔA in optical units ($\times 10^{-3}$). Curve 3—the CD spectrum of CLCD of (DNA–DAU) complex.

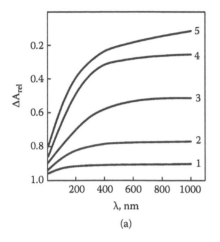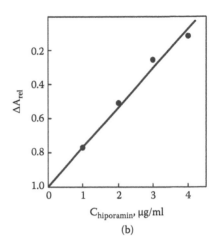

(a)   (b)

FIGURE 10.20 The dependences of the relative amplitude of the band (ΔA_{rel}) in the CD spectrum (λ = 515 nm) of DNA NaC on the time of hyporamin treatment (a) and on the concentration of hyporamin (b). 1—$C_{Hyporamin}$ = 0.494 µg/mL; 2—$C_{Hyporamin}$ = 0.990 µg/mL; 3—$C_{Hyporamin}$ = 1.978 µg/mL; 4—$C_{Hyporamin}$ = 2.966 µg/mL; 5—$C_{Hyporamin}$ = 3.953 µg/mL. C_{DNA} ~ 5.6 µg/mL/mL; C_{PEG} = 170 mg/mL, molecular mass of PEG = 4000 Da; C_{DAU} ~ 27.18 × 10^{-6} M; C_{Cu}^{2+} ~ 1 × 10^{-5} M; 0.3 M NaCl + 10^{-2} M Na$^+$-phosphate buffer.

of the abnormal band in the CD spectrum in the region of DAU chromophores' absorption, reflecting the existing nanobridges between DNA molecules in NaC. The biologically active components of hyporamin are hydrolyzed tannins [45], which form complexes with ions of bivalent metals. Hence, the formation of a more stable complex between hyporamin and Cu^{2+} ions results in the "extraction" of Cu^{2+} ions from nanobridges and their destruction.

Figure 10.20a represents the dependence of the change of relative amplitude of the abnormal band in the CD spectrum (λ = 515 nm) on the time of DNA NaC treatment with hyporamin. The data show that the change in the amplitude of the band in the CD spectrum of DNA NaC depends on the concentration of hyporamin in the solution. Figure 10.20b represents the dependence of the change of the relative band amplitude in the CD spectrum (λ = 515 nm) in the concentration of hyporamin. This dependence that corresponds to a 10-minute processing of DNA NaC with hyporamin is given in Figure 10.20 as an example. The arrow marks the minimum concentration of hyporamin that can be detected with the help of a biodetector based on DNA NaC. The direct proportionality between the relative amplitude of the band in the CD spectrum of DNA NaC and the concentration of hyporamin (in the interval from 0 to 4 µg/mL) permits using this dependence as the calibration line to detect a low (~0.5 µg/mL) concentration of hyporamin in the analyzed sample.

Consequently, NaC based on DNA molecules allows one to create biodetectors for a preliminary screening of phytogenotoxicants to detect the presence of these compounds and estimate relatively low concentrations of these drugs.

Regardless of their attractiveness, the main problem in the use of the "test-tube" form of DNA NaC as a biodetector is the insufficient temporal stability of their

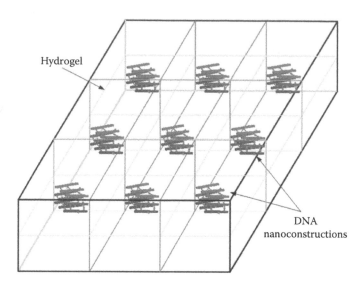

FIGURE 10.21   Scheme of a synthetic polymer matrix (hydrogel) that contains spatially separated particles of DNA NaC.

optical properties. This instability is caused by the fact that long storage (more than a week) results in the sedimentation of DNA NaC particles in the solution. Such a drop in concentration is followed by the uncontrollable change in the magnitude of the abnormal optical signal. To avoid this process, DNA NaCs (and, consequently, the optical signal generated by NaCs) were stabilized because immobilization in the content of a hydrogel does not distort their physicochemical properties. In this case, the biodetector is a synthetic polymer matrix that contains spatially separated particles of DNA NaC (Figure 10.21).

Such a biodetector, as noted earlier, is similar to the "membrane indicators" by the operation principle [2].

10.7   HYDROGELS CONTAINING DNA NACS AS NEW "FILM-TYPE" BIODETECTORS

The formation of nanobridges, containing DAU molecules and copper ions, between DNA molecules in CLCD particles, demonstrated in Section III, Chapter 9, allows one to develop a technology for the creation of hydrogels containing DNA NaCs.

Figure 10.22 presents the CD spectrum of DNA NaC in a hydrogel (the insert shown here is the CD spectrum recorded separately in the visible region of the spectra).

As shown above, an abnormal band in the region of DAU chromophores' absorption speaks in favor of the existence of nanobridges that contain alternating DAU molecules and Cu^{2+} ions in the content of rigid 3-D NaC. The hydrogel containing NaCs is optically isotropic, highly transparent, and maintains the abnormal optical activity of DNA NaC without keeping strict storage conditions (temperature,

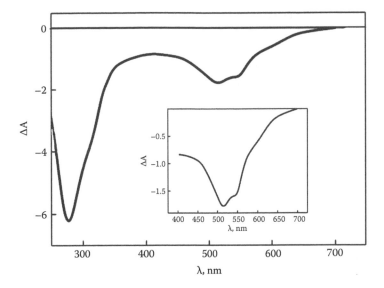

FIGURE 10.22 The CD spectra of a hydrogel containing DNA NaCs. $C_{DNA} = 39.33$ µg/mL; hydrogel size—6 × 20 × 2.67 mm. ΔA in optical units (×10⁻³); T = 22°C.

humidity) at least for several months after its preparation. This means that the operational stability of DNA NaC in a synthetic polymer matrix is quite high.

These results show that the approach that takes into account the physicochemical properties of a synthetic polymeric matrix and structural peculiarities of DNA CLCD opens a gateway for obtaining hydrogels containing spatially separated DNA NaCs with long-term stability of abnormal optical activity. Besides, low-molecular-mass compounds can easily diffuse into hydrogel.

Therefore, the combination of physicomechanical properties of the hydrogel and the presence of the abnormal band typical of DNA NaCs makes it possible to use such hydrogel as an integral "film-type" biodetector to determine BAC and drugs whose targets are the structural elements of the nanobridges [46].

As an example of a substance that is capable of diffusing into a hydrogel and disintegrating the structure of the DNA NaCs, a compound of medical importance, homocystein (HC), was tested. HC is produced in the human body during methionine metabolism. As the reverse transition of HC to methionine as a result of genetic defects of enzymatic systems that regulate the metabolism of methionine, or the lack of folic acid and vitamins B₆ и B₁₂ in the human body, HC begins to appear in the bloodstream, which causes the development of hyperhomocysteinemia. The long-term presence of HC in blood results also in the development of atherosclerosis, thrombovascular disease, pregnancy pathologies, and a number of other diseases [47].

The variety of clinical symptoms of hyperhomocysteinemia determines the necessity in the elaboration of the methods for a fast and accurate estimation of the HC concentration in blood plasma, which determines the timeliness and the efficiency of the therapy.

Currently, the HC detection in blood plasma is performed by means of different types of high-efficiency fluid chromatography [47]. Regardless of the relatively high accuracy and repeatability of these methods, they are characterized by the long time of analysis that may reach 4 hours, which is determined by the necessity of a special multistage preparation of the sample, use of expensive equipment, and, consequently, the employment of qualified personnel. The combination of these facts is the reason for the high cost of the analysis. This means that the development of a cheap, accurate, and simple method of detection of HC is an important challenge to modern biomedicine.

Considering the foregoing experimental data and a pronounced ability of HC to form strong chelate complexes with Cu^{2+} ions (the chemical formula of HC is shown in the insert in the Figure 10.23), it was interesting to test the hydrogel containing DNA NaCs as a "film-type" biodetector that can reflect the presence of HC in laboratory solutions. Figure 10.23 shows the CD spectra of a hydrogel containing NaCs in HC-containing solution. It is seen that the addition of HC decreases the amplitude of the negative band in the CD spectrum, which characterizes the presence of nanobridges between neighboring DNA molecules in NaCs. This means that HC molecules diffuse into the hydrogel, eliminate Cu^{2+} ions from the content of nanobridges, and induce the disintegration of the spatial structure of DNA NaC. This process is accompanied by a drop in the abnormal band in the CD spectrum and, ultimately, to its total disappearance. The obtained result shows that the DNA NaC, immobilized in a hydrogel, reacts easily to the presence of HC in the laboratory solution. Therefore, the attractive physicomechanical properties and high stability of the obtained hydrogels containing DNA NaC, combined with long-term conservation of the abnormal band in the CD spectra of these NaCs, allow them to be considered

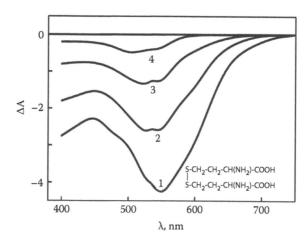

FIGURE 10.23 The CD spectra of DNA NaC in content of hydrogel placed in laboratory solution of Hc (C_{Hc} = 39.2 µg/mL) taken at various time: 1—0; 2—25; 3—45; 4—115 min. C_{DNA} = 19.9 µg/mL; C_{PEG} = 120 mg/mL; molecular mass of PEG = 35,000 Da; C_{DAU} = 26.94 × 10^{-6} M; C_{Cu}^{2+} = 9.82 × 10^{-6} M; 0.3 M NaCl + 10^{-2} M Na$^+$-phosphate buffer. Thickness of the hydrogel ~ 3 mm; T = 22°C. ΔA in optical units (×10^{-3}).

as reasonable background for the creation of integral "film-type" biodetectors that change their properties in response to the presence of chemical substances and BAC that destroy nanobridges in the analyzed sample.

10.8 SUMMARY

Various approaches to the design of sensing units (biodetectors) based on the consideration of properties of DNA molecules (or the properties of DNA complexes with various compounds) forming liquid-crystalline dispersions are described in this chapter. The intradisciplinary approach that takes into account achievements in the area of creation of nanoconstructions based on double-stranded DNA molecules, and achievements in polymer physical chemistry, microelectronics, and optics, allows one to design new types of biodetectors. The examples given previously show that biodetectors based on DNA nanoconstructions may well be introduced in biotechnology and medicine.

REFERENCES

1. Yevdokimov, Yu.M., Bundin, V.S., and Ostrovsky, M.A. Biosensors and sensory biology. *Sensory Syst.* (Russian Edition), 1997, vol. 11, p. 374–387.
2. *Biosensors: Fundamentals and Application.* A.P.F. Turner, I. Karube, G.S. Wilson, Ed. (Russian Edition). Moscow: Mir, 2002. p. 616.
3. Clark, L.C., Jr. and Lyons, C. Electrode systems for continuous monitoring in cardiovascular surgery. *Ann. N.Y. Acad. Sci.*, 1962, vol. 102, p. 29–45.
4. Cattrall, R.W. *Chemical Sensors.* New York: Oxford University Press, 1977, p. 143.
5. Brooks, S.L., Higgins, I.J., Newman, J.D., et al. Biosensors for process control. *Enzyme Microb. Technol.*, 1991, vol. 13, p. 946–955.
6. Cröregi, E., Gaspar, S., Niculescu, M., et al. Amperometric enzyme-based biosensors for application in food and beverage industry. *Physics and Chemistry Basis of Biotechnology.* De Cuyper, M., Bulte, J.M. Eds. London: Kluwer Acad. Publ., 2001, p. 105–129.
7. Dennison, M.J. and Turner, A.P.F. Biosensors for environmental monitoring. *Biotechnol. Adv.*, 1995, vol. 13, p. 1–12.
8. Wang, J. Amperometric biosensors for clinical and therapeutic drug monitoring. *J. Pharm. Biomed. Anal.*, 1999, vol. 19, p. 47–53.
9. Downs, M.E.A., Kobayashi, S., and Karube, I. New DNA technology and DNA biosensor. *Anal. Lett.*, 1987, vol. 20, p. 1897–1927.
10. Millan, K.M., Saraulo, A., and Mikkelsen, S.R. Voltammetric DNA biosensor for cystic fibrosis based on modified carbon paste electrode. *Anal. Chem.*, 1994, vol. 66, p. 2943–2948.
11. Yershov, G., Barsky, V., Belgovskij, A., et al. DNA analysis and diagnostics on oligonucleotide microchips. *Proc. Natl. Acad. Sci. USA*, 1996, vol. 93, p. 4913–4918.
12. Barsky, V.E., Kolchinsky, A.M., Lysov, Yu.P., and Mirzabekov, A.D. Biological microchips with hydrogel-immobilized nucleic acids, proteins and other compounds: Properties and applications in genomics. *Mol. Biol.* (Russian Edition), 2002, vol. 36, p. 563–584.
13. Zasedatelev, A.S. Biological microchips for medical diagnostics. *Science and Technologies in Industry* (Russian Edition), 2005, vol. 1, p. 18–20.

14. Lucarelli, F., Marrazza, G., Turner, A.P.F., et al. Carbon and gold electrodes as electrochemical transducers for DNA hybridization. *Biosens. Bioelectron.*, 2004, vol. 19, p. 515–530.
15. Gebhart, C.J., Ward, G.E., and Murtaugh, M.P. Species-specific cloned DNA probes for the identification of *Campylobacter hyointestinalis. J. Clin. Microbiol.*, 1989, vol. 27, p. 2717–2723.
16. Matthews, J.A. and Kricka, L.J. Analytical strategies for the use of DNA probes. *Anal. Biochem.*, 1986, vol. 169, p. 1–25.
17. Trukhan, E.M. Biosensors in systems for operative estimation environmental fitness. *Sensory Syst.* (Russian Edition), 1998, vol. 12, p. 135–144.
18. Kompanets, O.N. Portable optical biosensors for the determination of biologically active and toxic compounds. *Uspekhi Fizicheskikh Nauk* (Russian Edition), 2004, vol. 174, p. 686–690.
19. Skuridin, S.G., Dubinskaya, V.A., Irlyanov, D.I., et al. Identification of the plant genotoxicants by bioanalytical test-system based on the DNA liquid-crystalline biosensing units. *Sensory Syst.* (Russian Edition), 2006, vol. 20, p. 68–77.
20. Skuridin, S.G. and Yevdokimov, Yu.M. DNA liquid-crystalline dispersion particles as a basis for sensitive biosensor elements. *Biophysics* (Russian Edition), 2004, vol. 49, p. 468–485.
21. Yevdokimov, Yu.M., Skuridin, S.G., and Chernukha, B.A. Biosensing units based on liquid-crystalline dispersions of double-stranded nucleic acids. *Biotekhonologiya* (Russian Edition), 1992, No. 5, p. 103–109.
22. Yevdokimov, Yu.M., Salyanov, V.I., and Semenov, S.V. Analytical capacity of the DNA liquid-crystalline dispersions as biosensing units. *Biosens. Bioelectron.*, 1996, vol. 11, p. 889–901.
23. Akimenko, N., Cheltsov, P., Balcarova, Z., et al. A study of interactions of platinum(II) compounds with DNA by means of CD spectra of solutions and liquid crystalline microphases of DNA. *Gen. Physiol. Biophys.*, 1985, vol. 4, p. 597–608.
24. Yevdokimov, Yu.M., Skuridin, S.G., Salyanov, V.I., and Badaev, N.S. DNA liquid crystals—a possible system to test interactions between DNA and biologically active compounds of platinum(II) and some antibiotics. *Antibiotics Chemother.* (Russian Edition), 1988, vol. 33, p. 903–905.
25. Lippard, S.J. Chemistry and molecular biology of platinum anticancer drugs. *Pure Appl. Chem.*, 1987, vol. 59, p. 731–742.
26. Lippard, S.J. Platinum, gold and other metal chemotherapeutical agents. *ACS Symp.* (ser. 209), 1983, p. 5–311.
27. Yakovlev, D.Yu., Skuridin, S.G., Khomutov, A.R., et al. The reduction of thymine residues in DNA by combined action of UV light and hypophosphite. *J. Photochem. Photobiol. B: Biol.*, 1995, vol. 29, p. 119–123.
28. Akimenko, N.M., Burckhardt, G., Kadykov, V.A., et al. A compact form of methylated DNA in solutions containing poly(ethylene glycol). *Nucl. Acids Res.*, 1977, vol. 4, p. 3665–3676.
29. Nikogosyan, D.N., Repeyev, Yu.A., Yakovlev, D.Yu., et al. Photochemical alterations in DNA revealed by DNA-based liquid crystals. *Photochem. Photobiol.*, 1994, vol. 59, p. 269–276.
30. Yevdokimov, Yu.M., Skuridin, S.G., and Pozdnyakov, V.N. Liquid-crystalline biosensing unit to screen biologically active compounds and drugs interacting with double-stranded nucleic acid molecules. Russian Patent No. 1481974. 1991.

31. Yevdokimov, Yu.M. and Skuridin, S.G. Pseudocapsulated liquid crystals of nucleic acids. *Doklady of the USSR Acad. Sci.* (Russian Edition), 1988, vol. 303, p. 232–235.

32. Yevdokimov, Yu.M., Tokareva, L.G., and Skuridin, S.G. Biosensing unit to detect biologically active compounds interacting with double-stranded nucleic acid molecules. Russian Patent No. 2016888. 1994.

33. Kazanskii, K.S., Skuridin, S.G., Kuznetsova, V.I., and Evdokimov, Yu.M. Poly(ethylene oxide) hydrogels containing immobilized particles of an LC dispersion of desoxyribonucleic acid. *Polym. Sci., Series A* (Russian Edition), 1996, vol. 38, p. 875–883.

34. Yevdokimov, Yu.M., Kazanskii, K.S., Skuridin, S.G., and Varlamov, V.P. Polyfunctional liquid-crystalline composite material based on double-stranded nucleic acid and a novel method for its creation. Russian Patent No. 2224781, 2004.

35. Yevdokimov, Yu.M. and Skuridin, S.G. Biosensing units based on DNA liquid-crystalline dispersions. In *Resume of the Science and Techniques, Series Biotechnology* (Russian Edition), 1990, vol. 26, p. 134–161.

36. Skuridin, S.G. Biosensing units based on DNA liquid-crystalline dispersions. *Bulletin of the USSR Acad. Sci., Phys.* (Russian Edition), 1991, vol. 55, p. 1817–1824.

37. Skuridin, S.G., Yevdokimov, Yu.M., Efimov, V.S., et al. A new approach for creating double-stranded DNA biosensors. *Biosens. Bioelectron.*, 1996, vol. 11, p. 903–911.

38. Skuridin, S.G., Dubinskaya, V.A., Rebrov, L.B., et al. Regulation of the spatial organization of DNA molecules in particles of liquid-crystalline dispersions by means of biologically active compounds. *Russian Chem. Bull.* (Russian Edition), 2007, No. 12, p. 2405–2412.

39. Skuridin, S.G., Efimov, V.S., Nekrasov, A.V., et al. Method of heparin detection. Russian Patent No. 2123008. 1998.

40. Yevdokimov, Yu.M., Zakharov, M.A., and Skuridin, S.G. Nanotechnology based on nucleic acids. *Herald of the Russian Acad. Sci.* (Russian Edition), 2006, vol. 76, p. 712–721.

41. Yevdokimov, Yu.M., Skuridin, S.G., Nechipurenko, D.Yu., et al. Nanoconstructions based on double-stranded nucleic acids. *Int. J. Biol. Macromol.*, 2005, vol. 36, p. 103–115.

42. Yevdokimov, Yu.M., Salyanov, V.I., Mchedlishvily, B.V., et al. Molecular design based on double-stranded nucleic acids and synthetic polynucleotides for integral biosensing unit creation. *Sensory Syst.* (Russian Edition), 1999, vol. 13, p. 82–91.

43. Yevdokimov, Yu.M., Salyanov, V.I., Lortkipanidze, G.B., et al. Sensing biological effectors through the response of bridged nucleic acids and polynucleotides fixed in liquid-crystalline dispersions. *Biosens. Bioelectron.*, 1998, vol. 13, p. 279–291.

44. Skuridin, S.G., Dubinskaya, V.A., Zakharov, M.A., et al. Identification of the pharmaceutical substances with chelating properties by bioanalytical test-system based on DNA integral liquid-crystalline microchips. *Liquid Crystals and Their Application* (Russian Edition), 2005, No. 3–4, p. 64–74.

45. Karrer, P. *Organic Chemistry*. M.N. Kolosov, Ed. (Russian Edition). Leningrad: Goskhimizdat, 1960. p. 1216.

46. Skuridin, S.G., Dubinskaya, V.A., Lagutina, M.A., et al. Identification of the phytogenotoxicants by disposable biosensing units. *Biomedical Technologies and Radioelectronics* (Russian Edition), 2006, No 3, p. 38–43.

47. Skuridin, S.G., Amirova, S.R., Grigorenko, N.A., et al. Molecular constructions based on double-stranded nucleic acid liquid crystals: Formation, properties, practical application. *Liquid Crystals and Their Application* (Russian Edition), 2003, No. 3, p. 48–68.

SECTION III SUMMARY

NANO: The results of research performed by various authors represented in Section III of the book indicate the growth of a new branch of nanotechnology, nanobiotechnology, that is, an interdisciplinary science that combines the achievements of molecular biology, biotechnology, electronics, optics, and engineering. Though this science is still establishing its importance and is taking the first steps, it is obvious that the most important features of nanobiotechnology are molecules of biopolymers that are caused by their specific physicochemical properties—in particular, the molecules of nucleic acids or hybrid molecules based on nucleic acids. Using these molecules, spatial nanoconstructions with unique properties typical of the properties of metal nanoparticles—for instance, the dependence of melting temperature on size effect—can be formed. Such coincidence of the size effects for nanoconstructions formed by various initial materials has a principal meaning. The presence of size effects is the criterion that allows nanobiotechnology to be distinguished from other biological sciences where transformations of biopolymeric molecules with various sizes (including nanosized molecules) are investigated.

The approach based on the application of double-stranded nucleic acid molecules fixed in the spatial structure of liquid-crystal dispersion particles is particularly practical for nanodesign. The nanoconstructions obtained in this manner have unique properties such as abnormal optical activity, transition from "liquid" to "solid" state, dependence of melting temperature on the number of elements in the construction, change in the shape of the melting curve, depending on the number of structural elements, and so forth. Such properties are typical of nanoparticles of metals and semiconductors.

The nanoconstructions based on double-stranded NA molecules (DNA or RNA), regardless of the strategy of constructing, may be practically applied in various areas of science and technology.

1. Nanoconstructions based on double-stranded DNA CLCD with DNA concentration above 400 mg/mL may be used as carriers of genetic material or various biologically active/chemical compounds embedded into the structures. Application areas: health care, clinical biochemistry, biotechnology.
2. Nanoconstructions based on double-stranded DNA (RNA) CLCD or (DNA–Chi) complexes are sensing units of optical biosensors that can detect the presence of various BAC in different liquids. Application areas: health care, clinical biochemistry, ecology, biotechnology.
3. Nanoconstructions with controllable physicochemical properties immobilized in the structure of polymeric membranes (hydrogels) may be used in technology as molecular sieves, optical filters, and so forth.) Application areas: biotechnology, optics, electronics.

Section III of the book describes in detail various methods of creating sensing units based on nucleic acid molecules. In addition, much attention has been paid to the creation of the newest membrane biodetectors developed with the help of DNA nanoconstructions. There is hope that these biodetectors will allow significant modernization and simplification of the methods of detection of biologically significant compounds used in various areas of health care, biotechnology, and ecology.

Further research in the area of nanobiotechnology will not only contribute to the accumulation of knowledge about the properties of biological polymers and living cells but also foster the creation of new materials and devices with a wide range of practical abilities.

Index

251

9 781138 382145